# Communications in Computer and Information Science 1407

Editorial Board Members

Joaquim Filipe
*Polytechnic Institute of Setúbal, Setúbal, Portugal*
Ashish Ghosh
*Indian Statistical Institute, Kolkata, India*
Raquel Oliveira Prates
*Federal University of Minas Gerais (UFMG), Belo Horizonte, Brazil*
Lizhu Zhou
*Tsinghua University, Beijing, China*

More information about this series at http://www.springer.com/series/7899

Daniel Alejandro Rossit ·
Fernando Tohmé · Gonzalo Mejía Delgadillo (Eds.)

# Production Research

10th International Conference
of Production Research - Americas, ICPR-Americas 2020
Bahía Blanca, Argentina, December 9–11, 2020
Revised Selected Papers, Part I

*Editors*
Daniel Alejandro Rossit
Universidad Nacional del Sur
and CONICET
Bahía Blanca, Argentina

Gonzalo Mejía Delgadillo
Universidad de la Sabana
Chía, Colombia

Fernando Tohmé
Universidad Nacional del Sur
and CONICET
Bahía Blanca, Argentina

ISSN 1865-0929 ISSN 1865-0937 (electronic)
Communications in Computer and Information Science
ISBN 978-3-030-76306-0 ISBN 978-3-030-76307-7 (eBook)
https://doi.org/10.1007/978-3-030-76307-7

© Springer Nature Switzerland AG 2021
This work is subject to copyright. All rights are reserved by the Publisher, whether the whole or part of the material is concerned, specifically the rights of translation, reprinting, reuse of illustrations, recitation, broadcasting, reproduction on microfilms or in any other physical way, and transmission or information storage and retrieval, electronic adaptation, computer software, or by similar or dissimilar methodology now known or hereafter developed.
The use of general descriptive names, registered names, trademarks, service marks, etc. in this publication does not imply, even in the absence of a specific statement, that such names are exempt from the relevant protective laws and regulations and therefore free for general use.
The publisher, the authors and the editors are safe to assume that the advice and information in this book are believed to be true and accurate at the date of publication. Neither the publisher nor the authors or the editors give a warranty, expressed or implied, with respect to the material contained herein or for any errors or omissions that may have been made. The publisher remains neutral with regard to jurisdictional claims in published maps and institutional affiliations.

This Springer imprint is published by the registered company Springer Nature Switzerland AG
The registered company address is: Gewerbestrasse 11, 6330 Cham, Switzerland

# Preface

This CCIS volume includes selected articles presented at the 10th International Conference of Production Research - Americas (ICPR-Americas 2020), held in virtual form in Bahía Blanca, Argentina, during December 9–11, 2020. This conference was organized by the Americas chapter of the International Foundation of Production Research (IFPR). The aim of this conference was to exchange experiences and encourage collaborative work among researchers and practitioners from the Americas and the Caribbean region.

The first ICPR-Americas conference took place in November 2002 in St. Louis, Missouri, USA, under the general theme of "Production Research and Computational Intelligence for Designing and Operating Complex Global Production Systems". This conference generated very positive responses from the attendees. From then on, this ICPR regional meeting was held every other year. The second version of ICPR-Americas was held in August 2004 in Santiago, Chile, with the conference theme "Information and Communication Technologies for Collaborative Operations Management". The third edition of ICPR-Americas, held in Curitiba, Brazil, had the general theme of "Rethinking Operation Systems: New Roles of Technology, Strategy and Organization in the Americas' Integration Era". This event sought to promote a deep discussion about the role of production engineering in the Americas' integration process. In 2008, the venue of ICPR-Americas was Sao Paulo, Brazil, at the Universidade de São Paulo. The conference had the theme "The Role of Emerging Economies in the Future of Global Production: Creating New Multinationals". Bogotá, Colombia, hosted the fifth conference in 2010 on the subject of "Technologies in Logistics and Manufacturing for Small and Medium Enterprises". The sixth edition (2012) was organized around the topic "Production Research in the Americas Region: Agenda for the Next Decade" by the Universidad de Santiago de Chile. In Lima, Peru, the 2014 edition of ICPR-Americas addressed the general theme of "Towards Sustainable Eco-Industrialization through Applied Knowledge". The eighth edition was held in Valparaíso, Chile, in 2016, and more recently, in 2018, the ninth conference was held in Bogotá, Colombia, under the topic "Improving Supply Chain Management through Sustainability".

ICPR-Americas 2020 was the first edition held in virtual mode, due to the COVID-19 pandemic. Thanks to the participation and commitment of the attendees, the conference went on successfully, allowing many young researchers to participate in an international conference, in a year in which such opportunities were scarce. The ICPR-Americas meeting space provided them with the possibility of sharing their work, as well as exchanging ideas and points of view, all in the cordial atmosphere that characterizes the ICPR-Americas conferences.

ICPR-Americas 2020 was organized by a local committee from the Universidad Nacional del Sur, Argentina, supported by the university administration and the IFPR.

The conference received 280 submissions, all subjected to peer review. The refereeing process followed a single-blind procedure. The reviews were in the charge of a panel of experts and invited external reviewers (outside the Program Committee). Each submission had an average of three independent reviews and each reviewer was assigned, on average, two submissions. The best 53 articles were selected to be part of this CCIS volume.

March 2021

Daniel Alejandro Rossit
Fernando Tohmé
Gonzalo Mejía Delgadillo

# Organization

## General Chair

| | |
|---|---|
| Daniel Rossit | Universidad Nacional del Sur, Argentina |

## Program Committee Chair

| | |
|---|---|
| Fernando Tohmé | Universidad Nacional del Sur, Argentina |

## Organizing Committee

| | |
|---|---|
| Nancy López | Universidad Nacional del Sur, Argentina |
| Antonella Cavallin | Universidad Nacional del Sur, Argentina |
| Adrián Toncovich | Universidad Nacional del Sur, Argentina |
| Ernesto Castagnet | Universidad Nacional del Sur, Argentina |
| Diego Rossit | Universidad Nacional del Sur, Argentina |
| Mariano Frutos | Universidad Nacional del Sur, Argentina |
| Daniel Carbone | Universidad Nacional del Sur, Argentina |
| Luciano Sívori | Universidad Nacional del Sur, Argentina |
| Adrián Castaño | Universidad Nacional del Sur, Argentina |
| Martín Safe | Universidad Nacional del Sur, Argentina |
| Marisa Sánchez | Universidad Nacional del Sur, Argentina |
| Agustín Claverie | Universidad Nacional del Sur, Argentina |
| Fernando Nesci | Universidad Nacional del Sur, Argentina |
| Carla Macerates | CONICET, Argentina |

## Program Committee

| | |
|---|---|
| Adrián Toncovich | Universidad Nacional del Sur, Argentina |
| Mariano Frutos | Universidad Nacional del Sur and CONICET, Argentina |
| Antonella Cavallin | Universidad Nacional del Sur, Argentina |
| Marisa Analía Sánchez | Universidad Nacional del Sur, Argentina |
| Martín Safe | Universidad Nacional del Sur and CONICET, Argentina |
| Diego Gabriel Rossit | Universidad Nacional del Sur and CONICET, Argentina |
| Máximo Méndez Babey | Universidad de las Palmas de Gran Canaria, Spain |
| Héctor Cancela | Universidad de la República, Uruguay |
| Pedro Piñeyro | Universidad de la República, Uruguay |
| Sergio Nesmachnow | Universidad de la República, Uruguay |
| Adrián Ferrari | Universidad de la República, Uruguay |

| | |
|---|---|
| Marcus Ritt | Universidade Federal do Rio Grande do Sul, Brazil |
| Marcelo Seido Nagano | Universidade de São Paulo, Brazil |
| José Framiñan | Universidad de Sevilla, Spain |
| Rubén Ruiz | Universidad Politécnica de Valencia, Spain |
| Begoña González Landín | Universidad de las Palmas de Gran Canaria, Spain |
| Ricardo Aguasca Colomo | Universidad de las Palmas de Gran Canaria, Spain |
| Roger Ríoz-Mercado | Universidad de Nueva León, Mexico |
| Enzo Morosini Frazzon | Federal University of Santa Catarina, Brazil |
| Ciro Alberto Amaya Guio | Universidad de los Andes, Colombia |
| Cihan Dagli | Missouri University of Science and Technology, USA |
| Shimon Nof | Purdue University, USA |
| Bopaya Bidanda | University of Pittsburgh, USA |
| Sergio Gouvea | Pontifícia Universidade Católica do Paraná, Brazil |
| Edson Pinheiro | Pontifícia Universidade Católica do Paraná, Brazil |
| Fernando Deschamps | Pontifícia Universidade Católica do Paraná, Brazil |
| Cecilia Montt Veas | Pontifícia Universidade Católica de Valparaíso, Chile |
| Óscar C. Vásquez | Universidad de Santiago de Chile, Chile |
| Pedro Palominos | Universidad de Santiago de Chile, Chile |
| Luis Ernesto Quezada Llanca | Universidad de Santiago de Chile, Chile |
| Dusan Sormaz | Ohio University, USA |
| Gursel Suer | Ohio University, USA |
| José Ceroni | Pontifícia Universidade Católica de Valparaíso, Chile |
| Karen Y. Niño | Universidad Militar Nueva Granada, Colombia |
| Nubia Velasco | Universidad de los Andes, Colombia |
| Gonzalo Mejía Delgadillo | Universidad de la Sabana, Colombia |
| Jairo Rafael Montoya Torres | Universidad de la Sabana, Colombia |
| William Javier Guerrero Rueda | Universidad de la Sabana, Colombia |
| Leonardo Jose Gonzalez Rodriguez | Universidad de la Sabana, Colombia |
| Luis Alfredo Paipa Galeano | Universidad de la Sabana, Colombia |
| Alfonso Tullio Sarmiento Vasquez | Universidad de la Sabana, Colombia |
| Vícto Viana Céspedes | Universidad de la República, Uruguay |
| Diego Ricardo Broz | Universidad Nacional de Misiones, Argentina |
| Aníbal Blanco | CONICET, Argentina |
| Alberto Bandoni | Universidad Nacional del Sur, Argentina |
| José Luis Figueroa | Universidad Nacional del Sur, Argentina |
| Mauricio Miguel Coletto | Universidad Nacional de Río Negro, Argentina |
| Alejandro Olivera | Universidad de la República, Uruguay |
| Sandra Robles | Universidad Nacional del Sur, Argentina |
| Daniela Alessio | Universidad Nacional del Sur, Argentina |
| Frank Werner | Otto von Guericke University Magdeburg, Germany |
| María Clara Tarifa | Universidad Nacional de Río Negro and CONICET, Argentina |

| | |
|---|---|
| Fernanda Villarreal | Universidad Nacional del Sur, Argentina |
| Valentina Viego | Universidad Nacional de Río Negro and CONICET, Argentina |
| Lorena Brugnoni | Universidad Nacional del Sur and CONICET, Argentina |
| Jorge Lozano | Universidad Nacional del Sur and CONICET, Argentina |
| Guillermo Crapiste | Universidad Nacional del Sur and CONICET, Argentina |
| Facundo Iturmendi | Universidad Nacional de Río Negro, Argentina |
| Ana Maguitman | Universidad Nacional del Sur and CONICET, Argentina |
| Santiago Maiz | Universidad Nacional del Sur, Argentina |
| Héctor Chiacchiarini | Universidad Nacional del Sur and CONICET, Argentina |
| Gabriela Pesce | Universidad Nacional del Sur, Argentina |
| Susana Moreno | CONICET, Argentina |
| María Teresa González | Universidad Nacional del Sur and CONICET, Argentina |
| Ignacio Costilla | Universidad Nacional del Sur and CONICET, Argentina |
| Adrián M. Urrestarazu | Universidad Nacional del Sur, Argentina |
| Elda Monetti | Universidad Nacional del Sur, Argentina |
| Fabio Miguel | Universidad Nacional del Río Negro, Argentina |
| Diego Hernán Peluffo-Ordóñez | Yachay Tech, Ecuador |
| Pedro Ballesteros Silva | Universidad Tecnológica de Pereira, Colombia |
| Leandro Leonardo Lorente Leyva | Universidad Técnica del Norte, Ecuador |
| Katty Alicia Lagos Ortiz | Universidad Agraria del Ecuador, Ecuador |
| José Medina-Moreira | Universidad de Guayaquil, Ecuador |
| Miguel Heredia | Instituto Tecnológico de Gustavo A. Madero, Tecnológico de Nacional de México, Mexico |
| Israel David Herrera Granda | Universidad Técnica del Norte, Ecuador |
| Andrés Fioriti | CONICET, Argentina |
| Fernando Delbianco | Universidad Nacional del Sur and CONICET, Argentina |
| Claudio Delrieux | Universidad Nacional del Sur and CONICET, Argentina |
| Katyanne Farias | École des Mines de Saint-Etienne, France |
| Ana Carolina OIivera | Universidad Nacional de Cuyo and CONICET, Argentina |
| Francesco Pilati | University of Trento, Italy |
| Yanina Fumero | INGAR, Argentina |
| Gabriela Corsano | INGAR, Argentina |
| Marco Cedeño Viteri | Universidad Tecnológica de Chile INACAP, Chile |

| | |
|---|---|
| Marcela C. González Araya | Universidad de Talca, Chile |
| Leandro Rodriguez | Universidad Nacional de San Juan and CONICET, Argentina |
| Juana Zuntini | Universidad Nacional del Sur, Argentina |
| Natalia Urriza | Universidad Nacional del Sur, Argentina |
| María Angélica Viceconte | Universidad Nacional del Sur, Argentina |
| Marianela De Batista | Universidad Nacional del Sur, Argentina |
| Pau Fonseca i Casas | Universtitat Politècnica de Catalunya, Spain |
| Ariel Behr | Universidade Federal do Rio Grande do Sul, Brazil |
| José Fidel Torres Delgado | Universidad de los Andes, Colombia |
| Sepideh Abolghasem Ghazvini | Universidad de los Andes, Colombia |
| Alex Ricardo Murcia Cucaita | Universidad de los Andes, Colombia |
| Adriana Lourdes Abrego Perez | Universidad de los Andes, Colombia |
| Camil Martinez | Universidad de los Andes, Colombia |
| Jorge Luis Chicaiza Vaca | Dortmund Technical University, Germany |
| Fernando Daniel Mele | Universidad Nacional de Tucumán, Argentina |
| Humberto Heluane | Universidad Nacional de Tucumán, Argentina |
| Jorge Marcelo Montagna | INGAR, Argentina |
| Melisa Manzanal | Universidad Nacional del Sur and CONICET, Argentina |
| Gustavo Ramoscelli | Universidad Nacional del Sur, Argentina |
| José María Cabrera Peña | Universidad de Las Palmas de Gran Canaria, Spain |
| Francisco Javier Rocha Henríquez | Universidad de Las Palmas de Gran Canaria, Spain |
| Maarouf Mustapha | Universidad de Las Palmas de Gran Canaria, Spain |
| Carlos Hernández Hernández | Universidad de Las Palmas de Gran Canaria, Spain |
| Dagoberto Castellanos Nieves | Universidad de La Laguna, Spain |
| Martha Ramírez Valdivia | Universidad de la Frontera, Chile |
| Sri Talluri | Michigan State University, USA |
| Dmitry Ivanov | Berlin School of Economics and Law, Germany |
| Pietro Cunha Dolci | Universidade Santa Catarina do Sul, Brazil |
| Carlos Ernani Fries | Universidade Federal de Santa Catarina, Brazil |
| Claudemir Tramarico | Universidade Estadual Paulista, Brazil |
| Eduardo Ortigoza | Universidad Nacional de Asunción, Paraguay |
| Liang Gao | Huazhong University of Science and Technology, China |
| Carlos Contreras Bolton | Universidad de Concepción, Chile |
| Marcela Filippi | Universidad Nacional de Río Negro, Argentina |
| María Beatriz Bernabé Loranca | Benemérita Universidad Autónoma de Puebla, Mexico |
| Fernando Espinosa | Universidad de Talca, Chile |

| | |
|---|---|
| Patrick Hirsch | University of Natural Resources and Life Sciences, Austria |
| Erfan Babaee Tirkolaee | Mazandaran University of Science and Technology, Iran |
| Albert Ibarz | Universidad de Lerida, Spain |
| Darla Goeres | Montana State University, USA |
| Diane Walker | Montana State University, USA |
| Gabriela Vinderola | Universidad Nacional del Litoral and CONICET, Argentina |
| Pedro Rizzo | INTA, Argentina |
| Lorena Franceschinis | Universidad Nacional del Comahue and CONICET, Argentina |
| Vítor Alcácer | Instituto Politécnico de Setúbal, Portugal |
| Marcela Ibañez | Universidad de la República, Uruguay |
| Diego Passarella | Universidad de la República, Uruguay |
| Krzysztof Polowy | Poznan University of Life Sciences, Poland |
| Marta Glura | Poznan University of Life Sciences, Poland |
| Axel Soto | Universidad Nacional del Sur and CONICET, Argentina |
| Carlos Lorenzetti | Universidad Nacional del Sur and CONICET, Argentina |
| Rocío Cecchini | Universidad Nacional del Sur and CONICET, Argentina |
| Eduardo Xamena | Universidad Nacional de Salta and CONICET, Argentina |
| Diego Rodriguez | Universidad Nacional de Salta, Argentina |
| Silvio Gonnet | Universidad Tecnológica Nacional and CONICET, Argentina |
| Diego Pinto-Roa | Universidad Nacional de Asunción and CONACYT, Paraguay |
| Ángel Augusto Roggiero | Univesidad Nacional de Cuyo, Argentina |
| Claudia Noemí Zárate | Universidad Nacional de Mar del Plata, Argentina |
| Alejandra María Esteban | Universidad Nacional de Mar del Plata, Argentina |
| Carlos Vecchi | Universidad Nacional del Nordeste, Argentina |
| Ângela de Moura Ferreira Danilevicz | Universidade Federal do Rio Grande do Sul, Brasil |
| Germán Rossetti | Universidad Nacional del Litoral, Argentina |
| Marcelo Tittonel | Universidad Nacional de La Plata, Argentina |
| César Pairetti | Universidad Nacional de Rosario, Argentina |
| Victor Albornoz | Universidad Técnica Federico Santa María, Chile |
| Simone Martins | Universidade Federal Fluminense, Brazil |
| Antonio Mauttone | Universidad de la República, Uruguay |
| Débora Ronconi | Universidade de São Paulo, Brazil |
| Libertad Tansini | Universidad de la República, Uruguay |
| Carlos Testuri | Universidad de la República, Uruguay |

| | |
|---|---|
| Javier Dario Fernandez Ledesma | Universidad Pontificia Bolivariana, Colombia |
| Carlos Romero | Technical University of Madrid, Spain |
| Julio Arce | Universidade Federal do Paraná, Brazil |
| Pete Bettinger | University of Georgia, USA |
| Pascal Forget | Université du Québec à Trois-Rivières, Canada |
| Sergio Maturana Valderrama | Pontificia Universidad Católica de Chile, Chile |
| Gilson Adamczuk Oliveira | Universidade Tecnológica Federal do Paraná, Brazil |
| Juan José Troncoso Tirapegui | Universidad de Talca, Chile |
| Luis Diaz-Balteiro | Technical University of Madrid, Spain |
| Marcio Pereira da Rocha | Universidade Federal do Paraná, Brazil |
| Jamal Toutouh | Massachusetts Institute of Technology, USA |
| Maico Roris Severino | Universidade Federal de Goiás, Brazil |
| Joaquín Orejas | Universidad Nacional de Río Cuarto, Argentina |

# Contents – Part I

## Metaheuristics and Algorithms

## Industry 4.0 and Cyber-Physical Systems

# Contents – Part II

## Smart City



## Intelligent Systems and Decision Sciences

**Simulation**

**Machine Learning and Big Data**

# Optimization

# Optimal Reverse Supply Chain Design: The Case of Empty Agrochemical Containers

Glenda N. Yossen[1,2] and Gabriela P. Henning[1,2](✉)

[1] INTEC (Universidad Nacional del Litoral, CONICET), 3000 Santa Fe, Argentina
{gyossen,ghenning}@intec.unl.edu.ar
[2] Facultad de Ingeniería Química, Universidad Nacional del Litoral, 3000 Santa Fe, Argentina

**Abstract.** The growing use of agrochemicals added to the harm that contaminated containers cause to the environment and to the health of the population, make it necessary to establish a reverse supply chain for the recovery and correct treatment of empty containers. Current regulations establish that they must be triple-rinsed and sent to collection centers, where containers are consolidated and shipped to treatment/recycling plants. Containers can also be sent to these plants directly. Given this problem, this contribution proposes an MILP multi-period model to define the optimal configuration and operation of the reverse supply chain network of empty agrochemical containers along a given planning horizon. Provided a superstructure including farms and the location of the potential nodes (collection centers and plastic treatment plants), the model determines which facilities to install and when, their corresponding sizes, as well as how they would operate along the planning horizon. In addition, the model establishes the material flows among the various nodes in each planning period (network topology). The proposed model has been verified by addressing a realistic case study, which has been solved under different scenarios, exhibiting good computational performance. The obtained results allow reaching satisfactory conclusions in relation to model scalability and sensitivity.

**Keywords:** Reverse logistics · Supply chain design and planning · Empty agrochemical containers · Mixed-integer linear programming model

## 1 Introduction

Current supply chains are not limited to unidirectional product flows from manufactures to consumers. On the contrary, they need to deal with opposite material flows devoted to the reuse, recycling and/or final disposal of products, parts, as well as packaging and empty product containers. Managing these inverse flows confronts companies with novel challenges that need to be analyzed under the umbrella of reverse logistics. It can be defined as the process of planning, implementing and controlling backward flows of raw materials, in process inventory, packaging and finished goods, from a manufacturing, distribution or use point, to a point of recovery or point of proper disposal [1].

This contribution addresses the design of a reverse supply chain of agrochemical containers. It is intended to collect and thoroughly clean the plastic vessels, once they

© Springer Nature Switzerland AG 2021
D. A. Rossit et al. (Eds.): ICPR-Americas 2020, CCIS 1407, pp. 3–18, 2021.
https://doi.org/10.1007/978-3-030-76307-7_1

have been emptied and triple-rinsed, with the aim of recycling/using the plastic, decontaminating the washing water, while minimizing the associated environmental and health risks. The problem of configuring such network along a given planning horizon implies establishing: (i) the number and location of collection centers – intermediate points for collecting containers –, as well as plastic treatment plants to be installed, (ii) the time period in which each facility will be installed and the ones in which it will operate, (iii) the capacities of the various storage and processing facilities, (iv) for each time period, which are the flows of materials between farms, which generate the containers, and the collection/treatment points, as well as the flows between collection centers and plastic treatment plants. This problem is addressed by means of a multi-period MILP (Mixed Integer Linear Programming) model that is able to tackle the challenges previously presented while minimizing the total actual discounted cost of the network.

This contribution is organized as follows. Section 2 contains a brief literature review on the design of reverse supply chains. The mathematical model is presented in Sect. 3. In Sect. 4, a realistic case-study, which allows testing the formulation, is described. Computational results and a sensitivity analysis are presented in Sect. 5. The contribution ends with concluding remarks and a summary of future work.

## 2 Literature Review

Facility location decisions play a critical role in the design of supply chain networks, as pointed out in [2]. This review article presents a comprehensive analysis of several facility location approaches by identifying basic characteristics (number of products, number of periods of the planning horizon, type of network structure, uncertainty consideration, etc.) that models must capture in order to address different kinds of supply chain planning problems that are of practical interest. It considers both the typical node location-allocation decisions and their integration with other decisions that are relevant to problems of optimal design and planning of supply chain networks. This review paper shows that the supply chain configuration and planning problems have increasingly drawn the attention of the scientific community during the last two decades. In addition, sustainability aspects started playing a central role, thus impacting not only on the objectives to be pursued – economic, environmental and social – [3], but also on the reverse flows of materials to be considered [4].

Reverse logistics is defined as the process of planning, implementing and controlling backward flows of materials (raw materials, in-process inventory, packaging and finished goods), from a manufacturing, distribution or use point, to a point of recovery and/or proper disposal. Two main types of reverse logistics design and planning problems can be identified [2]: (i) The ones associated with closed-loop networks, where the reverse network is fully integrated with the forward chain, thus allowing the simultaneous definition of the optimal distribution and recovery networks, and; (ii) Recovery network problems, i.e. the ones that are limited to recovery and/or disposal activities. Regarding these last ones, [5] proposes a mathematical programming framework that accommodates several characteristics of practical relevance: a multi-period setting, modular capacities, capacity expansion of the facilities, reverse bills of materials, variable operational costs, among other features. Other generic mathematical programming formulations to address reverse network configuration problems were proposed in [3, 4].

In particular, [3] presents a multi-objective model that integrates the three dimensions of sustainability: economic, environmental, and social.

Despite the fact that the number of scientific publications related to the reverse logistics domain has been steadily growing, showing the increasing importance of the field due to the environmental awareness of the society, generic models cannot address all the situations that are found in practice yet. Even for the subfield of solid waste management, [6] have pointed out the various specific features that are to be considered, which may need to be further adapted under an extended producer responsibility approach.

Regarding the reverse supply chains of empty agrochemical containers, few contributions have been found. Lagarda-Leyva [7] addressed the reverse supply chain of agrochemical plastic containers associated with Roma tomato production in a Mexican region. The contribution proposed causal conceptual models that allow developing Forrester's dynamic models that, in turn, have been employed to simulate and assess the operation of the system under dynamic conditions within three different scenarios. Recently, [8] reported a preliminary approach to address the design of the reverse network of empty agrochemical containers in a region of Buenos Aires, Argentina. The proposal includes two MILP formulations. The first one is a single period facility location model that allows selecting, out of a set of potential places, the settings of temporary collection centers (CATs) and a main gathering operation site (OP), where the containers are get together. The potential facilities have fixed capacities. In addition, their operation and installation costs are omitted. The modeling of material flows and transportation costs is also very simple since the transportation from farms to collection centers is not taken into account (farms are not part of the problem). The second model, which must be applied to the solution of the first one, addresses the routing problem associated with the movement of empty containers from CATs to the main gathering point.

It can be concluded that the optimal configuration of the reverse supply chain network of empty agrochemical containers is still a challenging problem that deserves attention and will be addressed in the remaining sections of this contribution.

## 3 Mathematical Programming Formulation

In this Section a multi-period mixed-integer linear programming (MILP) model for the design of the reverse logistics network previously described is presented.

Given:

- a planning horizon of a certain length,
- the set of farms, each one generating a pre-specified number of empty containers in each period of the planning horizon,
- the superstructure that includes all the potential/existing collection centers and possible/existing plastic treatment plants, as well as the farms – i.e. all the nodes to be considered for the network design – and their interconnections – i.e. the arcs and their associated distance,
- the set of possible discrete sizes of collection centers/treatment plants,
- the sizes of the facilities already installed at the beginning of the planning horizon,
- maximum and minimum storage capacities for a collection center/treatment plant of a given size,

- maximum and minimum processing capacities of each treatment plant size,
- maximum capacity of each type of truck to be used,
- number of periods required by a new facility to start operating,
- installation and fixed operational cost of collection centers/treatment plants,
- variable processing cost of treatment plants,
- transportation cost per unit distance between the different types of nodes.

Determine:

- which collection centers and plastic treatment plants to install, and when to install them,
- capacity of each facility to be built,
- which collection centers and plastic treatment plants to operate along each period of the planning horizon,
- container flows between farms and collection centers, farms and treatment plants and collection centers and treatment plants during each planning period.

So as to minimize the total actual discounted cost of the network.

## 3.1 Model Assumptions

It is assumed that: (i) all the model parameters are deterministic, i.e. no uncertainty is considered, (ii) the containers that are generated in a given period must be collected and treated in the same period, (iii) collection centers and plastic treatment plants to be installed are of discrete capacities, (iv) variable operating costs are taken into account only for plastic treatment plants, (v) transportation costs only consider the distance traveled and omit the load of the vehicles.

## 3.2 Nomenclature

### Sets/indices

| | |
|---|---|
| $F/f$ | Farms |
| $J/j$ | Collection centers |
| $P/p$ | Plastic treatment plants |
| $J^i/P^i$ | Subset of transitory centers/treatment plants that are already installed at the beginning of the planning horizon |
| $M/m$ | Set of available sizes of collection centers |
| $N/n$ | Set of available sizes of plastic treatment plants |
| $T/t, t'$ | Time periods |

**Parameters**

| | |
|---|---|
| $g_{ft}$ | Number of containers to be generated by farm $f$ during period $t$ |
| $d_{fj}/d_{fp}/d_{jp}$ | Distance between nodes of different types |
| $k_m^{min}/k_n^{min}$ | Minimum capacity of a collection center/treatment plant of size *m/n* |
| $k_m^{max}/k_n^{max}$ | Maximum capacity of a collection center/treatment plant of size *m/n* |
| $nt^j/nt^p$ | Number of time periods required by a new collection center/treatment plant to start operating from the period it is installed |
| $c^{fj}/c^{fp}/c^{jp}$ | Transportation cost per distance unit between the different nodes of the network |
| $ir$ | Interest rate |
| $fc_m/fc_n$ | Fixed cost of operation of a collection center/treatment plant of size *m/n* |
| $vc$ | Variable production cost of plastic treatment plants |
| $cc_m^j/cc_n^p$ | Construction cost of a collection center/treatment plant of size *m/n* |
| $ec_n$ | Equipment cost of a treatment plant of size $n$ |
| $cap$ | Loading capacity of the trucks that travel between collection centers and treatment plants |
| $nf_{ft}$ | Number of trips required to transport the total generation of containers of farm $f$ during period $t$ |
| $nmax$ | $max\ (nt^j, nt^p)$ number of periods that must be considered prior to the start of the planning horizon for the installation of facilities |

**Decision variables**

| | |
|---|---|
| Binary variables | |
| $X_{jm}/X_{pn}$ | Equal to one if collection center $j$/treatment plant $p$ of size *m/n* is installed |
| $Y_{jt}/Y_{pt}$ | Equal to one if collection center j/treatment plant $p$ operates in period $t$ |
| $I_{jt}/I_{pt}$ | Equal to one if collection center j/treatment plant $p$ is installed in period $t$ |
| Positive variables | |
| $S_{fjt}/S_{fpt}/S_{jpt}$ | Number of containers sent between nodes of different types in period $t$ |

### 3.3 MILP Formulation

Using the above definitions, the model is formulated as follows.

$$\text{Min} \quad \sum_t \frac{Icc_t + Icp_t + Vcp_t + Ocj_t + Ocp_t + Tc_t^{fj} + Tc_t^{fp} + Tc_t^{jp}}{(1+ir)^{t-1}} \tag{1}$$

$$s.t. \quad g_{ft} = \sum_{\forall j \in J} S_{fjt} + \sum_{\forall p \in P} S_{fpt} \qquad \forall f \in F, \forall t \in T > nmax \tag{2}$$

$$\sum_{\forall f \in F} S_{fjt} = \sum_{\forall p \in P} S_{jpt} \qquad \forall j \in J, \forall t \in T > nmax \tag{3}$$

$$\sum_{\forall t', t' + nt^{j} \leq t} I_{jt'} \geq Y_{jt} \qquad \forall j \in J, j \notin J^{i}, \forall t \in T \tag{4}$$

$$\sum_{\forall t', t' + nt^{p} \leq t} I_{pt'} \geq Y_{pt} \qquad \forall p \in P, p \notin P^{i}, \forall t \in T \tag{5}$$

$$\sum_{\forall m \in M} X_{jm} = \sum_{\forall t \in T} I_{jt} \qquad \forall j \in J, j \notin J^{i} \tag{6}$$

$$\sum_{\forall n \in N} X_{pn} = \sum_{\forall t \in T} I_{pt} \qquad \forall p \in P, p \notin P^{i} \tag{7}$$

$$\sum_{\forall m \in M} X_{jm} \leq 1 \qquad \forall j \in J \tag{8}$$

$$\sum_{\forall n \in N} X_{pn} \leq 1 \qquad \forall p \in P \tag{9}$$

$$\sum_{\forall f \in F} S_{fjt} \leq k_{m = max}^{max} Y_{jt} \qquad \forall j \in J, \forall t \in T, t > nmax \tag{10}$$

$$\sum_{\forall f \in F} S_{fjt} \leq \sum_{\forall m \in M} k_{m}^{max} X_{jm} \qquad \forall j \in J, \forall t \in T, t > nmax \tag{11}$$

$$\sum_{\forall f \in F} S_{fpt} + \sum_{\forall j \in J} S_{jpt} \leq k_{n=max}^{max} Y_{pt} \qquad \forall p \in P, \forall t \in T, t > nmax \tag{12}$$

$$\sum_{\forall f \in F} S_{fpt} + \sum_{\forall j \in J} S_{jpt} \leq \sum_{\forall n \in N} k_{n}^{max} X_{pn} \qquad \forall p \in P, \forall t \in T, t > nmax \tag{13}$$

$$\sum_{\forall f \in F} S_{fjt} \geq k_{m=1}^{min} Y_{jt} + \sum_{\forall m \in M, m > 1} \left(k_{m}^{min} - k_{m=1}^{min}\right)\left(Y_{jt} + X_{jm} - 1\right) \qquad \forall j \in J, \forall t \in T, t > nmax \tag{14}$$

$$\sum_{\forall f \in F} S_{fpt} + \sum_{\forall j \in J} S_{jpt} \geq k_{n=1}^{min} Y_{pt} + \sum_{\forall n \in N, n > 1} \left(k_{n}^{min} - k_{n=1}^{min}\right)\left(Y_{pt} + X_{pn} - 1\right) \qquad \forall p \in P, \forall t \in T, t > nmax \tag{15}$$

$$Icc_{t} = \sum_{\forall j \in J, j \notin J^{i}} \sum_{\forall m \in M} cc_{m}^{j}\left(I_{jt} + X_{jm} - 1\right) \qquad \forall t \in T \tag{16}$$

$$Icp_t = \sum_{\forall p \in P, p \notin P^i} \sum_{\forall n \in N} (cc_n^p + ec_n)(I_{pt} + X_{pn} - 1) \qquad \forall t \in T \tag{17}$$

$$Vcp_t = \sum_{\forall p \in P} vc(\sum_{\forall f \in F} S_{fpt} + \sum_{\forall j \in J} S_{jpt}) \qquad \forall t \in T \tag{18}$$

$$Ocj_t = \sum_{\forall j \in J} \sum_{\forall m \in M} fc_m(Y_{jt} + X_{jm} - 1) \qquad \forall t \in T \tag{19}$$

$$Ocp_t = \sum_{\forall p \in P} \sum_{\forall n \in N} fc_n(Y_{pt} + X_{pn} - 1) \qquad \forall t \in T \tag{20}$$

$$Tc_t^{fj} = \sum_{\forall f \in F} \sum_{\forall j \in J} c^{fj} 2d_{fj} \frac{S_{fjt}}{g_{ft}} nf_{ft} \qquad \forall t \in T \tag{21}$$

$$Tc_t^{fp} = \sum_{\forall f \in F} \sum_{\forall p \in P} c^{fp} 2d_{fp} \frac{S_{fpt}}{g_{ft}} nf_{ft} \qquad \forall t \in T \tag{22}$$

$$Tc_t^{jp} = \sum_{\forall j \in J} \sum_{\forall p \in P} c^{jp} 2d_{jp} 1.2 \frac{S_{jpt}}{cap} \qquad \forall t \in T \tag{23}$$

The objective function, given by expression (1), comprises the different costs involved in the design and operation of the logistics network throughout the planning horizon: (i) installation costs of collection centers and treatment plants, (ii) variable costs of handling empty containers in treatment plants, (iii) fixed operation costs of facilities, and (iv) transportation costs between the different nodes of the network. In order to determine how much future projected cash flows are worth at present, these cost components are affected by an annual interest rate, according to the multi-period total discounted cost formula given by (1). The model includes constraints (2–23).

Equation (2) prescribes that all the containers generated by a farm in a certain period, must be sent to a collection center and/or to a treatment plant in the same planning period. Similarly, Eq. (3) states that all the containers that are received by a collection center in a given period must be sent to a plastic treatment plant in the same period.

Expression (4) assures that a collection center can operate in a certain period only if it has been installed in a previous one. Expression (5) forces treatment plants to obey the same restriction. Equation (6) states that if a collection center is installed in one of the periods, it must adopt one of the $m$ available sizes. In a similar way, Eq. (7) prescribes that if a plastic treatment plant is installed in one of the periods, it must be of one of the $n$ available sizes. Expressions (8) and (9) force that every collection center/treatment plant that is installed should adopt just one of the available sizes.

According to inequalities (10) and (11), the number of containers sent to a given collection center in each period, should not overpass the maximum storage capacity corresponding to the installed size. The maximum capacity of each treatment plant to be built is captured by means of expressions (12) and (13). They have the same rationale as (10) and (11). Following analogous ideas, the minimum capacity of each collection center and plastic treatment plant to be installed is given by expressions (14) and (15), respectively.

Expressions (16–23) capture the different cost terms participating in the objective function. Equation (16) defines the investment cost of new collection centers in each period, $Icc_t$. Similarly, Eq. (17) captures the investment cost of new treatment plants in each period. It takes into account both the construction cost and the equipment cost associated with the size of the plant to be installed. It should be noted that these costs are calculated by applying scale economies. Variable costs per period are only considered for plastic treatment plants, as shown in Eq. (18). They take into account unit processing cost and the total number of containers sent to each plant in each period. The fixed operation cost associated with collection centers in each period, $Ocj_t$, is given by Eq. (19). The $fc_m$ cost parameter depends on the adopted size $m$. Similarly, Eq. (20) establishes the fixed operation cost of treatment plants in each period, $Ocp_t$. The $fc_n$ cost depends on the size $n$ that is adopted for each treatment plant. The transportation costs from farms to collection centers for each one of the planning periods is captured by Eq. (21). As seen, round trips are taken into account by multiplying the transportation cost per distance unit from farm $f$ to collection center $j$, $c^{fj}$, by twice the distance between these nodes. In turn, this is multiplied by the proportion of the number of trips required to transport the total generation of containers of the farm that are just sent to the collection center. It should be noted that $nf_{ft}$ is calculated considering 1.5 times the actual number of trips that would be required if the complete vehicle capacity would have been employed. This assumption relies on the fact that most of the trips are made with less than the maximum allowed load. The proportion is calculated by dividing the number of containers sent from each farm to each collection center in a given period, by the container generation of the farm in such a period. The transportation cost from farms to treatment plants, for each planning period, is captured by Eq. (22) in the same way the transportation cost between farms and collection centers has been estimated. Finally, the transportation cost from collection centers to treatment plants, for each planning period, is calculated by Eq. (23). In this case, it is considered 1.2 times the total number of required trips, since a better container load consolidation can be carried out.

## 4 Case Study Description

The MILP model presented in the previous section has been implemented to establish the reverse logistic network configurations and operation policies associated with several scenarios of a realistic case study. Such case study corresponds to the handling and treatment of empty agrochemical containers in three northern departments of the Santa Fe province, Argentina. This province has one of the most significant agricultural productions in the country. Consequently, it has one of the largest generations of empty agrochemical containers, which are used to control unwanted pests and weeds of crops. As already presented, this reverse supply chain comprises three types of nodes: farms, transitory collection centers, and treatment plants. In these last facilities the plastic is thoroughly cleaned and subsequently pelletized. In addition, the water that is employed in the process is decontaminated before it is recycled, so as to minimize the use of fresh water. Treatment plants can also act as collection centers.

The operation of this recovery system is regulated by national law 27279, which is implemented in a very incipient way in Argentina. This law prescribes that empty agrochemicals containers must be triple-rinsed in farms, which are also responsible for their shipment to collection centers and/or treatment plants. Unfortunately, the system does not work well in practice, farmers fail to meet their obligations and there are not enough collection centers installed - which causes both high transportation costs and a large carbon footprint. Furthermore, government agencies do not have tools to assess how the law can be satisfactorily complied with, from an economic and environmental perspective, by establishing a proper recovery network. All these elements allow concluding that the problem deserves the attention of the scientific community.

Since this case study considers strategic and tactical decisions, it will be simplified in order to reduce the dimensionality of the resulting mathematical models. Therefore, aggregated farms of approximately 500 ha will be defined by grouping neighboring farms and assigning to them a fictitious location and an aggregate number of containers. Though it is a simplification, this assumption is pretty common in supply chain strategic decision making problems, which generally consider aggregated products, aggregated customers, or as in this case aggregated farms. The explicit consideration of farms is one of the main differences between the proposed approach and the first model reported in [8].

The number of containers generated per year by each farm has been obtained as a random parameter whose value lies between the minimum and maximum amounts of agrochemicals that are annually applied per hectare, multiplied by the number of hectares where agrochemicals are employed. These random values were also affected by a slight increasing tendency in order to represent that in this region open field livestock activities are being replaced by agricultural ones. The location of potential collection centers that has been considered coincides with the one of municipalities and counties located in the region of interest. With regard to the transportation of empty containers, according to current regulations in Argentina, the vehicles employed to make movements from farms to collection centers and/or treatment plants, should be pick-up vans, having a capacity limit that has been considered in the model. On the other hand, the transportation from collection centers to treatment plants is done by means of trailer trucks having a much greater capacity and lower transportation costs.

Initially, a simple base case has been tackled. It takes into account a super-structure composed of 300 farms, 20 potential collection centers, 8 potential treatment plants and no existing facilities that are already installed. Two different capacity modules for collection centers and treatment plants – i.e. two discrete sizes for each type of facility – have been allowed for. The planning horizon considers 6 years of operation of the collection-treatment system, plus an initial year for the installation of the first facilities. Thus, during the initial year the system does not operate, but the required facilities are being built. This base case has been further extended to test the scalability of the model and its sensitivity to changes in the number of farms, the number of potential collection centers/treatment plants, as well as the length of the planning horizon.

## 5 Computational Results

The model has been implemented in GAMS 23.6, running on an Intel Core i8 computer, with 3.4 GHz, 8 GB of RAM, and using CPLEX as a solver. Initially, the model – expressions (1–23) – was solved to assess its computational performance. The optimal solution corresponding to the base case was reached in 3283 s of CPU. Since this is the simplest case of those evaluated, and the CPU time was considerably high, a two-step relaxation approach was proposed to increase the computational efficiency. The proposed approach works as follows. In the first stage, the model composed of expressions (1–23) is solved as a linear formulation, relaxing all the binary variables. In a second stage, the original model is solved, but variables $I_{jt}$ and $I_{pt}$ are fixed to zero in those cases having such value in the relaxed model solution. The remaining binary variables – those having a value greater than zero in the relaxed model solution – are given freedom to take either one or zero values. This relaxation approach proved to be quite effective in all the examples that have been solved.

Figure 1 depicts the optimal solution of the base case, which was obtained in 1565 s. Thus, the same solution was reached with a reduction of 52.33% in the CPU time. During the initial year seven collection centers (red dots) and two treatment plants (black squares) are installed, and they operate along the whole planning horizon. Two collection centers are of the smallest size, with a maximum capacity of 5,000 containers per period. The remaining five correspond to the larger size, being able to handle up to 10,000 containers per period. One small and one large treatment plant are installed, with a maximum capacity of 60,000 and 100,000 empty containers, respectively.

**Fig. 1.** Optimal solution for the base case problem. Seven collection centers (red dots) and two treatment plants (black squares) are installed previous to the beginning of the planning horizon.

In order to analyze the model scalability and to test its response to changing conditions – i.e. to make a simple sensitivity analysis, – several additional scenarios have been solved using different data sets. Tables 1 and 2 summarize some of the scenarios that have been addressed. All the examples were solved to optimality if the optimal solution was found within 10000 s. Otherwise, a 5% gap was fixed as the stopping criterion, with the exception of one case in which a 10% gap was adopted. The sizes of the resulting models, as well as the computational results, are presented in Table 3. This table reports the statistics corresponding to the relaxation methodology. However, it should be remarked that in all the cases, the same solutions have been obtained when the speed-up approach is omitted, i.e. when solving the regular model. The following paragraphs present an analysis of the 10 examples that have been tackled.

Scenario 1 duplicates the number of farms with respect to the base case, but keeps constant the number of potential collection centers and plastic treatment plants. As shown

**Table 1.** Summary of some of the different dimension analyzed scenarios.

| | Farms | Possible collection centers | Possible treatment plants | Horizon length |
|---|---|---|---|---|
| Base case | 300 | 20 | 8 | 6 |
| Scen. 1 | 600 | 20 | 8 | 6 |
| Scen. 2 | 300 | 20 | 8 | 12 |
| Scen. 3 | 300 | 20 | 16 | 6 |
| Scen. 4 | 300 | 40 | 8 | 6 |

**Table 2.** Summary of some of the different analyzed scenarios disturbing parameters.

| | Disturbed parameter | Direction and intensity of disturbance |
|---|---|---|
| Scen. 5 | Containers generation | 50% reduction in period 3 |
| Scen. 6 | Containers generation | Generalized gradual decrease |
| Scen. 7 | Transportation costs | 10% increase |
| Scen. 8 | Transportation costs | 20% increase |
| Scen. 9 | Installation costs | 10% increase |
| Scen. 10 | Installation costs | 20% increase |

in Table 3, the number of binary decision variables does not increase, but the continuous ones almost duplicate, as well as the number of constraints. No optimal solution was obtained within 10000 s, but a solution with a 5% of relative gap was reached in 3099 s. The attained solution comprises the same number of collection centers as the base case solution, but has two more treatment plants. All the installed collection centers are of the large size. Regarding treatment plants, two small and two large are selected. In this way the capacity of the network is increased in order to serve a greater number of farms. All facilities are installed during the first period and operate along the whole planning horizon.

Scenario 2 takes into account the same superstructure as the base case, but doubles the length of the planning horizon. This leads to duplication in the number of continuous and binary variables, as well as the number of constraints. This more challenging scenario has been tackled with two different solution approaches. The first one considers the base case solution as given, and extends the horizon to evaluate potential new facilities. In this case, the optimal solution has been found in only 215 s. The second approach does not resort to any initial network and the 12 periods are taken into account from the beginning. A solution with a gap of 10% has been reached in 9758 s. The same solution (see Fig. 2) is obtained in both cases. It comprises twelve facilities: 9 collection centers and 3 treatment plants, all of which are installed during the initial period, with the exception of one treatment plant, which is built in period 5. The sizes of the various facilities are summarized in Table 4.

**Table 3.** Computational results corresponding to the different scenarios.

| | Total variables | Binary variables | Total constraints | Objective function | CPU s | Relative gap |
|---|---|---|---|---|---|---|
| Base case | 52,335 | 392 | 2,855 | 1,582,355.44 | 1,565 | 0.00 |
| Scen. 1 | 119,848 | 392 | 5,184 | 2,850,718.12 | 3,099 | 0.05 |
| Scen. 2 | 113,326 | 728 | 6,558 | 1,940,178.00 | 215* | 0.00 |
| Scen. 2 | 113,326 | 728 | 6,558 | 1,940,178.00 | 9,758# | 0.10 |
| Scen. 3 | 79,320 | 504 | 3,832 | 1,582,355.44 | 4,403 | 0.00 |
| Scen. 4 | 104,908 | 672 | 4,484 | 1,511,971.51 | 9,715 | 0.05 |
| Scen. 5 | 52,335 | 784 | 2,855 | 1,559,379.32 | 3,489 | 0.00 |
| Scen. 6 | 52,335 | 784 | 2,855 | 1,462,987.45 | 1,634 | 0.00 |
| Scen. 7 | 52,335 | 784 | 2,855 | 1,542,416.85 | 1,258 | 0.00 |
| Scen. 8 | 52,335 | 784 | 2,855 | 1,571,170.39 | 1,558 | 0.00 |
| Scen. 9 | 52,335 | 784 | 2,855 | 1,618,496.74 | 1,308 | 0.00 |
| Scen. 10 | 52,335 | 784 | 2,855 | 1,723,330.16 | 2,489 | 0.00 |

*Base Case solution taken for the six initial periods
#No solution given to the solver, 12 periods considered all at once.

**Fig. 2.** Optimal solution corresponding to scenario 2. Nine collection center (red dots) and three treatment plants (black squares) are installed prior to the beginning of the planning horizon.

A comparison of the solutions corresponding to the base case and scenario 2 shows that the effect of having a longer planning horizon is to increase the investment in the installation of facilities with the aim of reducing transportation costs, that otherwise would have been much higher. In fact, the base case solution has as its main cost components the installation and transportation ones, with a participation of 56.6% and 18.17%, respectively. In the case of scenario 2 solution, the participation of these components changes to 52.99% and 20.98%, respectively, despite the fact that three more facilities are installed.

Scenario 3 duplicates the number of potential treatment plants in relation to all the previous cases, but keeps constant the number of farms, collection centers and the

**Table 4.** Facilities to be installed in each one of the different scenarios.

| | Collection centers | Treatment plants | Size of collection centers | Size of treatment plants |
|---|---|---|---|---|
| Base case | 7 | 2 | 2 small, 5 large | 1 small, 1 large |
| Scen. 1 | 7 | 4 | All large | 2 small, 2 large |
| Scen. 2 | 9 | 3 | 5 small, 4 large | 2 small, 1 large |
| Scen. 3 | 7 | 2 | 2 small, 5 large | 1 small, 1 large |
| Scen. 4 | 12 | 2 | 9 small, 3 large | 1 small, 1 large |
| Scen. 5 | 7 | 2 | 2 small, 5 large | 1 small, 1 large |
| Scen. 6 | 6 | 2 | 1 small, 5 large | 1 small, 1 large |
| Scen. 7 | 7 | 2 | 2 small, 5 large | 1 small, 1 large |
| Scen. 8 | 7 | 2 | 2 small, 5 large | 1 small, 1 large |
| Scen. 9 | 7 | 2 | 2 small, 5 large | 1 small, 1 large |
| Scen. 10 | 7 | 2 | 2 small, 5 large | 1 small, 1 large |

length of the planning horizon with respect to the base case. The 16 possible treatment plants include the 8 ones associated with the base case. Provided the model has a bigger superstructure (504 versus 392 binary variables) with respect to the base case, it has the opportunity of finding a solution with a lower objective function value, or at least to reach the same solution as in the base case. The same solution as the base case one was reached in 4403 s, which indicates that in this case the new potential treatment plants do not lead to a solution improvement.

Along the same ideas, scenario 4 duplicates the number of potential collection centers in relation to all the previous cases, but keeps constant the number of farms, treatment plants and the length of the planning horizon with respect to the base case. As shown in Table 3, no optimal solution was obtained within 10000 s, but a solution having a 5% of relative gap has been reached in 9715 s. Given that the 40 potential collection centers include the 20 ones that correspond to the base case, a solution equivalent to the base case one, or better, was expected. In this case, the additional potential collection centers lead to an improvement in the base case solution. The new network comprises 5 extra collection centers, reducing the total discounted cost of the network due to savings in the transportation costs. In addition, the 5% gap indicates that further improvements can be obtained. Scenarios 3 and 4 allow concluding that the greater the number of potential nodes included in the superstructure, the greater the chances of obtaining good quality solutions. As expected, the increase in the number of binary variables demands an additional computational effort.

Scenarios 5 to 10 take into account the same superstructure as in the base case, but propose disturbances in different parameters to test the sensibility of the model to the changes. Scenarios 5 and 6 perturb the generation of empty containers from farms. In the first case, the generation of the base case is reduced by 50% in period 3, simulating an epoch with a reduced usage of agrochemicals due to changes in climatic conditions (drought). On the other hand, scenario 6 considers a gradual decrease in the generation of empty containers, in the same proportion for every farm, starting from period 3 up to the end of the planning horizon.

The optimal solution corresponding to scenario 5 was reached in 3489 s of CPU. It has the same facilities (see Fig. 1) identified in the base case solution. However, in this perturbed scenario two of the collection centers do not operate in period 3. The facilities that work in this period are the ones shown in Fig. 3. In turn, the optimal solution associated with scenario 6 was reached in 1634 s of CPU. It comprises six collection centers, one less than the base case solution. The reduction of facilities is due to the fact that fewer empty containers are generated along the planning horizon.

**Fig. 3.** Solution associated with scenario 6, showing the facilities that operate during period 3: Five collection centers (red dots) and two treatment plants (black squares).

Scenarios 7 and 8 disturb the transportation cost, by proposing an increase of 10% and 20%, respectively. The structure of the solutions that have been obtained in both cases is the same as the one reached for the base case. Nevertheless, the total discounted cost of the network is higher due to the rise in transportation costs. These results show that the increments in the transportation costs are not yet enough to produce a modification in the network topology.

Finally, scenarios 9 and 10 disturb the installation costs of both types of facilities, by proposing an increase of 10% and 20%, as compared to the ones of the base case, respectively. The optimal solution corresponding to scenario 9 was reached in 1308 s of CPU, whereas the one of scenario 10 required 2489 s of CPU. In both cases, the structure of the solution that was obtained is the same as the one associated with the base case. As expected, the total cost of the attained networks exhibit an increase, due to the rise in the installation costs of the facilities. Similarly to what happened in scenarios 7 and 8, the augmentations in the installation costs are not yet enough to produce a modification in the number and capacity of those facilities to be installed. As already presented, the base case solution has as its main cost components the installation and transportation ones, with a participation of 56.6% and 18.17%, respectively. In turn, the solution of scenario 9 has the following cost participation: 60.33% corresponding to installation costs and 17.8% to transportation ones. These percentages change to 61.4% and 16.7% in the case of the solution of scenario 10.

## 6 Concluding Remarks and Future Work

An MILP model for a multi-period reverse logistics network design problem has been presented. It takes into account several features that become relevant in supply chain configuration problems, such as: multi-period consideration, a three echelon network, discrete sizes for two facility types, minimum and maximum capacities for each of the available sizes of such facilities, as well as facilities that may have been already installed. In addition, the model considers variable costs per unit processed in the treatment plants, fixed operational costs and installation costs for all the available sizes of each facility type. Transportation costs only consider the distance traveled and omit the load of the vehicles, assuming that most of the trips are made with less than the maximum allowed load.

In order to reduce the computational load, a relaxation approach has been proposed. It rendered significant reductions in the CPU times without degrading, in any case, the quality of the attained solutions.

The model has been tested under various scenarios to analyze scalability and sensitivity issues. The obtained results have been satisfactory. Nevertheless, further testing will be done in the future to better study these aspects. For instance, a time horizon with periods of shorter duration and a superstructure with more nodes will be tackled. In addition, the solutions will be evaluated under more demanding conditions of change in the model parameters. This would allow to better understand the tradeoffs between installation and transportation costs. Finally, the incorporation of environmental and social objectives, in addition to the economic ones, will be pursued in the future. This will naturally lead to a multi-objective formulation.

## References


1. Dekker, R., Fleischmann, M., Inderfurth, K., Van Wassenhove, L.: Reverse Logistics Quantitative Models for Closed-Loop Supply Chains. Springer Verlag, Berlin (2004). https://doi.org/10.1007/978-3-540-24803-3
2. Melo, M., Nickel, S., Saldanha-da-Gama, F.: Facility location and supply chain management - a review. Eur. J. Oper. Res. **196**, 401–412 (2009). https://doi.org/10.1016/j.ejor.2008.05.007
3. Mota, B., Gomes, M., Carvalho, A., Barbosa-Povoa, A.: Towards supply chain sustainability: economic, environmental and social design and planning. J. Clean. Prod. **105**, 14–27 (2015). https://doi.org/10.1016/j.jclepro.2014.07.052
4. Salema, M.G., Barbosa-Povoa, A., Novais, A.: An optimization model for the design of a capacitated multi-product reverse logistics network with uncertainty. Eur. J. Oper. Res. **179**, 1063–1077 (2007). https://doi.org/10.1016/j.ejor.2005.05.032
5. Alumur, S., Nickel, S., Saldanha-da-Gama, F., Verter, V.: Multi-period reverse logistics network design. Eur. J. Oper. Res. **220**, 67–78 (2012). https://doi.org/10.1016/j.ejor.2011.12.045
6. Banguera, L., Sepúlveda, J., Ternero, R., Vargas, M., Vásquez, O.: Reverse logistics network design under extended producer responsibility: the case of out-of-use tires in the Gran Santiago city of Chile. Int. J. Prod. Econ. **205**, 193–200 (2018). https://doi.org/10.1016/j.ijpe.2018.09.006

7. Lagarda-Leyva, E.A., Morales-Mendoza, L.F., Ríos-Vázquez, N.J., et al.: Managing plastic waste from agriculture through reverse logistics and dynamic modeling. Clean Technol. Environ. Policy **21**, 1415–1432 (2019). https://doi.org/10.1007/s10098-019-01700-5
8. Sorichetti, A., Mammini, L., Savoretti, A., Bandoni, A.: Gestión de envases vacíos de agroquímicos, dos propuestas para el Sudoeste Bonaerense. Simposio Argentino de Informática Industrial e Investigación Operativa, pp. 55–70, Buenos Aires (2018)

# Two-Stage Stochastic Optimization Model for Personnel Days-off Scheduling Using Closed-Chained Multiskilling Structures

Orianna Fontalvo Echavez, Laura Fuentes Quintero, César Augusto Henao(✉), and Virginia I. González

Universidad del Norte, Barranquilla, Colombia
{oriannaf,lfuentesc,cahenao,vvirginia}@uninorte.edu.co

**Abstract.** This paper addresses a days-off scheduling problem that incorporates multiskilling decisions using a 2-chaining policy. The generated closed-chained multiskilling structures minimize the expected training and over/understaffing costs, while assigning rest days between working days to single-skilled and multiskilled employees. This research provides a suitable methodology for a wide range of industries where the main problem is to meet the demand requirements the seven days of the week, but where the employees cannot work daily. The methodology is structured in two steps. First, we develop a deterministic mixed integer linear programming model. Second, the deterministic model is reformulated as a two-stage stochastic optimization model in order to explicitly incorporate demand uncertainty. Our methodology was applied in a case study associated to a Chilean retail store. We use real data but also simulated data to represent various demand variability scenarios. Results showed that the model was able to do the days-off scheduling decisions cost-effectively for a planning horizon of two weeks. Additionally, it was found that the levels of understaffing and overstaffing can vary between weekdays (Monday to Friday) and weekend. Finally, we also observed that the 2-chaining policy was cost-effective in all scenarios; however, when there are very high levels of demand variability, it can be interesting to explore $k$-chaining with $k \geq 2$ and/or to hire more employees.

**Keywords:** Multiskilling · Chaining · Days-off scheduling · Workforce flexibility · Retail services

## 1 Introduction

The service sector currently contributes more than 60% of production and employment in the world and has a share in international trade of approximately 20% [1], so their systems should be optimized and highly studied. One of the processes with the greatest impact on a service organization is the personnel scheduling. However, in this sector is complex to establish effective personnel scheduling systems that ensure an adequate customer service level (CSL) and make a most efficient use of resources due to the high level of uncertain demand and unscheduled staff absenteeism. These phenomena

© Springer Nature Switzerland AG 2021
D. A. Rossit et al. (Eds.): ICPR-Americas 2020, CCIS 1407, pp. 19–32, 2021.
https://doi.org/10.1007/978-3-030-76307-7_2

produce periods of over- and under-staffing (i.e., staffing levels that are higher or lower than required, respectively) that can directly lead to an increase of labor costs and, in turn, deteriorate the customer service levels [2–4].

A source of labor flexibility, supported by the literature, that seeks to counteract the effects of staff shortage/surplus in a service company is to incorporate multiskilled employees who can work on multiple task types (or store departments) [2, 5]. In addition, chaining is a cost-effective policy that allows to transfer available multiskilled employees from overstaffed departments to understaffed departments. In this policy, a certain number of employees are trained to work in two or more departments, such that these allocation decisions are configured through a bipartite graph that involves different chain types. Particularly, the 2-chaining policy is based on the fact that certain employees are trained to work in a maximum of two departments. This policy has been shown to be one of the most cost-effective and recommended in the literature, because it does not lead to the over-training of the employees and optimizes the multiskilling approach [6].

Also, as mentioned above, the uncertainty in the staff demand is a relevant phenomenon in the context of the personnel scheduling problem in the service companies. Thus, to make decisions under uncertain demand, is useful to apply a two-stage stochastic optimization (TSSO) model that explicitly considers such uncertainty.

Given the above, the intended contribution of this paper to the personnel scheduling problem in the service sector is a model that determines the multiskilling levels that minimize the expected costs associated with training and over/understaffing. The proposed methodology is structured in two steps. First, a days-off scheduling problem is initially formulated as a deterministic mixed integer linear programming model (MILPM), which assigns rest days between working days to single-skilled and multiskilled employees over a two-week planning horizon. Note that, since 2-chaining policies have shown to be very effective in terms of providing flexibility to address uncertain demands, the training plan for employees will be structured through closed chains. Second, the MILPM is reformulated as a TSSO model in order to explicitly incorporate demand uncertainty. Then, our methodology is applied to a case study of a Chilean retail store by using a commercial software. The performance of the optimization approach is assessed for various demand variability scenarios in order to measure the benefits of the closed-chain multiskilling structures to face uncertainty.

## 2 Literature Review

There is a wide range of authors who proposed different mathematical models to solve the personnel scheduling problems. Table 1 characterizes a series of articles published in recent decades to obtain an accurate contextualization of the state of the art. The studies are described by the following characteristics:

(i) *Personnel scheduling problem* (PSP): based on the classification made by Henao et al. [7], we indicate the type of problem addressed in each study. May be: (a) staffing (S); (b) shift scheduling (SS); (c) days-off scheduling (DOS); (d) tour scheduling (TS); or (e) assignment (A).
(ii) *Multiskilling* (MS): this criterion indicates whether the multiskilling was modeled through a parameter (Par), i.e. as model input, or a decision variable (Var).

(iii) *Chaining* (CH): according to the classification presented in Henao et al. [7], the chain types used in each study are indicated. These could be: (a) closed long chain (CLC), one multiskilled employee per task type forming one chain that connects all task types; (b) closed short chain (CSC), a subset of multiskilled employees that connects just a subset of task types; (c) open chain (OC), a long chain which it was not closed; and (d) all-for-one (AxO), where all multiskilled employees are trained for assignment on a single task type.

(iv) *Workforce type* (WFT): May be: (a) homogeneous workforce (HO), which means that the productivity of an employee does not change if he or she is assigned different task types; and (b) heterogeneous workforce (HT), which means that, if employees increase the number of task types they are able to work, then productivity can decrease.

(v) *Uncertainty* (U): Indicates whether or not the problem posed by study considers: (a) demand uncertainty (D); and/or (b) supply variability (O) (i.e., unscheduled staff absenteeism).

(vi) *Methods* (ME): Indicates the solution method used. May be: (a) simulation (S); (b) heuristics (H); (c) traditional optimization (OPT); (d) analytic (AN); and (e) metaheuristics (M).

(vii) *Uncertainty modeling approach* (UMA): Indicates the modeling approach that was used to explicitly incorporates uncertainty into an optimization model. May be: (a) two-stage stochastic optimization (TSSO); (b) robust optimization (RO); and (c) closed-form equation (CF).

(viii) *Application (APP):* We have found different fields in which the authors do their research, such as hospitals (H), retail (R), airports (AI), cyber security (CS), services (S), and manufacturing (MA).

Table 1 shows that multiskilling is a source of labor flexibility very used to address the personnel scheduling problems. Complementing it with chaining undoubtedly has a high impact to minimize the levels of over/understaffing. However, even though the chaining represents a growing trend according to the state of the art, there is still a need for more work to prove its effectiveness. On the other hand, the authors focus on shift scheduling or tour scheduling, so days-off scheduling represents a gap in the literature. This represents a field of interest, since days-off scheduling is a need in a wide range of industries where the main problem is to meet the demand requirements the seven days of the week, but where the employees cannot work daily. Examples include restaurants, power plants, hospitals, and retail stores. In addition, few studies explicitly incorporated the demand uncertainty into the mathematical formulation.

The contribution of this work is twofold. First, we evaluate the impact of closed-chained multiskilling structures on days-off scheduling in the service sector, particularly for the retail industry. Second, we propose a TSSO approach to make the uncertain demand explicit. It is important to note that our modeling will consider multiskilling as a decision variable, thus, the model will decide how many employees will be multiskilled and in which store departments, while at the same time will produce a set of closed chains. This will allow to consider a much wider range of assignment possibilities and providing a much more efficient way to satisfy the staff demand.

**Table 1.** Characteristics of published studies related to the personnel scheduling problem.

| Reference | PSP | MS | CH | WFT | U | ME | UMA | APP |
|---|---|---|---|---|---|---|---|---|
| Brusco and Johns [8] | S + SS | Par | CLC,CSC | HT | – | OPT | – | S |
| Alvarez-Valdes et al. [9] | TS | – | – | HO | – | OPT + H | – | AI |
| Inman et al. [5] | S + A | Par | CLC,CSC, OC, AxO | HO | O | S | – | MA |
| Jordan et al. [10] | A | Par | CLC | HT | D | S | – | MA |
| Kabak et al. [11] | S + TS | – | – | – | – | OPT + S | – | R |
| Campbell [12] | DOS | Par | – | HT | D | OPT + H | TSSO | – |
| Paul and McDonald [13] | S + A | Var | CLC, CSC | HT | D | OPT + M | CF | H |
| McDonald [14] | A | Par | OC | HT | – | OPT | – | MA |
| Henao et al. [15] | A | Var | CLC,CSC | HO | D | OPT + H | RO | R |
| Ying and Tsai [16] | A | Par | – | HO | – | OPT + H | – | MA |
| Restrepo et al. [17] | TS | Par | – | HO | D | OPT + M | TSSO | – |
| Altner et al. [18] | S + SS | – | – | – | D | OPT + H | TSSO | CS |
| He et al. [19] | S + SS | – | – | HO | D | OPT | TSSO | H |
| Felberbauer et al. [20] | S + A | Par | – | HT | D | OPT + M | TSSO | MA |
| Altner et al. [21] | DOS | Var | – | HO | D | OPT + M | TSSO | CS |
| Henao et al. [7] | A | Var | CLC, CSC | HO | D | AN + H | CF | R |
| Porto et al. [2] | S + TS | Var | CLC,CSC | HO | D | OPT | – | R |
| *This paper* | ***DOS*** | ***Var*** | ***CLC, CSC*** | ***HO*** | ***D*** | ***OPT*** | ***TSSO*** | ***R*** |

## 3 Problem Description

The problem we propose to study consists in generating a training plan for a retail store, and then, obtaining a schedule of resting and working days for each employee over a two-week planning horizon. It is assumed that each department has hired a set of employees initially single-skilled and therefore trained to work in a single department. In more specific terms, our problem will be to solve a biweekly DOS that simultaneously decides how many single-skilled employees should be multiskilled, in which second department to form closed-chained multiskilling structures, and how to assign working and resting days to each employee. In addition, each employee can work maximum five consecutive days and must work a given number of weekly hours as specified by his/her contract. The objective is to minimize the expected costs associated with training and over/understaffing. Table 2 shows the notation used in the formulation of the problem.

**Table 2.** Sets, parameters, and variables of the days-of scheduling problem.

| Sets | |
|---|---|
| $I$ | Store employees, indexed by $i$ |
| $L$ | Store departments, indexed by $l$ |
| $I_l$ | Employees under contract in department $l$, indexed by $i$, $\forall l \in L$ |
| $S$ | Set of weeks, indexed by $s$ |
| $D$ | Days of the biweekly planning horizon, indexed by $d$, $\{M1, Tu1, W1, Th1, F1, S1, Su1, M2, Tu2, W2, Th2, F2, S2, Su2\}$ |
| $D_s$ | Set of days of the week $s$ |
| $D5_d$ | Set of consecutive days after the day $d$ |
| $D^+$ | Set of maximum consecutive days, indexed by $d$, $\{M1, Tu1, W1, Th1, F1, S1, Su1, M2, Tu2\}$ |
| Parameters | |
| $r_{ld}$ | Number of daily hours demanded in department $l$ on day $d$, $\forall l \in L, d \in D$ |
| $h$ | Number of weekly hours an employee must work according to his/her contract |
| $t$ | Number of working hours an employee must work per day |
| $m_i$ | Department for which employee $i$ was initially trained, $\forall i \in I$ |
| $\mu$ | Staff shortage cost per hour (equivalent to the expected cost of lost sales) |
| $o$ | Staff surplus cost per hour (equivalent to the opportunity cost of idle employee-hours) |
| $c$ | Training cost of an employee to work in any department |
| Variables | |
| $x_{il}$ | Equal to 1 if employee $i$ is trained for department $l$, otherwise 0, $\forall i \in I, l \in L$ |
| $v_i$ | Equal to 1 if employee $i$ is multiskilled, otherwise 0, $\forall i \in I$ |
| $\omega_{ild}$ | Number of working hours assigned to employee $i$ in department $l$ in day $d$, $\forall i \in I, l \in L, d \in D$ |
| $y_{id}$ | Equal to 1 if employee $i$ works in day $d$, otherwise 0, $\forall i \in I, d \in D$ |
| $k_{ld}$ | Staff shortage in hours in department $l$ on day $d$, $\forall l \in L, d \in D$ |
| $\delta_{ld}$ | Staff surplus in hours in department $l$ in day $d$, $\forall l \in L, d \in D$ |
| $z_i$ | Equal to 1 if employee $i$ has days off in the first weekend (S1, Su1), and equal to 0 if employee has days off in the second weekend (S2, Su2), $\forall i \in I$ |

Our formulation incorporates the following assumptions: (1) Employees can be trained in a maximum of two store departments. (2) Employees have the same level of productivity for each department (i.e., homogeneous workforce). (3) Employees may work a maximum of five consecutive days. (4) The planning horizon has 14-days (two weeks). (5) All employees must work exactly five days per week with 9-h work shifts per working day. (6) Unscheduled staff absenteeism is not considered. (7) The daily hours demanded in each department is a random parameter. (8) All employees are initially trained in single department. (9) The cost of staff shortage is the same for all departments, which is a common assumption on the field. (10) The cost of staff surplus is an opportunity cost for idle staff, and it is also equal for all departments. (11) Training costs are equal for each department.

## 4 Methodology

In this section we present a deterministic version of our mixed integer linear programming model for solving the days-off scheduling problem with chaining as just described above. Later, we will introduce the necessary modifications to transform the model into a two-stage stochastic formulation incorporating demand uncertainty for solving the same problem.

### 4.1 Deterministic Model

The deterministic model can now be formulated as follows:

$$Min \sum_{i \in I} \sum_{l \in L: l \neq m_i} c x_{il} + \sum_{d \in D} \sum_{l \in L} (\mu k_{ld} + o \delta_{ld}) \quad (1)$$

$s.t.$

$$\sum_{i \in I} \omega_{ild} + k_{ld} - \delta_{ld} = r_{ld} \; \forall l \in L, d \in D \quad (2)$$

$$\omega_{ild} \leq t x_{il} \; \forall i \in I, l \in L, d \in D \quad (3)$$

$$\sum_{l \in L} \omega_{ild} = t y_{id} \; \forall i \in I, d \in D \quad (4)$$

$$\sum_{d \in D_S} y_{id} = 5 \; \forall i \in I, s \in S \quad (5)$$

$$y_{id} + \sum_{\bar{d} \in D5_d} y_{i\bar{d}} \leq 5 \; \forall i \in I, d \in D^+ \quad (6)$$

$$y_{id} \leq 1 - z_i \; \forall i \in I, d \in D\{S1, Su1\} \quad (7)$$

$$y_{id} \leq z_i \; \forall i \in I, d \in D\{S2, Su2\} \quad (8)$$

$$x_{il} = 1 \; \forall i \in I, l \in L : l = m_i \quad (9)$$

$$v_i = \sum_{l \in L: l \neq m_i} x_{il} \; \forall i \in I \quad (10)$$

$$v_i \leq 1 \; \forall i \in I \quad (11)$$

$$\sum_{i \in I_l} v_i = \sum_{i \in \{I - I_l\}} x_{il} \ \forall l \in L \tag{12}$$

$$y_{id} \in \{0, 1\} \ \forall i \in I, d \in D \tag{13}$$

$$x_{il} \in \{0, 1\} \ \forall i \in I, l \in L \tag{14}$$

$$z_i \in \{0, 1\} \ \forall i \in I \tag{15}$$

$$\omega_{ild} \geq 0 \ \forall i \in I, l \in L, d \in D \tag{16}$$

$$k_{ld} \geq 0 \ \forall l \in L, d \in D \tag{17}$$

$$\delta_{ld} \geq 0 \ \forall l \in L, d \in D \tag{18}$$

$$v_i \geq 0 \ \forall i \in I \tag{19}$$

In this formulation the objective function (1) minimizes the following three biweekly costs: personnel training, staff shortage, and staff surplus. Constraints (2) along with the constraints (17) and (18) yield the levels of staff shortage/surplus associated with each department. Constraints (3) ensure that individual employees be assigned to work in, and only in, departments they have been trained for. Constraints (4) ensure that employees work exactly the number of hours required in each working day. Constraints from (5) to (8) implement days off scheduling constraints developed by Altner et al. [21]. Constraints (5) indicate that each employee must work five days per week. Note that, constraints (4) and (5) ensure that employees work exactly the weekly number of hours required by the employment contract, i.e., $h$. Constraints (6) indicate that each employee can work a maximum of 5 consecutive days. Constraints (7) and (8) ensure that each employee has at least one weekend off. Constraints (9) determine the department each employee was originally trained for. Constraints (10) determine whether an employee is multiskilled or single-skilled. Constraints (11) ensure that an employee is trained in a maximum of two departments. Constraints (12) indicate that the number of employees who have been trained in the department $l$ (from other departments) should be equal to the number of multiskilled employees belonging to the department $l$. Constraints (11) and (12) ensure that employees are assigned to closed-chained multiskilling structures. Finally, constraints (13)–(19) define the domain of each problem variable.

## 4.2 Two-Stage Stochastic Optimization Model

The two-stage stochastic optimization model (TSSO) arises when full knowledge about the distribution of uncertain demand is available and is defined with two variable types: first-stage and second-stage. The first-stage variables are related to decisions that are made before the realization of the uncertain parameters become available, that is, decisions such as the training of employees in a second department. The second-stage variables are related to decisions that are made after the realization of daily demand is known, that is, decisions such as personnel assignment to minimize the levels of over/under-staffing. The second-stage variables depend on the first-stage variables and the uncertain parameter.

To solve this problem the sample average approximation (SAA) approach will be implemented. This approach creates one set of second-stage variables for every possible scenario of demand. The quality of the SAA is related to how large is the set of samples. The set of scenarios comes from the continuous distribution of the random parameters $\xi$ represented by $\xi \in \Xi$, where $\Xi$ is the set of scenarios generated, which forms the SAA problem. Below, we present the TSSO version of the deterministic formulation (1)–(19), written as follows:

$$Min \sum_{i \in I} \sum_{l \in L: l \neq m_i} c x_{il} + \frac{1}{|\Xi|} \left[ \sum_{\xi \in \Xi} \sum_{d \in D} \sum_{l \in L} \mu k_{ld}(\xi) + o \delta_{ld}(\xi) \right] \quad (20)$$

$s.t.$

$$\sum_{i \in I} \omega_{ild}(\xi) + k_{ld}(\xi) - \delta_{ld}(\xi) = r_{ld}(\xi) \; \forall l \in L, d \in D, \xi \in \Xi \quad (21)$$

$$\omega_{ild}(\xi) \leq t x_{il} \; \forall i \in I, l \in L, d \in D, \xi \in \Xi \quad (22)$$

$$\sum_{l \in L} \omega_{ild}(\xi) = t y_{id} \; \forall i \in I, d \in D, \xi \in \Xi \quad (23)$$

$$\sum_{d \in D_S} y_{id} = 5 \; \forall i \in I, s \in S \quad (24)$$

$$y_{id} + \sum_{\bar{d} \in D5_d} y_{i\bar{d}} \leq 5 \; \forall i \in I, d \in D^+ \quad (25)$$

$$y_{id} \leq 1 - z_i \; \forall i \in I, d \in D\{S1, Su1\} \quad (26)$$

$$y_{id} \leq z_i \; \forall i \in I, d \in D\{S2, Su2\} \quad (27)$$

$$x_{il} = 1 \; \forall i \in I, l \in L : l = m_i \quad (28)$$

$$v_i = \sum_{l \in L: l \neq m_i} x_{il} \; \forall i \in I \quad (29)$$

$$v_i \leq 1 \; \forall i \in I \quad (30)$$

$$\sum_{i \in I_l} v_i = \sum_{i \in \{I - I_l\}} x_{il} \; \forall l \in L \quad (31)$$

$$y_{id} \in \{0, 1\} \; \forall i \in I, d \in D \quad (32)$$

$$x_{il} \in \{0, 1\} \; \forall i \in I, l \in L \quad (33)$$

$$z_i \in \{0, 1\} \; \forall i \in I \quad (34)$$

$$\omega_{ild}(\xi) \geq 0 \; \forall i \in I, l \in L, d \in D, \xi \in \Xi \quad (35)$$

$$k_{ld}(\xi) \geq 0 \; \forall l \in L, d \in D, \xi \in \Xi \quad (36)$$

$$\delta_{ld}(\xi) \geq 0 \; \forall l \in L, d \in D, \xi \in \Xi \quad (37)$$

$$v_i \geq 0 \; \forall i \in I \quad (38)$$

The objective function (20) aims to minimize the training cost (first-stage decisions) and the expected biweekly cost of staff shortages and surpluses (second-stage decisions). In this TSSO model the decision about which employees will be multiskilled is done in the first stage and this will affect the levels of staff shortage/surplus in the second stage. The main objective in the TSSO is to choose the first-stage variables in order to minimize the sum of first stage costs and the expected value of the random second stage costs. Constraints (21)–(38) are accordingly extended from (2)–(19) to consider the uncertainty demand and scenario-dependent second-stage variables.

## 5 Case Study, Results, and Discussion

This section presents the case study used to validate our methodology, as well as its computational implementation.

### 5.1 Case Study

The case study is based on real and simulated data, which is derived from a retail store in Chile. This retail store currently has employees hired in specific departments. Each department currently constitutes a store business unit and at that level is where the store's training and employee scheduling decisions are made.

The real data corresponds to the information related to the store size, the number of departments, the number of employees hired in each department, employment-contract characteristics, mean values on the daily demand in each department, as well as costs related to the model. Regarding the simulated data, these include information about the random parameter of daily hours demand in each store department, which corresponds to the uncertain parameter in our model. The following is a detailed description of the data.

**Real Data.** The case study considers a retail store with 6 departments and 30 employees, that is, $|L| = 6$ and $|I| = 30$. The employment contract stipulates that an employee must comply with 45 h per week ($h = 45$) and 9 h per working day ($t = 9$). Regarding the cost parameters, there is a minimal training cost value per employee ($c = US\$1 biweek/employee$), which has been estimated according to Henao et al. [7, 15, 22]. This minimal training cost allows interpreting the results as an upper bound on the potential contribution of multiskilling to store performance [2, 23]. Finally, the costs related to under/over-staffing are estimated on an hourly basis and are $u = US\$60/h$ and $s = US\$15/h$, respectively.

**Simulated Data.** Since the TSSO model (20)–(38) is solved by a SAA method, this implies that we use a Monte Carlo simulation to randomly generate 60 scenarios for the parameters of daily hours demand in each department $r_{ld}(\xi), \forall l \in L, \forall d \in D, \xi \in \Xi$ such that $\Xi = 60$. We consider that the probability density function (pdf) for the demand of each department follows a normal distribution and it is zero-truncated to prevent generating negative demand values. Finally, in the simulation Monte Carlo, we consider three representative variability levels for demand in a department: coefficient of variation $CV = 10\%, 30\%, 50\%$.

## 5.2 Results and Discussion

This section presents the results of the TSSO model presented in Sect. 4.2 according to the conditions presented in the case study described in Sect. 5.1. The model was coded in AMPL and solved using the commercial software CPLEX 12.9.0 running through NEOS-server for 8 h (restrictive). Considering the running conditions, we obtained optimality gaps of less than 0.3%.

The results and discussion of our methodology are divided into the following three subsections: models' size and computational times; multiskilling results; and days-off scheduling results. The last two subsections only present results associated with the TSSO model.

**Model's Size and Computational Times.** As it was mentioned before, the run time of the TSSO model was restricted to 8 h. Thus, for each coefficient of variation of the staff demand ($CV = 10\%, 30\%$ and $50\%$) the same computational time was maintained. Regarding the deterministic model, the running time was of approximately 14.3 s for each CV. Table 3 presents the differences in model size between the deterministic model (DT) and the associated TSSO model.

**Table 3.** Comparison between the deterministic and stochastic models.

| Element | TSSO | DT |
|---|---|---|
| Constraints | 156 738 | 3 102 |
| Binary variables | 600 | 600 |
| Continuous variables | 161 322 | 2 730 |

**Multiskilling results.** The purpose of this subsection is to evaluate the multiskilling requirements associated with each CV. Thus, for each CV, these requirements were measured through three metrics and are presented in Table 4. First, we present the number of multiskilled employees required in each store department. Second, we calculate the percentage of multiskilled employees ($\%ME$), which represents the number of multiskilled employees required, in relation to the total number of hired employees. Third, the percentage of total multiskilling ($\%TM$) is calculated, representing the number of

**Table 4.** Multiskilling results and metrics with different demand uncertainty levels.

| | Store department | | | | | | | |
|---|---|---|---|---|---|---|---|---|
| | 1 | 2 | 3 | 4 | 5 | 6 | | |
| *%CV* | No. multiskilled employees | | | | | | *%ME* | *%TM* |
| **10** | 3 | 2 | 2 | 2 | 2 | 3 | 47 | 9 |
| **30** | 5 | 5 | 2 | 3 | 3 | 5 | 77 | 15 |
| **50** | 7 | 5 | 3 | 3 | 4 | 8 | 100 | 20 |

additional skills trained, in relation to the theoretical maximum number feasible (at the store level).

We observed that the number of multiskilled employees per store department, %ME, and %TM increase as the %CV of demand increases. This result is intuitive because the greater the variability in demand, there will be greater opportunities to transfer multiskilled employees from overstaffed departments to understaffed departments.

In detail, when demand variability is the lowest (i.e., CV = 10%), the %ME is almost equal to 50%. In turn, when demand variability is equal to 30%, the %ME is equal to 77%. Finally, when demand variability is the highest (i.e., CV = 50%), the optimal number of multiskilled employees per department is equal to its total staff, i.e., ME = 100%. This result is interesting since it indicates that when there are very high levels of demand variability, a 2-chaining policy cannot be sufficient to minimize the costs of over/understaffing. That is, it can be interesting to evaluate two new strategies: (1) to explore *k*-chaining policies where certain employees can be trained to work in two or more departments [24]; and (2) to hire more employees.

**Days-off scheduling results.** For the second week of our planning horizon and each CV in the demand, Fig. 1 shows a comparison between the number of employees assigned and the over/understaffing hours in each store department. A plot of the first week, not shown in the figure, would indicate the same general tendency. Figure 1 shows that those days where there is more staff demand (i.e., weekend), a greater number of employees are assigned.

In detail, when $CV = 10\%$, we observed that from Monday to Friday, less employees are assigned in comparison to the weekend; this is reasonable given that on Saturday and Sunday there are higher personnel requirements. But as $CV$ increases, the number of employees assigned to weekdays (Monday to Friday) versus weekends become similar. This result is not surprising, since at higher levels of variability in demand, the model decides to assign almost the same amount of staff every day, which is a conservative decision. In addition, this makes sense since the over/understaffing costs do not vary according to the day of the week.

Regarding of the demand satisfaction, Fig. 1 shows that as CV increases, the levels of over/understaffing are also greater. This is also a predictable behavior, since, if there is greater demand uncertainty, it is more difficult to adjust the staff supply to the staff demand. Finally, the high levels of staff shortage during the weekend suggest that it is

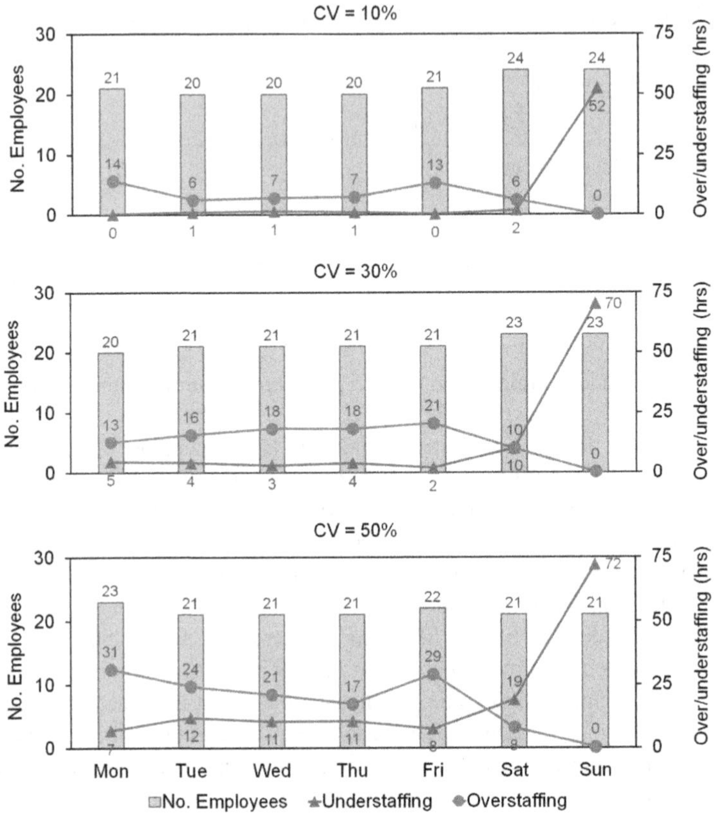


**Fig. 1.** Comparison between the number of employees assigned and the over/understaffing hours in each department for different levels of uncertainty in the demand.

convenient to hire more employees and/or to allow multiskilled employees to be trained in more than one additional department.

## 6 Conclusions

This research presents a new methodology to determine the multiskilling levels that minimize the expected costs associated with training and over/understaffing. In a novel way we ensure the training plans using the approach 2-chaining. The problem consists in assigning rest days between working days to single-skilled and multiskilled employees over a two-week planning horizon. The proposed methodology initially formulates a deterministic mixed integer linear programming model, and secondly, presents a stochastic version of the model (TSSO). The TSSO model is validated through a case study associated to a Chilean retail store. The performance of the TSSO model is assessed for various demand variability scenarios in order to measure the benefits of the closed-chain multiskilling structures to face uncertainty.

Regarding the results associated with the model's size and its computational times, it was found that the stochastic version of the model led to an increase in the computational times. However, the TSSO times are acceptable since personnel training decisions are tactical-strategic and, thus, made in medium and long-term planning horizons.

Regarding the results associated with the multiskilling requirements, we observe an intuitive result, since the required number of multiskilled employees grows with the demand coefficient of variation. In addition, we found that when there are very high levels of demand variability, it can be interesting to explore $k$-chaining policies where certain employees can be trained to work in two or more departments.

Regarding the results associated with the days-off scheduling problem, we observe (as expected) that those days with the highest demand also have a greater number of assigned employees. Thus, there is a greater staff supply on weekends. In addition, we can also observe that when there are very high levels of demand variability, increases in multiskilling are not enough, since increases in capacity are also required to minimize the levels of staff shortage. That is, it is necessary to hire more employees.

As regards future research, we think it would be valuable to include into the model staffing decisions, where the amount of personnel to be hired is decided. Also, it can be interesting to evaluate $k$-chaining policies where certain employees are trained to work in $k$ departments with $k \geq 2$. Finally, the TSSO model could also explicitly incorporate the uncertainty in the staff supply, that is, unscheduled staff absenteeism.

## Reference


1. Kinfemichael, B., Morshed, A.M.: Unconditional convergence of labor productivity in the service sector. J. Macroecon. **59**, 217–229 (2019)
2. Porto, A.F., Henao, C.A., López-Ospina, H., González, E.R.: Hybrid flexibility strategic on personnel scheduling: retail case study. Comput. Indus. Eng. **133**, 220–230 (2019)
3. Mac-Vicar, M., Ferrer, J.C., Muñoz, J.C., Henao, C.A.: Real-time recovering strategies on personnel scheduling in the retail industry. Comput. Indus. Eng. **113**, 589–601 (2017)
4. Álvarez, E., Ferrer, J.C., Muñoz, J.C., Henao, C.A.: Efficient shift scheduling with multiple breaks for full-time employees: a retail industry case. Comput. Indus. Eng. **150**, 106884 (2020)
5. Inman, R.R., Jordan, W.C., Blumenfeld, D.E.: Chained cross-training of assembly line workers. Int. J. Product. Res. **42**(10), 1899–1910 (2004)
6. Wang, X., Zhang, J.: Process flexibility: a distribution-free bound on the performance of k-chain. Oper. Res. **63**(3), 555–571 (2015)
7. Henao, C.A., Muñoz, J.C., Ferrer, J.C.: Multiskilled workforce management by utilizing closed chains under uncertain demand: a retail industry case. Comput. Indus. Eng. **127**, 74–88 (2019)
8. Brusco, M., Johns, T.: Staffing a multiskilled workforce with varying levels of productivity: an analysis of cross-training policies. Decis. Sci. **29**(2), 499–515 (1998)
9. Alvarez-Valdes, R., Crespo, E., Tamarit, J.M.: Labour scheduling at an airport refueling installation. J. Oper. Res. Soc. **50**(3), 211–218 (1999)
10. Jordan, W., Inman, R., Blumenfeld, D.: Chained cross-training of workers for robust performance. IIE Trans. **36**(10), 953–967 (2004)
11. Kabak, O., Ulengin, F., Aktas, E., Onsel, S., Topcu, I.: Efficient shift scheduling in the retail sector through two-stage optimization. Eur. J. Oper. Res. **184**(1), 76–90 (2008)
12. Campbell, G.M.: A two-stage stochastic program for scheduling and allocating cross-trained workers. J. Oper. Res. Soc. **62**, 1038–1047 (2011)

13. Paul, J., MacDonald, L.: Modeling the benefits of cross-training to address the nursing shortage. Int. J. Product. Econ. **150**, 83–95 (2014)
14. McDonald, T.: Analysis of a worker assignment model with skill chaining. In: Proceedings of IIE Annual Conference, Norcross (2015)
15. Henao, C.A., Ferrer, J.C., Muñoz, J.C., Vera, J.: Multiskilling with closed chains in a service industry: A robust optimization approach. Int. J. Product. Econ. **179**, 166–178 (2016)
16. Ying, K., Tsai, Y.: Minimizing total cost for training and assigning multiskilled workers in seru production systems. Int. J. Product. Res. **55**(10), 2978–2989 (2017)
17. Restrepo, M.I., Gendron, B., Rousseau, L.M.: A two-stage stochastic programming approach for multiactivity tour scheduling. Eur. J. Oper. Res. **262**(2), 620–635 (2017)
18. Altner, D.S., Rojas, A.C., Servi, L.D.: A two-stage stochastic program for multi-shift, multi-analyst, workforce optimization with multiple on-call options. J. Sched. **21**(5), 517–531 (2018)
19. He, F., Chaussalet, T., Qu, R.: Controlling understaffing with conditional value-at-risk constraint for an integrated nurse scheduling problem under patient demand uncertainty. Oper. Res. Perspect. **6**, (2019)
20. Felberbauer, T., Gutjahr, W.J., Doerner, K.F.: Stochastic project management: multiple projects with multi-skilled human resources. J. Sched. **22**, 271–288 (2019)
21. Altner, D.S., Mason, E.K., Servi, L.D.: Two-stage stochastic days-off scheduling of multi-skilled analysts with training options. J. Comb. Optim. **38**(1), 111–129 (2019)
22. Henao, C.A., Muñoz, J.C., Ferrer, J.C.: The impact of multi-skilling on personnel scheduling in the service sector: a retail industry case. J. Oper. Res. Soc. **66**(12), 1949–1959 (2015)
23. Porto, A.F., Henao, C.A., López-Ospina, H., González, E.R., González, V.I.: Dataset for solving a hybrid flexibility strategy on personnel scheduling problem in the retail industry. Data Brief **32**, 106066 (2020)
24. Abello, M.A., Ospina, N.M., De la Ossa, J.M., Henao, C.A., González, V.I.: Using the k-chaining approach to solve a stochastic days-off-scheduling problem in a retail store. In: Rossit, D.A., Tohmé, F., Mejía, G. (eds.) Production Research. ICPR-Americas 2020. Communications in Computer and Information Science, vol. 1407, Springer Nature Switzerland AG (2021)

# An Optimization Model for University Course Timetabling. A Colombian Case Study

Jaén Suárez-Rodríguez, Juan C. Piña, Laura Malagón-Alvarado, Valentina Blanco, Melissa Correa, Laura De La Rosa, Mariana Lopera, Juan Valderrama, and Carlos A. Vega-Mejía(✉)

Semillero Logística Empresarial, Operations and Supply Chain Management Research Group, Escuela Internacional de Ciencias Económicas y Administrativas, Universidad de La Sabana, Chía, Colombia
carlos.vega6@unisabana.edu.co

**Abstract.** Course Timetabling at universities can be defined as the scheduling of different courses in specific time slots. To carry out this process it is necessary to take into account key factors such as: available courses, the distribution of time slots and lecturers' availability. In addition, there are resources that are essential for the model performance. For instance, the available classrooms and their capacity to completely satisfy the specific demand of students. In this study, a mixed integer linear programming model is presented to find the optimal schedule of the courses. Results show that the proposed model provides feasible solutions where all the proposed factors are considered.

**Keywords:** University course timetabling · Scheduling · Mixed integer linear programming · Classroom programming

## 1 Introduction

Scheduling has been widely used in educational institutions due to its importance for optimizing physical and human resources [1]. Institutions such as universities, which have a high demand for subjects and students, must organize each of these resources in the best way to comply with all the quality conditions they have offer and are accountable for. Not having a systematic methodology to generate feasible schedules efficiently may result in the use of human resources to carry out the task manually. This situation worsens when considering all the different conditions given to determine the viability of schedules to be used.

For educational institutions, organizing academic schedules is not an easy task. Moreover, it is not unusual that budget for these processes is scarce, meaning that an efficient planning schedule must be carried out by qualified staff and with minimum cost. Lawrie [2] and Tripathy [3] took that into consideration and defined the number of hours per week that were necessary for a course to prevent

© Springer Nature Switzerland AG 2021
D. A. Rossit et al. (Eds.): ICPR-Americas 2020, CCIS 1407, pp. 33–46, 2021.
https://doi.org/10.1007/978-3-030-76307-7_3

courses from overlapping, as main requirements. These generates the fulfillment of the estimated hours and the assignment of teachers without conflict. During the following years, different authors adapted these first models, adding specific conditions for their institutions or some other general ones that could be applied to other studies. An example of this is the study by Schimmelpfeng [4], where the application of a Mixed Integer Programming (MIP) model was carried out at the German School of Economics and Administration, adding elements such as the capacity of the available rooms and establishing as a objective function the minimization of violation of planning constraints.

The aim of this paper is to propose a MIP model for the course timetabling in a Colombian University. In this case study, an academic program and its core courses are considered. The academic program consists in 10-semesters, 17 core subjects, each with a certain number of subgroups determined by the expected demand of students. The model focuses mainly on avoiding the overlapping of subjects from the same semester, guaranteeing that students can enroll the courses of the semester they are. The proposed MIP model also considers the classroom capacity, the assignment of professors to courses according to a subject ability and the type of contract to determine the maximum number of hours each professor can teach per week.

The rest of this paper is organized as follows. Section 2 presents a bibliometric analysis of related studies. Section 3 shows the MIP model formulation. Section 4 presents the computational results. Lastly, Sect. 5 drawns some conclusions of the study and provides some interesting avenues for future research.

## 2 Bibliometric Analysis

In order to delve deeper into schedule and timetabling problems, a bibliometric analysis was carried out. Bibliometric analysis helps one to understand who are the authors more interested on a specific topic, as well as which are the journals, countries and institutions that have worked more on it. (Sierra-Henao et al., 2019) [5]. In addition, citation and co- citation analysis help identify the most cited authors, articles and prevailing topics. According to Noyons et al. (1999) [6], bibliometrics combine two main procedures: scientific mapping and performance analysis. The importance of this type of analysis is to estimate how well the reviewed topics have been studied and draw conclusions from there, as mentioned by Solano, Castellanos, Rodríguez and Hernández [7].

Table 1 shows a summary of the methodology followed for the analysis. The bibliometric analysis was based on the literature collected from Scopus database (www.scopus.com) as it is defined as "the largest abstract and citation database, with over 1.4 billion cited references indexed from high-quality of peer-reviewed literature as scientific journals, books and conference proceedings" (Elsevier, 2020) [8] (Fig. 1).

After having collected the search results from the query string, a software called VOSviewer was used. This software is a tool for building and visualizing bibliometric networks for better understanding and detect non relevant repeating

**Table 1.** Research method.

| Unit of analysis | Published |
|---|---|
| Period of analysis | 2010 to 2020 |
| Search engine | Scopus |
| Query string | TITLE-ABS-KEY(( course OR class* ) AND timetable AND schedul* AND optim*) AND ( LIMIT-TO ( PUBYEAR , 2020 ) OR LIMIT-TO ( PUBYEAR , 2019 ) OR LIMIT-TO ( PUBYEAR , 2018 ) OR LIMIT-TO ( PUBYEAR , 2017 ) OR LIMIT-TO ( PUBYEAR , 2016 ) OR LIMIT-TO ( PUBYEAR , 2015 ) OR LIMIT-TO ( PUBYEAR , 2014 ) OR LIMIT-TO ( PUBYEAR , 2013 ) OR LIMIT-TO ( PUBYEAR , 2012 ) OR LIMIT-TO ( PUBYEAR , 2011 ) OR LIMIT-TO ( PUBYEAR , 2010 ) ) AND ( LIMIT-TO ( DOCTYPE , "ar" ) OR LIMIT-TO ( DOCTYPE , "cp" ) OR LIMIT-TO ( DOCTYPE , "re" ) ) AND ( LIMIT-TO ( LANGUAGE , "English" ) ) AND ( LIMIT-TO ( SUBJAREA , "COMP" ) OR LIMIT-TO ( SUBJAREA , "ENGI" ) OR LIMIT-TO ( SUBJAREA , "MATH" ) OR LIMIT-TO ( SUBJAREA , "DECI" ) ) |
| Number of evaluated articles | 1,059 |

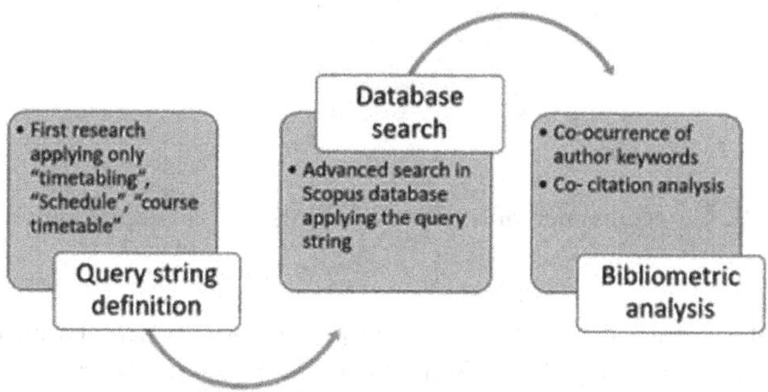


**Fig. 1.** Summary of methodology used.

terms to excluded them from the list. VOSviewer was developed by Van Eck and Waltman [9] where the software has the capacity to distinguish two types of bibliographic maps, one of them related between the investigated elements and the other facilitates the identification of groups of related elements.

First of all, an analysis of author keywords was performed. Figure 2 shows the 85 most frequent keywords with at least three occurrences for easier reading. Keywords are represented with circles, where the largest circles represent the keywords that have been used the most. As shown in Fig. 2, eight different clusters are separated by distinct colors. The green cluster identified as "timetabling", represents the prescriptive approaches for solving this type of problems.

Keywords related with "scheduling problem", such as "combinatorial optimization" and "course timetabling problem" appear in the yellow cluster. Other words such as "timetabling problem", "genetic algorithm", "high school timetabling problem", "exam timetabling" and "particle swarm optimization" appear in the purple, blue, orange, red and light blue clusters, respectively. In addition, the brown subgroup are linked with "robustness" and "course timetabling" topics.

Additionally, in order to obtain a deeper insight into the gained results, Fig. 3 shows the timeline visualization overlay of the co-occurrence of author keywords on the evaluated contributors. In other words, it exhibits the evolution of the different keywords through the years. Considering what was mentioned above, new or trending concepts can be highlighted: university course timetabling, educational timetabling, bi-criteria optimization, and integer linear programming. The latter has the highest average publication year among 85 keywords between

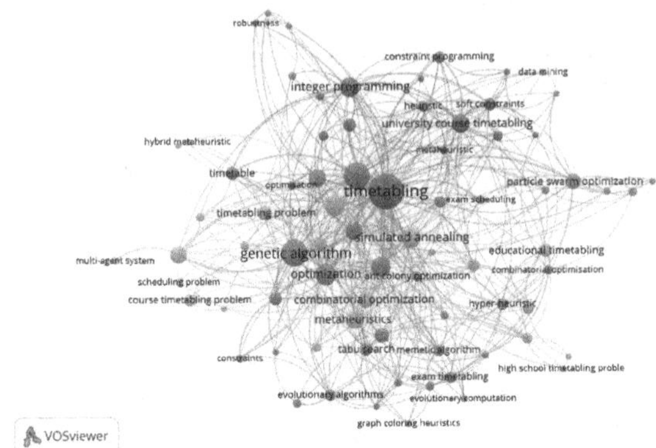


**Fig. 2.** Co-occurrence of author keywords. (Color figure online)

2018. At the other end of the spectrum, keywords as "course timetabling problem" and "school timetabling problem" have been less popular recently with an average publication year between 2012.

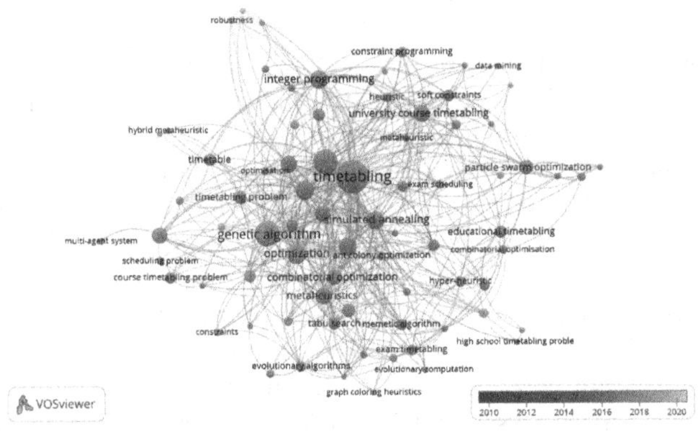


**Fig. 3.** Evolution of co-occurrence of author keywords from 2010 to 2020.

For the co-citation of authors analysis, a minimum of 20 citations per author was considered. Out of the 8,702 authors within the analysis, 218 of them met the threshold. As shown in Fig. 3, authors were classified in 5 different clusters evaluating the strength of correlations between co-citations. Authors Burke, E.K, Mccollum, B. and Qu, R, are the most relevant and most cited for timetabling and scheduling related topics. Table 2 shows a summary on the top 10 most cited authors, their total citations and the references to which citations are associated.

**Table 2.** Top 10 most cited authors

| Rank | Authors | Year | Title | TC |
|---|---|---|---|---|
| 1 | Burke and Bykov [10] | 2016 | An adaptive flex-deluge approach to university exam timetabling | 1,204 |
| 2 | Abdul Rahman et al. [11] | 2014 | Adaptive linear combination of heuristic ordering in constructing examination timetables | 735 |
| 3 | Qu et al. [12] | 2015 | Hybridising heuristics within an estimation distribution algorithm for examination timetabling | 491 |
| 4 | Abdullah and Turabieh [13] | 2012 | On the use of multi neighbourhood structures within a tabu-based memetic approach to university timetabling problems | 433 |
| 5 | Burke et al. [14] | 2010 | Hybrid variable neighbourhood approaches to university exam timetabling | 391 |
| 6 | McCollum et al. [15] | 2010 | Setting the research agenda in automated timetabling: the second international timetabling competition | 373 |
| 7 | Abuhamdah et al. [16] | 2014 | Population based local search for university course timetabling problems | 315 |
| 8 | Lewis and Thompson [17] | 2015 | Analysing the effects of solution space connectivity with an effective metaheuristic for the course timetabling problem | 218 |
| 9 | Muklason et al. [18] | 2017 | Fairness in examination timetabling: student preferences and extended formulations | 217 |
| 10 | Oner et al. [19] | 2011 | Optimization of university course scheduling problem with a hybrid artificial bee colony algorithm | 192 |

## 3 Problem Formulation

This section presents a MIP model to schedule a specific number of subjects, considering institutional policies and physical resources characteristics. The objective function and operational constraints aim to minimize the number of missed hours to fulfill professors workload while assuring that all classes are programmed and avoiding the overlapping of subjects from the same semester.

Therefore, the model considers students demand, professor availability, classrooms capacity, days and hours where classes could be programmed, and type of contract (full or part time) of the professors, seeking to generate optimal schedules for both students and professors. The following sections present the sets, parameters, decision variables, objective function and operational constraints of the model.

### 3.1 Indices

The following indices were defined for the proper development of the model:

$i$ = timeslots (T1, ..., T21).
$j$ = days (Monday, ..., Friday).
$k$ = courses (C1, ..., C59).
$p$ = professors (1, ..., 25).
$t$ = contract type (full time, part time).
$s$ = semester (1, ..., 10).
$a$ = available classrooms (S1, ..., S10).
$m$ = number of days in which the class is taught. That is, a class needs either one day a week or two.

### 3.2 Parameters

The following parameters were defined for the development of the model:

$Hours_k$ = Week hours for course $k$.
$Maxdays_m$ = Days $m$ per week required for course. Either one or two.
$Numberhours_i$ = Hours for timeslot $i$.
$Capacity_a$ = Student capacity of classroom $a$.
$Students_k$ = Expected demand of students for course $k$.
$Contract_{p,t}$ = 0–1 Matrix indicating if professor $p$ has contract type $t$.
$Semester_{k,s}$ = 0–1 Matrix indicating if course $k$ is offered in semester $s$.
$Timeslots_{i,i'}$ = 0–1 Matrix indicating if timeslot $i$ does not overlap with timeslot $i'$.
$Courseprof_{k,p}$ = 0–1 Matrix indicating if course $k$ can be taught by professor $p$.
$Courseday_{k,m}$ = 0–1 Matrix indicating if course $p$ requires $m$ days per week.
$Disp_{p,k,i,j}$ = 0–1 Matrix indicating if professor $p$ is available for teaching course $k$ during timeslot $i$ on day $j$.
$Minhoursparttime$ = Maximum number of teaching hours per week for part-time lecturers.
$Maxhoursfulltime$ = Minimum number of teaching hours per week for full-time professors.
$Maxhoursfulltime$ = Maximum number of teaching hours that can be allocated per day for each semester.
$BigM$ = Very large integer value.

### 3.3 Decision Variables

The following decision variables were defined for the model:

$X_{p,k,i,j}$ = 1 if professor $p$ teaches course $k$ in timeslot $i$ of day $j$, 0 otherwise.
$Y_{k,i,j}$ = 1 if course $k$ is taught in timeslot $i$ of day $j$, 0 otherwise.
$N_{a,i,k,j}$ = 1 if course $k$ is taught in timeslot $i$ of day $j$ in classroom $a$, 0 otherwise.

$W_{p.k}$ = If professor $p$ teaches course $k$.
$G_{k,j}$ = If course $k$ is taught on day $j$.
$DELTA_p$ = Hours professor $p$ misses to fulfill his workload.
$DELTA2_{p,k,i,j}$ = Hours missed to assign all courses.

### 3.4 Objective Function

The model is aimed at minimizing the hours professor $p$ misses to fulfill his workload:

$$min\ Z = BigM \cdot \sum_{p} DELTA_p \tag{1}$$

### 3.5 Constraints

The following constraints were considered for the problem:

$$\sum_{i}\sum_{j} Y_{k,i,j} \cdot Numberhours_i = Hours_k,\ \forall k \tag{2}$$

$$\sum_{i} Y_{k,i,j} \leq 1,\ \forall k, j | Courseday_{k,m} = 2 \tag{3}$$

$$G_{k,j} + G_{k,j+1} \leq 1,\ \forall k, j | j \neq Friday \tag{4}$$

$$\sum_{j} G_{k,j} \cdot Courseday_{k,m} \leq Maxdays_m,\ \forall k, m \tag{5}$$

$$\sum_{k} Y_{k,i,j} \cdot Semester_{k,s} \leq 1,\ \forall i, j, s \tag{6}$$

$$Y_{k,i,j} = 0,\ \forall k, i = \{T4, T5, T6, T10, T11, T14, T15, T18, T20, T21\}, j = Friday \tag{7}$$

$$\sum_{k}\sum_{i} Y_{k,i,j} \cdot Semester_{k,s} \leq Maxhoursday,\ \forall k, s \tag{8}$$

$$\sum_{p} W_{p,k} \leq 1,\ \forall k \tag{9}$$

$$\sum_{p} X_{p,k,i,j} \leq 1,\ \forall k, i, j \tag{10}$$

$$\sum_{k} X_{p,k,i,j} + \sum_{k} \sum_{i' | i' \neq i \wedge Timeslots_{i,i'} = 0} X_{p,k,i',j} \leq 1,\ \forall p, i, j \tag{11}$$

$$X_{p,k,i,j} \leq Courseprof_{p,k},\ \forall p, k, i, j \tag{12}$$

$$\sum_k \sum_i \sum_j X_{p,k,i,j} \cdot Contract_{p,t} \geq Minhoursfulltime - DELTA_p,\ \forall p, t = Fulltime \tag{13}$$

$$\sum_k \sum_i \sum_j X_{p,k,i,j} \cdot Contract_{p,t} \leq Maxhoursparttime,\ \forall p, t = PartTime \tag{14}$$

$$\sum_p X_{p,k,i,j} = Y_{k,i,j}\ \forall k, i, j \tag{15}$$

$$X_{p,k,i,j} \leq disp_{p,i,j} + DELTA2_{p,k,i,j},\ \forall p, k, i, j \tag{16}$$

$$\sum_a N_{a,i,k,j} \leq 1,\ \forall i, k, j \tag{17}$$

$$\sum_k N_{a,i,k,j} + \sum_k \sum_{i'|i' \neq i \wedge Timeslots_{i,i'}=0} N_{a,i',k,j} \leq 1,\ \forall i, j, a \tag{18}$$

$$\sum_k N_{a,i,k,j} \cdot Students_k \leq Capacity_a,\ \forall a, i, j \tag{19}$$

$$\sum_k N_{a,i,k,j} = Y_{k,i,j},\ \forall i, j, k \tag{20}$$

$$DELTA2_{p,k,i,j} \leq 1,\ \forall p, k, i, j \tag{21}$$

$$W_{p,k} \leq 1,\ \forall p, k \tag{22}$$

$$G_{k,i} \leq 1,\ \forall k, j \tag{23}$$

$$X_{p,k,i,j} \leq W_{p,k},\ \forall p, k, i, j \tag{24}$$

$$Y_{k,i,j} \leq G_{k,j}\ \forall k, i, j \tag{25}$$

$$X_{p,k,i,j}, Y_{k,i,j}, N_{a,i,k,j} \in (0,1), \forall i, j, k, p, a \tag{26}$$

$$W_{p,k}, G_{k,j}, DELTA_p, DELTA2_{p,k,i,j} \geq 0, \forall j, k, p \tag{27}$$

Constraints 2 guarantee that all the hours of each class are scheduled. Constraints 3 guarantee that, if a course has more than one session per week (i.e., requires two or more days of the week), the sessions of the class cannot be scheduled on the same day. Constraints 4 apply to those subjects that require days between their sections, in this case at least one day in between. Constraints 5 guarantee that a class cannot be programmed in more days than it is permitted to. In addition, it is necessary to ensure that classes that belong to the same semester do not overlap, this is done by constraints 6. Considering the university, Friday afternoons are reserved for postgraduate courses, meaning that constraints 7 guarantee that no undergraduate classes are assigned during said timeslots. Constraints 8 limit the maximum number of hours a day that can be taught to each semester.

It is also considered that each of the courses may be taught by a single professor, just as they may teach only one class at a time. This is ensured with constraints 9, 10 and 11. Constraints 12 establish that professors will only be able to teach courses that they are able to teach. Additionally, according to the contract type, full-time professors must teach a minimum 6 h of class per week, while part-time professors may dictate maximum 21 h. This is done in constraints 13 and 14 respectively. Constraints 15 establish that professors can only be assigned to courses-slots that have already been assigned. Part-time professors have a certain availability according to their needs. Constraints 16 contemplate the availability matrix of all professors, both full-time and part-time professor, in order to assign them to courses that are available in time slots to teach the class.

For the last resource (i.e., classrooms), constraints 17 ensure that each class may be assigned to only one timeslot on a single day. On the other hand, constraints 18 guarantee that classrooms may have only one course allocated on each timeslot. Constraints 19 establish that all courses will be assigned to classroom that can support the expected student demand. Lastly, the number of classrooms assigned must correspond to the number of courses in each of the time slots of each day. This is given by constraints 20. Constraints 21 to 25 establish relationship conditions between variables, and constraints 24 and 25 indicate the nature of the variables of the model.

## 4 Results

The proposed model was coded and solved using GAMS (General Algebraic Modeling System) version 28.2.0 and solved using CPLEX on a laptop computer equiped with an Intel Core i5 processor, 8 GB RAM and a 64-bit Windows 10 operating system.

For this particular project, it is important to consider, within the result analysis, different scenarios that provide useful information for decision makers. Despite the foundations of the model being exclusively on course timetabling, it is ideal to develop ideas that revolve on the impact that the model may have on other areas, such as economics or the optimization of resource utilization.

The structure of the scenarios considered is based on two variables: teachers' availability and courses' demand. Based on information shared by the University, the code states an initial expected students demand for each course and a specific amount of groups for dividing students in order to meet classrooms maximum capacity and a homogeneous distribution. For the problem analyzed in this paper, only ten classrooms of the university were considered, seven of them can accommodate 40 people while the other three have a capacity for 30. The academic program has 20 teachers, 12 of them are part-time, while the rest are full-time teachers.

For evaluating different scenarios, eight indicators were selected, depending on classrooms, courses, teachers and time frames. The following indicators were defined: classroom usage hours per day (%HU), classroom average occupied capacity (%AVG CAP), scheduled and unscheduled courses (%CA and %UC, respectively), assigned full- and part-time professors, critical time slots, and number of classrooms used.

Three scenarios were examined. The first one with the normal expected student demand. The second one consideres a 10% demand increase. It is worth mentioning, that in this scenario classes still have a maximum number of 40 students, as this is the largest classroom capacity. The third one evaluates a 10% demand decrease. In addition, each scenario considers five possible professor availability (100%, 75%, 60%, 50% and 40%).

**Table 3.** Results with expected student demand.

| | | 100% | 75% | 60% | 50% | 40% |
|---|---|---|---|---|---|---|
| Classrooms | %HU | 33.82% | 33.82% | 33.82% | 33.27% | 32.73% |
| | %AVG CAP | 91.66% | 90.06% | 83.55% | 88.98% | 91.70% |
| Courses | %CA | 100% | 100% | 100% | 98% | 97% |
| | %UC | 0% | 0% | 0% | 2% | 3% |
| Professors used | Full-Time | 8 | 8 | 8 | 8 | 8 |
| | Lecturers | 10 | 12 | 11 | 12 | 12 |
| Critical hour | | 17–18 | 8–9 | 9–10 | 8–9 | 9–10 |
| Classrooms used | | 10 | 10 | 9 | 10 | 10 |
| Used resources indicators | | 100% | 75% | 60% | 50% | 40% |
| Classrooms | %HU | 33.82% | 33.82% | 37.58% | 33.27% | 32.73% |
| | %AVG CAP | 91.66% | 90.06% | 92.84% | 88.98% | 91.70% |

This first scenario, represented in Table 3, is based on the usual expected demand for each course. It is very important to highlight the fact that with the 40% and 50% professors' availability the total courses were not assigned. Despite the latter, the percentage of average capacity used for the 40% scenario, is the best within this frame, mainly because of the use of classrooms 7 and 10, whose capacities were entirely used by one course in a time frame, this is why these spaces have the lowest percentage of hours used. It is important to point out that the teacher 19 is the only entitled to teach for 6 groups due to its subject speciality; therefore when the availability is less than 60%, the professor is not able to teach all of them, in the 40% he matches 4 out of 6 and 5 out of 6 in the scenario with 50% of professor availability.

Once it is possible to teach all the courses due to professors' availability, an interesting phenomenon appears: the proposed schedule distributed the hours through the classrooms in a very similar way, resulting in an equal percentage of used hours. Nevertheless, it is interesting to see that the average capacity used increases as the availability does, stating the fact that when teachers have better availability, it is more likely to use resources better.

However, in the scenario with 60% of professor availability, there is a fact that needs to be taken into account: not all the classrooms are used, that is, there is at least one empty classroom which increases the usage percentage of the classrooms. That is the reason why the second part of Table 3 is important, because it takes into account only used resources. Therefore, analyzing the behavior of the results, the 60% scenario is the best possible, regarding average capacity used of the classrooms and the hours used. This is explained by the fact that, the model focuses mainly on avoiding the overlapping of subjects from the same semester and ensuring certain amount of hours for part-time professors.

**Table 4.** Results with a student demand increase of 10%.

| Students demand increases 10% | | 100% | 75% | 60% | 50% | 40% |
|---|---|---|---|---|---|---|
| Classrooms | %HU | 33.82% | 33.27% | 33.82% | 33.09% | 33.09% |
| | %AVG CAP | 74.89% | 82.40% | 82.22% | 80.46% | 80.35% |
| Courses | %CA | 100% | 100% | 100% | 98% | 97% |
| | %UC | 0% | 0% | 0% | 2% | 3% |
| Professors used | Full-Time | 8 | 8 | 8 | 8 | 8 |
| | Lecturers | 9 | 10 | 10 | 11 | 11 |
| Critical hour | | 17–18 | 8–9 | 8–9 | 9–10 | 8–9 |
| Classrooms used | | 8 | 9 | 9 | 9 | 9 |
| Used resources indicators | | 100% | 75% | 60% | 50% | 40% |
| Classrooms | %HU | 42.27% | 36.97% | 37.58% | 36.77% | 36.77% |
| | %AVG CAP | 93.61% | 91.55% | 91.35% | 89.40% | 89.27% |

The results for the second scenario are presented in Table 4. It can be observed that the best used hours percentage is achieved with a 100% professor availability. Basically, when demand increased there were only 2 courses that could be taught in less than 30 people capacity classrooms, each of them with 22 students demand. The model scheduled these two classes in one classroom, leaving low demands concentrated in one space and not distributed, saving the possibility of lowering the average of the rooms with a capacity of 40 people. This scenario is the best possible for all indicators: hours used, average classroom capacity, classrooms used and teachers scheduled. Nevertheless, all the instances (i.e., 75%, 60%, 50% and 40%) for this scenario have a good performance in terms of average capacity used.

**Table 5.** Results with a student demand decrease of 10%.

| Students demand decreases 10% | | **100%** | **75%** | **60%** | **50%** | **40%** |
|---|---|---|---|---|---|---|
| Classrooms | %HU | 33.82% | 33.82% | 34.18% | 33.27% | 32.73% |
| | %AVG CAP | 84.04% | 84.34% | 81.99% | 83.96% | 81.89% |
| Courses | %CA | 100% | 100% | 100% | 98% | 97% |
| | %UC | 0% | 0% | 0% | 2% | 3% |
| Professors used | Full-Time | 8 | 8 | 8 | 8 | 8 |
| | Lecturers | 10 | 12 | 11 | 12 | 12 |
| Critical hour | | 9–10 | 12–13 | 17–18 | 11–12 | 12–13 |
| Classrooms used | | 10 | 10 | 10 | 10 | 10 |

For the last scenario, when demand decreases, the model tends to distribute equally the classes within the classrooms as it can be concluded after analysing Table 5. For the different variants of teachers' availability, the 10 classrooms were used; all of them with less than 85% capacity used. This is due to the use of all classrooms where the low and high students courses are mixed and the fact the demand itself decreased. It is really interesting the fact that when demand is lower the model tends to distribute almost equally teachers and classrooms. This could be explained by considering that a demand decrease would generate smaller groups and, therefore, a better usage could be achieved for the classrooms with capacity for 30 students.

## 5 Conclusions and Future Research

This paper studied a university course timetabling problem. The objective was to schedule the set of courses of an academic program while professors and classrooms were assigned. A set of data from a Colombian University was taken as a sample to carry out a mathematical model that meets the conditions of schedules, crossings of subjects from the same semester, the maximum number of hours that a professor can dictate and keeping in mind the demand of students for each of the courses to allocate classrooms.

The problem is solved and provides to the researchers a tool to assess and evaluate different scenarios. The computational results show that it is possible adapt the model to different educational scenarios that require the optimal use of resources. The initial conditions are very favourable for the University as classrooms and teachers are enough. Nevertheless, numbers may highlight the underuse of available classrooms and the possibility of leaving these spaces for another purpose; for all scenarios at least 57% hours are free in the classrooms, meaning that other programs could take advantage of this spaces.

The results also show that the model can comply with the guidelines established by the board of directors and the quality standards necessary in the academic offer. This is clear as the scheduling conditions can be modified to meet specific requirements and make the assignment process more complete, while maintaining high levels of service for both students and teachers and for the institution.

As future research, the model can be considered to be applied with all academic programs of the university to coordinate efficiently the manage of resources and to deliver one system that take control of them. Also, the inclusion of distances between classrooms can be contemplated as part of a new optimization criterion. The route that students and teachers must take could be an objective to take into account when there are two or more consecutive classes, so the less distance they travel the more time they will have to arrive. Also, integrating other scenarios with alternate students demand. In addition, as in the result analysis, assigning an opening cost for each classroom and each time frame, could transform the programming into an optimization tool as well.

It is also possible to add preference conditions at specific times, determined by the institutions to give priority to other academic activities. This should contain a matrix with weights for each of the time slots where it will be possible to indicate which of them are more pertinent to specific interests. Finally, new ways of addressing the problem can be designed, such as heuristic methods, that allow finding feasible solutions comparable with the results obtained in this research and obtaining conclusions about the performance of the models and the management of resources for their optimization.

## References

1. Rudová, H., Müller, T., Murray, K.: Complex university course timetabling. J. Sched. **14**(2), 187–207 (2011). https://doi.org/10.1007/s10951-010-0171-3
2. Lawrie, N.L.: An integer linear programming model of a school timetabling problem. Comput. J. **12**(4), 307–316 (1969)
3. Tripathy, A.: A Lagrangean relaxation approach to course timetabling. J. Oper. Res. Soc. **31**(7), 599–603 (1980)
4. Schimmelpfeng, K., Helber, S.: Application of a real-world university-course timetabling model solved by integer programming. OR Spectrum **29**(4), 783–803 (2007). https://doi.org/10.1007/s00291-006-0074-z
5. Sierra-Henao, A., Muñoz-Villamizar, A., Solano-Charris, E., Santos, J.: Sustainable development supported by industry 4.0: a bibliometric analysis. In: Borangiu, T., Trentesaux, D., Leitão, P., Giret Boggino, A., Botti, V. (eds.) SOHOMA 2019. SCI, vol. 853, pp. 366–376. Springer, Cham (2020). https://doi.org/10.1007/978-3-030-27477-1_28
6. Noyons, E.C., Moed, H.F., Luwel, M.: Combining mapping and citation analysis for evaluative bibliometric purposes: a bibliometric study. J. Am. Soc. Inform. Sci. **50**(2), 115–131 (1999)
7. López, E.S., Quintero, S.C., del Rey, M.L.R., Fernández, J.H.: La bibliometría: Una herramienta eficaz para evaluar la actividad científica postgraduada. MediSur **7**(4), 59–62 (2009)

8. Elsevier. How Scopus works: information about Scopus product features. https://www.elsevier.com/solutions/scopus/how-scopus-works
9. Van Eck, N.J., Waltman, L.: Software survey: VOSviewer, a computer program for bibliometric mapping. Scientometrics **84**(2), 523–538 (2010). https://doi.org/10.1007/s11192-009-0146-3
10. Burke, E.K., Bykov, Y.: An adaptive flex-deluge approach to university exam timetabling. INFORMS J. Comput. **28**(4), 781–794 (2016)
11. Rahman, S.A., Bargiela, A., Burke, E.K., Özcan, E., McCollum, B., McMullan, P.: Adaptive linear combination of heuristic orderings in constructing examination timetables. Eur. J. Oper. Res. **232**(2), 287–297 (2014)
12. Qu, R., Pham, N., Bai, R., Kendall, G.: Hybridising heuristics within an estimation distribution algorithm for examination timetabling. Appl. Intell. **42**(4), 679–693 (2014). https://doi.org/10.1007/s10489-014-0615-0
13. Abdullah, S., Turabieh, H.: On the use of multi neighbourhood structures within a tabu-based memetic approach to university timetabling problems. Inf. Sci. **191**, 146–168 (2012)
14. Burke, E.K., Eckersley, A.J., McCollum, B., Petrovic, S., Qu, R.: Hybrid variable neighbourhood approaches to university exam timetabling. Eur. J. Oper. Res. **206**(1), 46–53 (2010)
15. McCollum, B., et al.: Setting the research agenda in automated timetabling: the second international timetabling competition. INFORMS J. Comput. **22**(1), 120–130 (2010)
16. Abuhamdah, A., Ayob, M., Kendall, G., Sabar, N.R.: Population based local search for university course timetabling problems. Appl. Intel. **40**(1), 44–53 (2014). https://doi.org/10.1007/s10489-013-0444-6
17. Lewis, R., Thompson, J.: Analysing the effects of solution space connectivity with an effective metaheuristic for the course timetabling problem. Eur. J. Oper. Res. **240**(3), 637–648 (2015)
18. Muklason, A., Parkes, A.J., Özcan, E., McCollum, B., McMullan, P.: Fairness in examination timetabling: student preferences and extended formulations. Appl. Soft Comput. **55**, 302–318 (2017)
19. Oner, A., Ozcan, S., Dengi, D.: Optimization of university course scheduling problem with a hybrid artificial bee colony algorithm, pp. 339–346 (2011)

# Determination of Contribution Rates in the Retirement System of an Institution of Higher Education Through an Optimization Model

Marco A. Montufar Benítez[1](✉), José Raúl Castro Esparza[2], Eva Hernández Gress[3], José Luis Mota Reyes[1], Hector Rivera Gomez[1], and Octavio Castillo Acosta[1]

[1] Universidad Autónoma del Estado de Hidalgo, 42184 Pachuca, Hidalgo, Mexico
montufar@uaeh.edu.mx
[2] Universidad de Guadalajara, Guadalajara, Jalisco, Mexico
[3] Tecnológico de Monterrey, 42083 Pachuca, Hidalgo, Mexico

**Abstract.** Retirement and pension systems in the world have undergone important changes due to governmental impositions in several aspects. In this research, we develop a linear optimization model to determine the optimal contribution percentages in a retirement system designed to support high education institution teachers. The parameters considered by the model, which is developed in the LINGO language, are the interest rate generated by the investment, the inflation rate, the total number of years a worker saved, the years in which coverage is desired, and the salary of the workers. The results show the optimal investment percentages concerning the worker salary, assuming that coverage is desired for a certain number of years in the future and the total amount saved for some time. At the end we show a sensitivity analysis making the mentioned parameters take values in certain reasonable ranges.

**Keywords:** Higher education · Linear programming · Optimization · Retirement

## 1 Introduction

The purpose of pension systems is to provide workers with an economic income during their post-employment life so that they can meet their living expenses and cope with different situations that may arise. A pension plan consists of an arrangement of regular payments, generally, during a worker's working life, that will allow him to guarantee savings for old age [1, 2].

While countries around the world are experiencing unprecedented increases in life expectancy and fertility rates, there are changes in accounting standards, reductions in contributions, and low financial returns. Added to this is the high volatility in the financial markets, which directly impacts the purchasing power of workers. All these elements have led to a downturn in the financial systems and an increase in interest rates.

© Springer Nature Switzerland AG 2021
D. A. Rossit et al. (Eds.): ICPR-Americas 2020, CCIS 1407, pp. 47–55, 2021.
https://doi.org/10.1007/978-3-030-76307-7_4

Pension funds in the UK and USA, which have traditionally had relatively high capital allocations, have been hit hard [3].

In Mexico, both the Mexican Social Security Institute (IMSS) and the Institute of Security and Social Services for State Workers (ISSSTE) approach social security by establishing two differentiated regimes: the mandatory regime and the voluntary regime. Traditionally created to provide workers with a dignified retirement, Mexico's pension systems face enormous challenges due to a complex scenario. The laws of both institutions establish that social security in the mandatory regime must provide for retirement, unemployment at an advanced age, and old age for workers. Historically, as it can be seen, the Mexican state has updated the pension plans. However, the traditional pension model has undergone several changes due to multiple problems that have arisen both globally and locally. The financial liberalization model that has prevailed in Mexico for several decades led to changes in the pension system. In Mexico, the IMSS retirement pension system was reformed, which changed from a public system to a private system of individual contributions in 1997 [4]. At the same time, in 2007, the ISSSTE law was reformed [5], so that the new law allows the creation of a pension system that currently operates as a Retirement Fund Administrator (AFORE).

As it can be seen, the creation of retirement savings funds is of great importance from every perspective: social, economic, political, and finally, human. Ensuring a decent standard of living for retired workers who have contributed their labor force during a lifetime of work is a problem of the utmost importance. The improvement in medical treatments for chronic degenerative diseases that afflict older adults has led to a significant increase in life expectancy, with the consequent impact on their quality of life. In other words, longevity can have important consequences on workers' pension plans. This is an issue that legislators, actuaries, and financial engineers need to address from their respective disciplines.

According to data provided by the IMSS, ISSSTE, and the National Institute of Statistics, Geography, and Informatics (INEGI), the proportion of active workers contributing to pension systems for the economically active population (EAP) in employment fluctuated around 37% between 2000 and 2009 [6]. Currently, close to 60% of the employed EAP is not contributing to access a retirement pension when they reach old age. This, of course, leads to different questions about the social and economic functions that pension systems should fulfill [3].

For all of the above reasons, it is essential to address the issue of retirement savings to mitigate the conditions of helplessness in which a large part of the Mexican population finds itself when they reach an advanced age. More than twenty years after the reform of the IMSS Law, it is necessary to provide viable alternatives that allow workers to have a pension plan for retirement.

Lin [7] makes use of a stochastic Ito process, considering three state variables: the crediting rate, the risk-free interest rate, and the wage growth rate. This process is studied through a Monte Carlo Least Squares (LMS) simulation model to fix the prices of the options where the benefit obtained is taken from the maximum between the defined benefits and the defined contributions. The model assumes that the maximum age to remain in the pension system is 65 years. However, it also considers that there are two possibilities where the worker can retire with less than that age: (1) the worker has more than 15 years of experience and more than 55 years of age, and (2) the worker has more than 25 years of experience.

This paper develops a retirement savings model for academic workers of a public higher education institution in Mexico, whose objective is to mitigate the economic consequences that the current retirement plans (AFORES) are not going to cover. The study is approached from two perspectives: (1) The first one makes an analysis of the coverage in time that the savings can cover making use of the contributions suggested by the educational institution, and (2) Under an optimization perspective, the minimum percentage of salary that is recommended to contribute to having a certain coverage in years, under certain years of work, is determined.

## 2 Materials and Methods

### 2.1 Description of the Problem

In 2019, the Mexican Institute of Social Security [8] published a new law to regulate retirements, this new law arises for different reasons, among them we can find the average increase in life expectancy of Mexicans, according to the [9] life expectancy has increased considerably; in 1930 people lived on average 34 years; 40 years later in 1970 this indicator stood at 61; in the year 2000 it was 74, in 2016 it was 75. 2 years and especially for the year 2016 in the state of Hidalgo life expectancy was 74.6 years. Another important aspect is the origin of the economic resource obtained for retirements since, in the old Social Security law, the resource granted to retirees came from economically active people. If we observe data provided by the INEGI [9] we will see that the birth rate was reduced by 50% between the years 1970 and 1990, this causes that if there are no economically active people to be able to subsidize the retirement of retired people there would be an economic problem of great dimensions when trying to grant this resource.

This law mentions that people who have a formal job and start working after July 1997 will only receive the money saved by themselves in their retirement fund (Afore), which will be granted in installments according to their life expectancy at that time. Taking into account that the economic rewards given by these retirement schemes (AFORES) are low, for example, as is shown in [10], a 43-year-old person today, with a monthly salary of 20,000 MXN and an accumulated savings of 100,000 MXN up to this moment, would be receiving approximately 3,600 MXN per month at the age of 60, almost one-fifth of his current salary, without considering inflation. In light of this not-so-favorable scenario, some institutions have opted to create their retirement system to support the worker. In this paper, we developed an optimization model that could support decision-making in a higher education institution to implement an internal retirement savings system. Specifically, our model determines the percentage of the worker's salary that should be invested in a trust to achieve certain levels of future economic quality of life. The following sections develop the optimization model, as well as the input data used in its analysis.

The savings and retirement process consists of teachers of different ages and salaries who wish to save a percentage of their salary for a retirement fund which, over a period of time determined by them, is increased by the bank interest produced by a trust fund in a banking institution. When the teacher decides to retire, the money collected over the years of savings will be given to him/her so that he/she can enjoy his/her old age (Fig. 1).

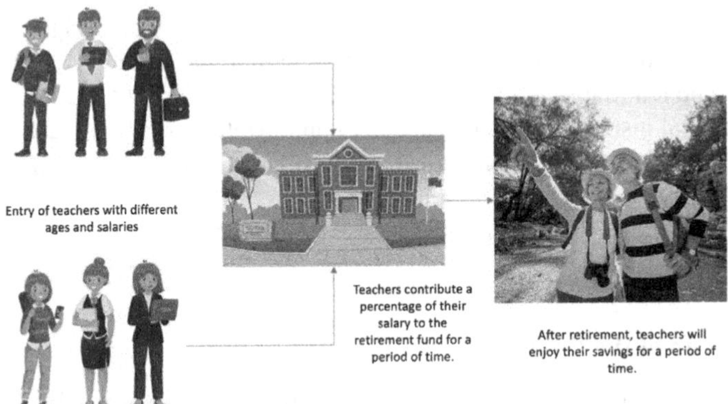

Entry of teachers with different ages and salaries
Teachers contribute a percentage of their salary to the retirement fund for a period of time.
After retirement, teachers will enjoy their savings for a period of time.


**Fig. 1.** General savings and retirement process

## 2.2 Data Collection

The educational institution where we gathered information established that the benefited teachers should have some important characteristics to be able to enter their retirement program. As a primary characteristic, the worker must belong to the academic union, to receive fairly and legally all the benefits established in their collective bargaining agreement. An important element in the model is the salary of the worker, since this is not the same for all professors, because in higher education institutions there are different types of positions with which the personnel is hired, due to their experience, level of studies, publications, etc., among other factors that define their hiring category.

No less important in this optimization model is the percentage of salary that both the teachers and the institution itself would contribute to the retirement system; initially, it was believed that 5% of the monthly salary would not mean a considerable reduction in their standard of living. Given that the money contributed by the workers and the institution would be invested in some financial instrument, it was considered that these would yield an approximate annual interest rate of 6% (Personal communication).

Also, our model takes into account inflation, which is handled as another parameter, because if a person plans to live for 10 years after retirement, with the money he/she can live on today, it is necessary to inflate his/her current salary to bring it to future value. In this case, an inflation rate of 3.33% is considered, since according to the Bank of Mexico [11], this is the annual inflation rate that the country suffers. As part of this reasoning, and to improve salaries, the institution grants an approximate salary increase of 3% per year in an attempt to counteract such inflation.

## 2.3 Model Formulation

Next, we define through the following notation the variables and parameters of the optimization model:

Sets:

$i$: Set of years saving, $i = \{1, ..., N\}$
$j$: Set of years with coverage, $j = \{N + 1, ..., n\}$

Parameters:

$N$: Number of years the person saves
$n$: Number of years the person receives a pension (coverage)
$\alpha$: Rate of salary increase
$\beta$: Inflation rate
$\kappa$: Interest rates paid by the trust
$\theta$: Interest rate on bank savings account
$S_0$: Starting salary

Variables:

$\gamma$: Percentage of contribution (decision variable)
$R_i$: Retirement in year $j$
$A_i$: Contribution in year $i$
$S_i$: Salary in year $i$
$B_i$: Balance in year $i$
$B_j$: Balance in year $j$

With these definitions, the mathematical model is as follows:

$$\text{Min } Z = \gamma$$

S.T.

$$A_i = S_i(2 * \gamma),\ i = 1, ..., N \tag{1}$$

$$A_j = 0,\ j = N + 1, ..., N + n \tag{2}$$

$$S_i = S_{i-1}(1 + \alpha),\ \ i = 1, ..., N \tag{3}$$

$$S_j = 0, j = N + 1, ..., N + n \tag{4}$$

$$B_i = (B_i - 1 + A_i)(1 + \kappa), i = 1, ..., N \tag{5}$$

$$R_j = S_0(1 + \beta)^j, j = N + 1, ..., N + n \tag{6}$$

$$B = (B - R)(1 + \theta), jj - 1j \tag{7}$$

$$j = N + 1, ..., N + n \tag{8}$$

$$R_i = 0, i = 1, ..., N \tag{9}$$

$$BN + n \geq 0 \tag{10}$$

$$\gamma \geq 0 \tag{11}$$

A similar model is proposed by [12], in which the contribution is made year after year with an incremental way and considers changes in the interest rate.

## 3 Results and Sensitivity Analysis

The model was run with the following parameter values, which were explained in Sect. 2.2: $\boldsymbol{N = 30}$, $\boldsymbol{n = 10}$, $\boldsymbol{\alpha = 3\%}$, $\boldsymbol{\beta = 3.3\%}$, $\boldsymbol{\kappa = 6\%}$, $\boldsymbol{\theta = 3\%}$, $\boldsymbol{S_0 = 240,000}$.

The contribution percentage obtained was $\gamma = 13\%$, which is more than twice as high as that proposed (5%). For the workers and the Institution to perceive the behavior of the contribution percentage, we analyzed two scenarios, considering the years worked or saved (***N***) and the years of life after retirement that are desired to be covered (***n***). The first scenario considers that the person saves a total time of 30 and wishes to live between 10 and 20 years with the money collected after retirement (see Fig. 2).

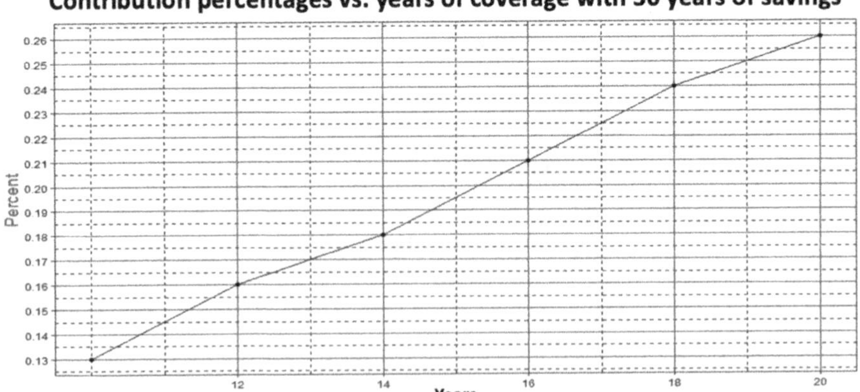


**Fig. 2.** Contribution percentages according to years of coverage at the moment of retirement.

As it can be seen in Fig. 2, there is an approximately linear relationship between the years of coverage after retirement and the contribution percentages, since if the worker wishes to live more years with the generated money, the annual contribution percentage should be higher.

In the second scenario we analyzed the case of a retired teacher who intends or expects to live 10 years with the money collected, but on this occasion, the time of saving was considered as an independent variable. In this scenario we varied the savings time from 16 years to 30 years, obtaining the following results (see Fig. 3).

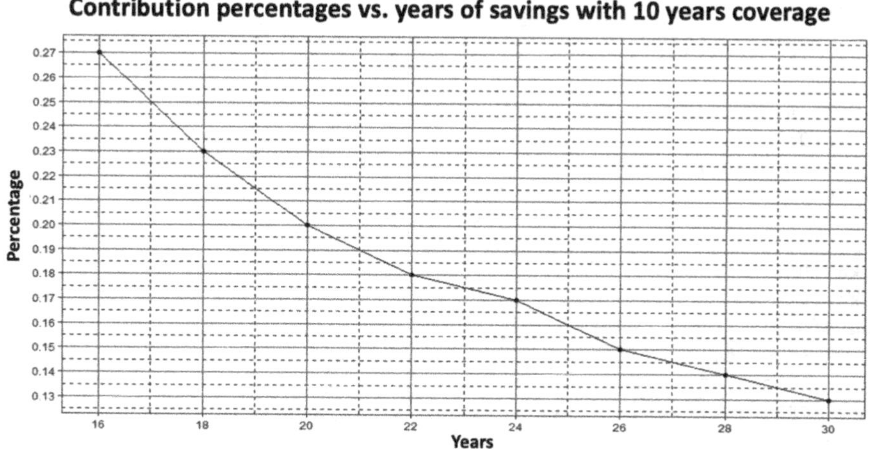


**Fig. 3.** The ratio of years of savings and contribution percentages for retirement with 10 years of financial coverage.

In Fig. 3, we can see that the curve is descending, therefore, as we save more time the contribution percentage is lower to live ten years of retirement.

For the moment this model does not incorporate the effect of the life expectancy of the teacher, since it is clear, in any case, that even if the worker wishes to enjoy savings for a certain number of years, this may be truncated by the death of the individual. Future studies could include a type of stochastic programming model, including the factor of a lifetime or the randomness of interest rates, even changing the approach to a simulation model.

## 4 Conclusions

The model developed allowed the Educational Institution to define the best salary contribution percentage, considering the opinions of the teachers after they were informed of the analysis generated with the model. What was evident from this analysis is that a 5% salary contribution percentage is insufficient to cover the future living expenses of most workers. The model turns out to be useful for all those companies that require a quantitative analysis of the different scenarios in their retirement plans. The authors are currently working on a stochastic simulation model to explore more realistic scenarios in the behavior of the system.

# A Appendix

```
Lingo program

SETS:
  IVNA/1..20/:vna;
  IVNC/1..10/:vnc;
ENDSETS
SETS:
  Ahorro:Sa,Aa,Ba;
  Cobertura:Rc,Bc;
ENDSETS
DATA: na=40;
  nc=20;
  Ahorro=1..na;
  cobertura=1..nc;
  alfa=.03;
  beta=.033;
  ka=.06;
  kc=.03;
ENDDATA
SUBMODEL BASE:
   !funcion objetivo;
   [AportationRate]Min= gamma;
   !RESTRICCIONES;
   So=240000;
   Sa(1)=So;
@FOR(Ahorro(t)|(t #GE# 2 #AND#
t#LE#na):Sa(t)=Sa(t1)*(1+alfa);!salarios;
@FOR(Ahorro(t)|(t #GT# na): Sa(t)=0);salarios;
@FOR(Ahorro(t)|(t #AND# t #LE#
na):AA(t)=Sa(t)*(2*gamma));!aportaciones durante ahorro;
@FOR (Ahorro (t)|(t #GT# na):AA(t)=0);!aportaciones durante cobertura;
Ba(1)=(So*(2*gamma))*(1+ka);!Balance final año 1; @FOR (ahorro(t)|(t #GE# 2
#AND# t #LE# na):Ba(t)=(Ba(t- 1)+Aa(t))*(1+ka));!balance
ahorro;
@For(Cobertura(t) |(t#GE# 2 #AND# t #LE#
nc):Rc(t)=Sa(1)*(1+beta)^(na+t));!retiros cobertura;
Bc(1)=Ba(na)-Rc(1))*(1+kc);
@FOR cobertura(t) |(t#GE# 2 #AND# t #LE# nc):Bc(t)=(Bc(t-
1)-
Rc(t))*(1+kc));!balance durante cobertura; Bc(nc)>=0;
ENDSUBMODEL
```

# References

1. Blanchard, O., Fischer, S.: Macroeconomics Annual 1989. National Bureau of Economie Research, The MIT Press, Cambridge (1989)
2. Rofman, R.: Social Security Coverage in Latín America. The World Bank, Washington (2005)

3. Marco, M., Gregoriou, G.N., Masala, G.: Pension Fund Risk Management Financial and Actuarial Modeling. CRC Press/Taylor & Francis Group, Florida (2010)
4. IMSS: Ley del Seguro Social, Instituto Mexicano del Seguro Social, Diario Oficial de la Federación (1997)
5. ISSSTE: Ley del Instituto de Seguridad y Servicios Sociales de los Trabajadores del Estado 2007, published on el Diario Oficial de la Federación on 31 March 2007, México (2007)
6. Murillo, S., Venegas, F.: Cobertura de los sistemas de pensiones y factores asociados al acceso a una pensión de jubilación en México. Papeles de Población **17**(67), 209–25 (2011)
7. Lin, C.-G., Yang, W.-N., Chen, S.-C.: Analysis of retirement benefits with options. Econ. Model. **36**, 130–135 (2014). https://doi.org/10.1016/j.econmod.2013.09.025
8. IMSS: Ley del Seguro Social (2019). https://www.imss.gob.mx/sites/all/statics/pdf/leyes/LSS.pdf. Accessed 28 Nov 2019
9. INEGI: Cuentame.inegi (2019). https://cuetame.inegi.org.mx/poblacion/esperanza.aspx?tema=p. Accessed 29 June 2019
10. Finauta (2019). https://app.finauta.com/retiro. Accessed 19 June 2019
11. Banxico Banco de México (2020). https://www.banxico.org.mx. Accessed 5 July 2020
12. Wayne, L.W., Albright, S.C., Broadie, M.: Practical Management Science. Cengage Learning, Boston (2019)

# Optimizing Traceability in the Meat Supply Chain

Bárbara V. Schmidt and M. Susana Moreno(✉)

Planta Piloto de Ingeniería Química – PLAPIQUI, Universidad Nacional del Sur - CONICET, 8000 Bahía Blanca, Argentina
smoreno@plapiqui.edu.ar

**Abstract.** Over the past decades, several measures have been taken to solve the different public health problems caused by food. Among them, the traceability is one of outstanding importance, which allows tracking and tracing products efficiently. The objective of this work is to minimize the total dispersion of production batches in the supply chain of meat products in order to optimize their traceability. To do so, a mixed integer linear programming (MILP) model is developed, whose decision variables are employed to determine the existence of batches and their number at each node, their quantities, and the involved actors. The results show that an optimal number of batches can be produced and their dispersion diminished, which in turn allows minimizing the costs of possible product recalls in the case of a food safety incident. In addition, the formulation allows tracing the product batches at each node of the supply chain, identifying the batches of animals and raw materials utilized in their production.

**Keywords:** Meat supply chain · Batch dispersion · Traceability · MILP · Optimization

## 1 Introduction

Due to the different public health problems caused by food in the last decades, consumers are demanding more and more information about its quality and safety. This growing concern over food safety has generated greater pressure not only on food producers and distributors but also on the competent authorities for establishing systems and regulations that allow the identification, control, and tracking a food from its origin until the end of its commercialization [1]. It is in this context that traceability systems have gained notoriety, constituting a key tool for managing the food safety and quality in different agri-food Supply Chains (SC), in general, and in the Meat SC (MSC), in particular.

Although there exist several definitions for traceability, according to the *Codex Alimentarius* [2], it is defined as "the ability to follow the movement of a food through specific stage(s) of production, processing and distribution". Thus, establishing a traceability system is an effective method to guarantee the food safety, proving that the food is healthy and reliable such that, in the event of any problem, the information is available to limit its risk, allowing, if necessary, to withdraw the products from the market. However,

© Springer Nature Switzerland AG 2021
D. A. Rossit et al. (Eds.): ICPR-Americas 2020, CCIS 1407, pp. 56–70, 2021.
https://doi.org/10.1007/978-3-030-76307-7_5

this system does not reduce the quantity of recalled products in those cases where the production is carried out by mixing batches as happens, for example, in the production of meat products.

In the food industry, raw material batches from several suppliers, with different qualities attributes and prices, are often mixed to obtain batches of finished products. This is known as batch dispersion problem, in which disassembly and assembly processes occurs during production [3]. The mixture of raw materials batches in different product batches increases not only the complexity of their backward traceability or tracing but also the probability of contamination of the food which could cause product recalls.

At plant level, the work by Dupuy et al. [4] is one of the pioneers in addressing the batch dispersion problem from a mathematical modeling approach. These authors developed a mixed integer linear programming (MILP) model to optimize traceability by minimizing batch dispersion during the production of sausages, controlling the mixture of production batches to limit size and, consequently, cost and media impact of product recalls in case of problem. Later, Tamayo et al. [5] used genetic algorithms to solve the same problem proposed in [4]. Hu et al. [6] proposed a mathematical model based on dynamic programming to improve the traceability system in a Chinese dumpling factory by minimizing the product batches withdrawn from the market due to possible food safety crises. Dabbene et al. [7] posed a new MILP formulation to optimize traceability where an interconnected graph is used to represent the production process. These authors adopt the worst-case recall cost as performance criterion of the model and apply their approach to the same example proposed in [4] in order to demonstrate its effectiveness.

Regarding traceability in the MSC, Vélez Cervantes [8] developed a linear programming (LP) model for planning a fresh fish SC located at Esmeraldas (Ecuador) where a "traceability management system" is implemented by considering the quality of the shellfish around each node of the SC. On the other hand, Mohammed et al. [9] developed a fuzzy multi-objective MILP model for the design and product distribution planning in the MSC that incorporates traceability by calculating the costs for using a radio frequency identification (RFID) system.

To the best of the authors' knowledge, the integration of traceability and batch dispersion in MSC has not been implemented yet. For this reason, this work presents a MILP model for the optimization of traceability by reducing the batch dispersion throughout the MSC. Moreover, the problem allows to determine the actors involved in the production of meat products and calculate the size and number of both raw material and finished product batches obtained at the processing facilities and the batches produced in the slaughterhouses, identifying those that will be used in the production of each final product, thus achieving its traceability through the whole supply chain.

## 2 Problem Definition

The problem under study is intended to optimize the traceability system of an MSC composed by three echelons, as is shown in Fig. 1. Each slaughterhouse $f \in \{1,\ldots, F\}$ slaughters one or more animal species $a \in \{1,\ldots,A_f\}$, producing batches $h \in \{1,\ldots,H_{af}\}$ for each animal. The animals are sold in the form of half carcasses to the processing plants $e \in \{1,\ldots, E\}$ where they are dismembered in $m \in \{1,\ldots, M_e\}$ raw materials making batches $i \in \{1,\ldots, I_{me}\}$. In the following stage, these raw materials are cut to obtain the components $c \in \{1,\ldots, C_e\}$ which, according to a given recipe, are then mixed in the appropriate proportions to obtain the batches $k \in \{1,\ldots, K_{pe}\}$ of each product $p \in \{1,\ldots, P_e\}$ produced in each plant $e$. Finally, the products $p$ are distributed to each retailer $r \in \{1,\ldots, R\}$ in order to satisfy its demand $D_{pr}$ in a time horizon $HT$.

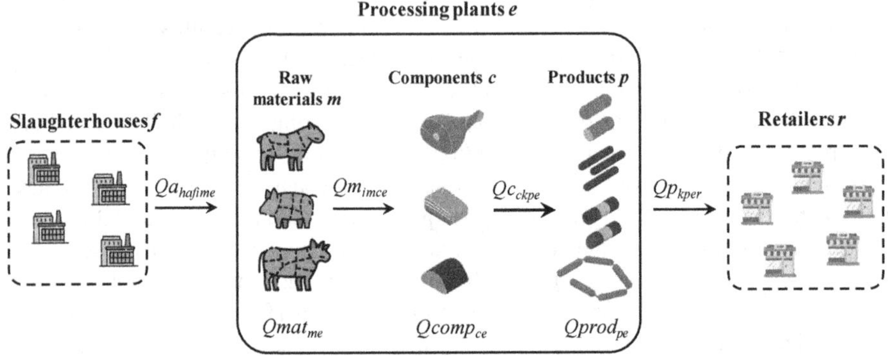


**Fig. 1.** Structure of the MSC under study.

In this problem it is assumed that, the specification of the different primal cuts $m$ that can be retrieved from each half carcass of animal $a$, the cutting pattern applied to raw material $m$ to obtain each component $c$, and the production recipe specifying the components $c$ utilized in the production of final product $p$ are known in each processing plant $e$. The following parameters represent these data:

- $Pa_{am}$: proportion of raw material $m$ obtained from animal $a$.
- $Pm_{mc}$: proportion of component $c$ obtained from raw material $m$.
- $Pc_{cp}$: proportion of component $c$ used in product $p$.

Also, the amount of product $p$ demanded by each retailer $r$, $D_{pr}$, and the maximum and minimum possible batch sizes for the batches of animal $a$ in each slaughterhouse $f$ ($bt_{af}^{\max}$, $bt_{af}^{\min}$) and, for each batch of raw material $m$ ($b_{me}^{\max}$, $b_{me}^{\min}$) and product $p$ ($bs_{pe}^{\max}$, $bs_{pe}^{\min}$) in every processing plant $e$ are model parameters. These latter bounds are based on the maximum and minimum processing or storage capacity of the corresponding echelon or stage. Note that, in this work, the maximum number of batches of animal, raw material, and product, respectively, is proposed (that is, the maximum number of elements of sets $H_{af}$, $I_{me}$, and $K_{pe}$). These numbers have been calculated based on the minimum possible size of the corresponding batch and the demands.

The objective function of the problem consists in minimizing the batch dispersion during the production of meat products, that is, from the purchase of the animal to the slaughterhouse until obtaining the final product, considering each stage of the production process.

In summary, the problem consists in determining the following simultaneously:

a) the supplier slaughterhouse, the quantity of required animal and the size of its batches, identifying the batch that is sent to each processing plant,
b) the number and size of raw material batches obtained at each processing plant and which ones are used for producing each product,
c) the number and size of product batches obtained at each plant,
d) the plant that supplies each retailer, identifying the product and the batch that each one receives.

Therefore, this model allows to identify the batch of animal (from a given slaughterhouse) that produces the product batch that is received by the retailer, thus guaranteeing the traceability of the product.

## 3 Model Formulation

This section describes the posed MILP formulation for minimizing the batch dispersion in a three-echelon MSC.

### 3.1 Slaughterhouse Constraints

First, in order to determine the batches produced in the slaughterhouses, the following binary variable is defined:

$$x_{haf} = \begin{cases} 1 \text{ if batch } h \text{ of animal } a \text{ is produced in slaughterhouse } f \\ 0 \text{ otherwise} \end{cases}$$

$$BT_{haf} \leq bt_{af}^{\max} \cdot x_{haf} \qquad \forall h \in H_{af}, a \in A_f, f \tag{1}$$

$$BT_{haf} \geq bt_{af}^{\min} \cdot x_{haf} \qquad \forall h \in H_{af}, a \in A_f, f \tag{2}$$

Equations (1) and (2) ensure that the size of batch $h$ of animal $a$ at slaughterhouse $f$, $BT_{haf}$, vary within its upper and lower bounds ($bt_{af}^{\max}$, $bt_{af}^{\min}$), respectively. Also, these constraints impose that the batch size is zero if it is not produced ($x_{haf} = 0$).

$$x_{h+1,af} \leq x_{haf} \quad \forall\, 1 \leq h < H_{af}, a \in A_f, f \tag{3}$$

Equation (3) establishes that batches $h$ in each slaughterhouse $f$ are consecutively produced in ascending numerical order, i.e. batch $h + 1$ can be produced only if batch $h$ has already been produced.

$$Qan_{af} = \sum_{h \in H_{af}} BT_{haf} \qquad \forall a \in A_f, f \tag{4}$$

Equation (4) defines the total amount of animal $a$ processed in slaughterhouse $f$, $Qan_{af}$, as the sum of the batches $h$ produced.

$$Pa_{am} \cdot BT_{haf} = \sum_{e} \sum_{i \in I_{me}} Qa_{hafime} \qquad \forall h \in H_{af}, a \in A_f, f, m \in MR_{ae} \tag{5}$$

Equation (5) expresses that each batch $h$ of meat animal carcasses $a$ sent to the processing plants $e$, is cut up to obtain the raw material cuts $m$, according to the known proportion $Pa_{am}$. The continuous variable $Qa_{hafime}$ represents the quantity of animal $a$ in batch $h$ sent from the slaughterhouse $f$ to the processing plant $e$ that is used in the batch $i$ of raw material $m$. Here, for each plant $e$, the set $MR_{ae}$ is defined containing the subset of raw materials $m$ that are obtained from animal $a$.

$$\sum_{e} \sum_{m \in MR_{ae}} \sum_{i \in I_{me}} Qa_{hafime} \leq Qa_{af}^{\max} \cdot x_{haf} \qquad \forall h \in H_{af}, a \in A_f, f \tag{6}$$

If the batch $h$ of animal $a$ in slaughterhouse $f$ exists (i.e., $x_{haf} = 1$), Eq. (6) ensures that the quantity $Qa_{hafime}$ is less than the maximum amount of animal that can be processed at the slaughterhouse $f$, $Qa_{af}^{\max}$.

$$BT_{h+1,af} \leq BT_{haf} \quad \forall\, 1 \leq h < H_{af}, a \in A_f, f \tag{7}$$

In order to avoid alternative optimal solutions with the same objective function value, Eq. (7) is used to enforce that batch sizes follow a numerical sequence.

## 3.2 Processing Plant Constraints

**Raw Materials**

Let $w_{ime}$ be the binary variable defined by:

$$w_{ime} = \begin{cases} 1 \text{ if batch } i \text{ of raw material } m \text{ is produced in processing plant } e \\ 0 \text{ otherwise} \end{cases}$$

$$\sum_{e} \sum_{m \in MR_{ae}} w_{ime} \leq \sum_{f} \sum_{h \in H_{af}} x_{haf} \quad \forall a \in A_f, i \in I_{me} \tag{8}$$

Constraint (8) ensures that if the batches $h$ of animal $a$ are not produced, then none of the batches $i$ of raw materials $m$ belonging to that animal are elaborated at any plant $e$.

$$B_{ime} \leq b_{me}^{\max} \cdot w_{ime} \quad \forall i \in I_{me}, m \in M_e, e \tag{9}$$

$$B_{ime} \geq b_{me}^{\min} \cdot w_{ime} \quad \forall i \in I_{me}, m \in M_e, e \tag{10}$$

In case batch $i$ of raw material $m$ is elaborated in the processing plant $e$, i.e. $w_{ime} = 1$, Eqs. (9) and (10) require the batch size $B_{ime}$ to be between its upper $b_{me}^{\max}$ and lower $b_{me}^{\min}$ bound, respectively.

$$w_{i+1,me} \leq w_{ime} \quad \forall 1 \leq i < I_{me}, m \in M_e, e \tag{11}$$

In order to reduce the number of alternative solutions, Eq. (11) specifies that the production of raw material batches $i$ is made in ascending numerical order.

$$B_{ime} = \sum_{f} \sum_{a \in A_f} \sum_{h \in H_{af}} Qa_{hafime} \quad \forall i \in I_{me}, m \in M_e, e \tag{12}$$

Due to the fact that a batch $i$ of raw material $m$ in each plant $e$ can be obtained from different batches $h$ of animal $a$ supplied by different slaughterhouses $f$, Eq. (12) establishes that its size, $B_{ime}$, is equal to the sum of the quantities that come from all of them.

$$\sum_{f}\sum_{a\in A_f}\sum_{h\in H_{af}} Qa_{hafime} \leq Qa_{af}^{\max}\cdot w_{ime} \qquad \forall i\in I_{me}, m\in M_e, e \tag{13}$$

If the batch $i$ of raw material $m$ exists in processing plant $e$ ($w_{ime} = 1$), then, as enforced by Eq. (13), the quantity of animal $a$ in batch $h$ sent from slaughterhouse $f$ to plant $e$ used in batch $i$ of raw material $m$, $Qa_{hafime}$, must not exceed its upper limit, $Qa_{af}^{\max}$.

$$Qmat_{me} = \sum_{i\in I_{me}} B_{ime} \qquad \forall m\in M_e, e \tag{14}$$

The total amount of raw material $m$ available in each processing plant $e$, $Qmat_{me}$, is defined by the sum of the batches $i$ produced (Eq. 14).

$$Pm_{mc}\cdot B_{ime} \geq Qm_{imce} \qquad \forall i\in I_{me}, m\in M_e, c\in CS_{me}, e \tag{15}$$

Equation (15) represents the disassembly of the raw material $m$ to obtain the components or meat cuts $c$, according to the known proportion $Pm_{mc}$. Here, the continuous variable $Qm_{imce}$ is the amount of component $c$ that is obtained from the batch $i$ of raw material $m$ and the set $CS_{me}$ contains the subset of components $c$ that come from raw material $m$ in the processing plant $e$.

$$Qmat_{me} \geq \sum_{c\in CS_{me}}\sum_{i\in I_{me}} Qm_{imce} \quad \forall m\in M_e, e \tag{16}$$

Equation (16) requires the summation of the quantities of all components $c$ obtained from all the raw material batches $i$ cannot be greater than the total amount of raw material $m$ from which they are obtained.

$$Qm_{imce} \leq Qm_{me}^{\max}\cdot w_{ime} \qquad \forall i\in I_{me}, m\in M_e, c\in CS_{me}, e \tag{17}$$

Equation (17) states that the amount of component $c$ obtained from batch $i$ of raw material $m$ in the plant $e$, must be less than or equal to the maximum amount of raw material that can be processed at the plant, $Qm_{me}^{\max}$.

$$B_{i+1,me} \leq B_{ime} \quad \forall 1\leq i < I_{me}, m\in M_e, e \tag{18}$$

To avoid symmetric solutions, Eq. (18) is added to guarantee that the assignment of raw material batch sizes is carried out in numerical sequence.

**Components**

$$Qcomp_{ce} = \sum_{m\in MS_{ce}}\sum_{i\in I_{me}} Qm_{imce} \quad \forall c\in C_e, e \tag{19}$$

Equation (19) establishes that the quantity of component $c$ produced in plant $e$, $Qcomp_{ce}$, comes exclusively from the summation of all batches $i$ of raw materials $m$ processed in the plant. The set $MS_{ce}$ contains the subset of raw materials $m$ that can produce the component $c$.

$$Qcomp_{ce} = \sum_{p \in PS_{ce}} \sum_{k \in K_{pe}} Qc_{ckpe} \quad \forall c \in C_e, e \tag{20}$$

Besides, Eq. (20) defines that $Qcomp_{ce}$ is equal to the summation of the amount of component $c$ used in the elaboration of all batches $k$ of finished products $p$ in the processing plant $e$. Here, the set $PS_{ce}$ contains the subset of products $p$ that use the component $c$ in the plant $e$.

**Products**

Let $z_{kpe}$ be the binary variable defined by:

$$z_{kpe} = \begin{cases} 1 \text{ if batch } k \text{ of product } p \text{ is produced in processing plant } e \\ 0 \text{ otherwise} \end{cases}$$

$$BS_{kpe} \leq bs_{pe}^{\max} \cdot z_{kpe} \quad \forall k \in K_{pe}, p \in P_e, e \tag{21}$$

$$BS_{kpe} \geq bs_{pe}^{\min} \cdot z_{kpe} \quad \forall k \in K_{pe}, p \in P_e, e \tag{22}$$

Equations (21) and (22) require the size of batch $k$ of product $p$ in plant $e$, $BS_{kpe}$, to be between the maximum and minimum possible batch sizes ($bs_{pe}^{\max}$ and $bs_{pe}^{\min}$), respectively.

$$z_{k+1,pe} \leq z_{kpe} \quad \forall 1 \leq k < K_{pe}, p \in P_e, e \tag{23}$$

As in the previous sections, in order to reduce the search space, it is assumed that in each processing plant the product batches are produced in ascending order, which is enforced through Eq. (23).

$$Qprod_{pe} = \sum_{k \in K_{pe}} BS_{kpe} \quad \forall p \in P_e, e \tag{24}$$

Equation (24) shows that the total quantity of product $p$ produced in the processing plant $e$, $Qprod_{pe}$, is defined as the summation of all product batches $k$.

$$Pc_{cp} \cdot BS_{kpe} = Qc_{ckpe} \quad \forall c \in CR_{pe}, k \in K_{pe}, p \in P_e, e \tag{25}$$

Equation (25) calculates the amount of component $c$ used in the batch $k$ of product $p$ in plant $e$, $Qc_{ckpe}$ by considering the proportion $Pc_{cp}$ given in the production recipe. In this expression, $CR_{pe}$ denotes the set of components that can be used for producing finished product $p$ in the plant $e$.

$$Qc_{ckpe} \leq Qc_{ce}^{\max} \cdot z_{kpe} \quad \forall c \in CR_{pe}, k \in K_{pe}, p \in P_e, e \tag{26}$$

Equation (26) forces the amount of component $c$ used in batch $k$ of product $p$ in the plant $e$ to be lower than the maximum amount of component that can be processed, $Qc_{ce}^{\max}$. This upper limit only applies if there exists the batch $k$ of product $p$ in the plant $e$.

$$Qprod_{pe} = \sum_{c \in CR_{pe}} \sum_{k \in K_{pe}} Qc_{ckpe} \qquad \forall p \in P_e, e \tag{27}$$

The mass balance in (27) establishes that the amount of product $p$ is equal to the sum of all components $c$ mixed (assembled) to produce it in plant $e$.

$$BS_{k+1,pe} \leq BS_{kpe} \qquad \forall 1 \leq k < K_{pe}, p \in P_e, e \tag{28}$$

In order to improve the solution efficiency, Eq. (28) removes equivalent solutions by forcing that, for each product $p$, the size of batch $k$ must be greater than or equal to the size of batch $k + 1$.

### 3.3 Retailers Constraints

$$D_{pr} = \sum_{e} \sum_{k \in K_{pe}} Qp_{kper} \qquad \forall p \in P_e, r \tag{29}$$

Let $Qp_{kper}$ be a nonnegative variable that represents the amount of finished product $p$ belonging to batch $k$ produced in the plant $e$ and delivered to the retailer $r$. Equation (29) ensures the demand satisfaction in the different retailers. This equation states that the total demand of product $p$ in each retailer $r$ has to be fulfilled using as many batches $k$ of product $p$ as necessary that may come from one or more plants $e$.

$$\sum_{r} Qp_{kper} \leq BS_{kpe} \qquad \forall k \in K_{pe}, p \in P_e, e \tag{30}$$

Equation (30) states that the quantity of product $p$ belonging to batch $k$ produced in plant $e$ sent to all retailers $r$ must be less than or equal to the batch size, $BS_{kpe}$.

$$\sum_{k \in K_{pe}} \sum_{r} Qp_{kper} \leq Qprod_{pe} \qquad \forall p \in P_e, e \tag{31}$$

In turn, Eq. (31) expresses that the total amount of product $p$ delivered to all retailers must be less than or equal to the amount of product $p$ available in the plant $e$, $Qprod_{ep}$.

### 3.4 Constraints on the Use of Batches

The following binary variables are introduced to express the relationships between the different types of batches along the MSC:

$$v_{hafime} = \begin{cases} 1 & \text{if batch } h \text{ of animal } a \text{ from slaughterhouse } f \text{ is used in batch } i \text{ of raw} \\ & \text{material } m \text{ in processing plant } e \\ 0 & \text{otherwise} \end{cases}$$

$$y_{imckpe} = \begin{cases} 1 & \text{if batch } i \text{ of raw material } m \text{ produces the component } c \text{ utilized in} \\ & \text{batch } k \text{ of finished product } p \text{ in processing plant } e \\ 0 & \text{otherwise} \end{cases}$$

$$v_{hafime} \leq x_{haf} \qquad \forall h \in H_{af}, a \in A_f, f, i \in I_{me}, m \in MR_{ae}, e \tag{32}$$

$$v_{hafime} \leq w_{ime} \qquad \forall h \in H_{af}, a \in A_f, f, i \in I_{me}, m \in MR_{ae}, e \tag{33}$$

If production of both batch $h$ of animal $a$ in the slaughterhouse $f$ ($x_{haf} = 0$) and batch $i$ of raw material $m$ in the plant $e$ ($w_{ime} = 0$) is null, then the binary variable $v_{hafime}$ takes value zero. This condition is guaranteed by Eqs. (32) and (33).

$$Qa_{hafime} \leq Qa_{af}^{\max} \cdot v_{hafime} \qquad \forall h \in H_{af}, a \in A_f, f, i \in I_{me}, m \in MR_{ae}, e \tag{34}$$

Equation (34) establishes that the quantity $Qa_{hafime}$ is limited by the maximum amount of animal that can be processed in the slaughterhouse.

$$y_{imckpe} \leq w_{ime} \qquad \forall i \in I_{me}, m \in MS_{ce}, c \in CR_{pe}, k \in K_{pe}, p \in P_e, e \tag{35}$$

$$y_{imckpe} \leq z_{kpe} \qquad \forall i \in I_{me}, m \in MS_{ce}, c \in CR_{pe}, k \in K_{pe}, p \in P_e, e \tag{36}$$

Also, if the production of both batch $i$ of raw material $m$ ($w_{ime} = 0$) and batch $k$ of product $p$ in the processing plant $e$ ($z_{kpe} = 0$) is null, then binary variable $y_{imckpe}$ takes value zero. Inequalities (35) and (36), respectively, are used to enforce this condition.

$$Qmcp_{imckpe} \leq Qcomp_{ce}^{\max} \cdot y_{imckpe} \quad \forall i \in I_{me}, m \in MS_{ce}, c \in CR_{pe}, k \in K_{pe}, p \in P_e, e \tag{37}$$

$$\sum_{p \in PS_{ce}} \sum_{k \in K_{pe}} Qmcp_{imckpe} = Qm_{imce} \qquad \forall i \in I_{me}, m \in MS_{ce}, c \in C_e, e \tag{38}$$

$$\sum_{m \in MS_{ce}} \sum_{i \in I_{me}} Qmcp_{imckpe} = Qc_{ckpe} \qquad \forall c \in CR_{pe}, k \in K_{pe}, p \in P_e, e \tag{39}$$

Equations (37)–(39) relates the continuous variable $Qmcp_{imckpe}$ defined as the amount of the batch $i$ of raw material $m$ used in the batch $k$ of finished product $p$ in plant $e$ with the binary variable $y_{imckpe}$, the quantity of raw material $m$ used to obtain component $c$, and the amount of component $c$ used to produce product $p$.

### 3.5 Objective Function

$$FO = \sum_{e} \sum_{m \in MS_{ce}} \sum_{i \in I_{me}} \sum_{c \in CR_{pe}} \sum_{p \in P_e} \sum_{k \in K_{pe}} y_{imckpe} + \sum_{f} \sum_{a \in A_f} \sum_{h \in H_{af}} \sum_{e} \sum_{m \in MR_{ae}} \sum_{i \in I_{me}} v_{hafime} \tag{40}$$

The objective function of the posed problem consists in minimizing the batch dispersion in the entire MSC.

## 4 Study Case

The application of the proposed MILP model will be illustrated with a study case that considers an MSC consisting of three slaughterhouses ($F = 3$), two processing plants ($E = 2$) and three retailers ($R = 3$). In addition, there are 2 types of animals ($A = 2$) and 12 different products ($P = 12$), namely, pure pork chorizo (p1), salami (p2), second-quality or mixture chorizo (p3), mortadella (p4), Vienna sausage (p5), boiled ham (p6), cooked blood sausage (p7), pork cheese (p8), longaniza (p9), bondiola (p10), pork loin (p11), and bacon (p12). For producing these products, seven raw materials ($M = 7$) corresponding to primal cuts of both types of animals and thirteen components ($C = 13$) referring to the raw material cuts are considered.

The slaughterhouse f1 can slaughter pigs (a1), f2 cows (a2), and f3 both species (a1 and a2). In turn, for each slaughterhouse the number of batches is upper bounded by $H_{f1} = 5$ for f1, $H_{f2} = 3$ for f2 and $H_{f3} = 4$ for f3. Each processing plant can produce only a subset of products. Namely, plant e1 can elaborate p1, p2, p3, p4, p7, p8 and p9 ($P_{e1} = 7$) whereas plant e2 can produce p2, p3, p4, p5, p6, p10, p11 and p12 ($P_{e2} = 8$). In both plants, the maximum number of raw material batches and finished products batches are $I = 5$ and $K = 4$, respectively.

Data on the proportions of raw materials given by each animal and the components given by each raw material are shown in Table 1 while the proportion of components used in the production of each product and the monthly product demands of the retailers are detailed in Table 2.

**Table 1.** Data for the disassembly of animals and raw materials.

| $m$ | $Pa_{am}$ | | $Pm_{mc}$ | | | | | | | | | | | | |
|---|---|---|---|---|---|---|---|---|---|---|---|---|---|---|---|
| | a1 | a2 | c1 | c2 | c3 | c4 | c5 | c6 | c7 | c8 | c9 | c10 | c11 | c12 | c13 |
| m1 | 0.18 | | 0.25 | 0.75 | | | | | | | | | | | |
| m2 | 0.13 | | | | 0.50 | | | | | | 0.42 | 0.08 | | | |
| m3 | | 0.17 | | | | 0.30 | 0.70 | | | | | | | | |
| m4 | 0.07 | | | | | | | 0.20 | 0.80 | | | | | | |
| m5 | 0.14 | | | | | | | | | 0.38 | | | | | 0.62 |
| m6 | 0.17 | | | | | | | | | | | | 0.07 | 0.33 | 0.60 |
| m7 | 0.13 | | | | | | | | | | | | | | 1.00 |

The presented MILP model was implemented and solved in GAMS 31.1.1 on a PC Intel (R) Core i7-10700, 2.9 GHz using CPLEX 12.10 solver. The formulation involves 6,527 equations with 3,924 continuous variables and 1,685 binary variables. The resolution time is 13.58 CPU seconds with a 3% margin of optimality. In the optimal solution, the value of the objective function for the batch dispersion is 100, which means that 10 batches $h$ for both types of animal obtained in the slaughterhouses f1 and f3 are dispersed in 30 raw material batches $i$ which, in turn, are part of 19 batches $k$ of finished product in both processing plants.

**Table 2.** Data for the assembly of each product and their demands.

| $p$ | $Pc_{cp}$ | | | | | | | | | | | | | $D_{pr}(kg)$ | | |
|---|---|---|---|---|---|---|---|---|---|---|---|---|---|---|---|---|
| | c1 | c2 | c3 | c4 | c5 | c6 | c7 | c8 | c9 | c10 | c11 | c12 | c13 | r1 | r2 | r3 |
| p1 | 0.35 | 0.40 | 0.25 | | | | | | | | | | | 100 | 0 | 0 |
| p2 | 0.30 | 0.50 | 0.20 | | | | | | | | | | | 200 | 100 | 100 |
| p3 | | 0.45 | 0.25 | 0.30 | | | | | | | | | | 300 | 100 | 200 |
| p4 | 0.20 | | 0.10 | 0.30 | 0.40 | | | | | | | | | 2000 | 1000 | 800 |
| p5 | 0.15 | 0.30 | 0.35 | | 0.20 | | | | | | | | | 3000 | 1700 | 0 |
| p6 | 0.20 | 0.60 | 0.20 | | | | | | | | | | | 2000 | 1000 | 900 |
| p7 | | | | | | | 0.10 | | 0.50 | | | | 0.40 | 1000 | 0 | 0 |
| p8 | | | | | | | | | 0.50 | | | | 0.50 | 0 | 0 | 750 |
| p9 | | | | | | 0.25 | | | | 0.25 | | | 0.50 | 0 | 0 | 800 |
| p10 | | | | | | | | 1.00 | | | | | | 0 | 3500 | 0 |
| p11 | | | | | | | | | | | 1.00 | | | 0 | 1200 | 0 |
| p12 | | | | | | | | | | | | 1.00 | | 0 | 1500 | 0 |

Figure 2 illustrates the results obtained in the optimal solution giving a detail of the numbers of batches of each animal obtained in the selected slaughterhouses and the number of raw material batches and product batches produced in both processing plants, as well as their dispersion throughout the MSC. Table 3 shows a detail of the quantities of each batch of animal sent to each plant to form the different raw material batches whereas Tables 4 and 5 summarize the amount of each product batch delivered from plants 1 and 2 to each retailer, respectively.

As mentioned previously, the developed model allows identifying the batch of animal used in each finished product. For example, it can be seen from Fig. 2 that the total amount of second-quality chorizo (p3) requested by the three retailers is met by producing a single batch (k1) in processing plant e2. This batch is composed of 270 kg of c2 coming from the batch i1 of m1, 150 kg of c3 obtained from batch i1 of m2, and 180 kg of c4 coming from the batch i1 of m3. Moreover, in Table 3, it can be seen that batch i1 of m1 comes from the batch h3 of pork (a1) obtained in f1 and the batch h4 obtained in f3, the batch i1 of m2 comes from the batch h2 of f3, and the batch i1 of m3 comes from batch h2 of beef (a2) obtained in f3. Although the batch i1 of m2 could be obtained by using batches from the same slaughterhouse, this situation occurs because decisions are made in order to minimize dispersion and generate the exact amounts of components. It should be pointed out that, in this problem, total costs are not considered, therefore, choosing batches coming from one slaughterhouse or another is indistinct, since the unique performance criteria pursued by the model is to minimize batch dispersion.

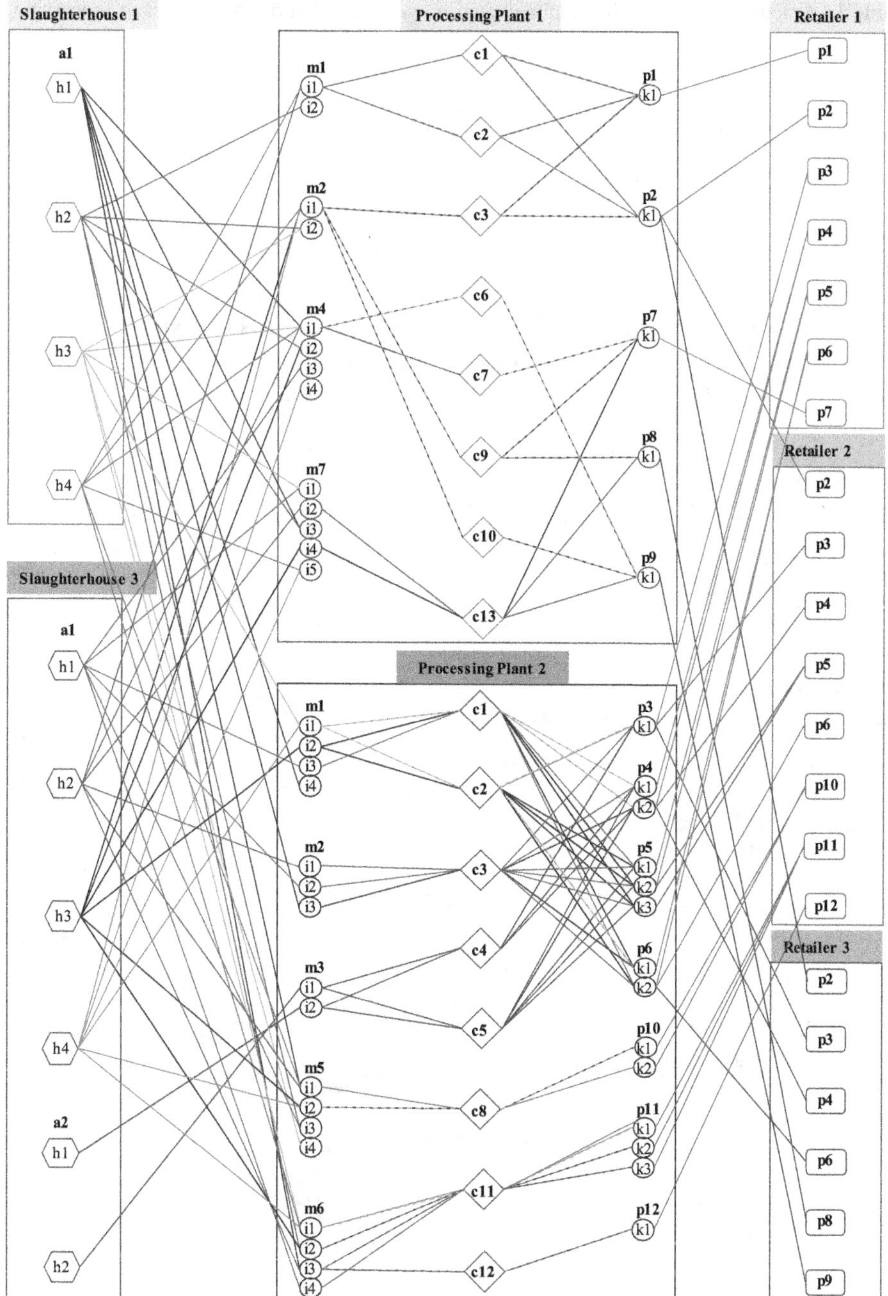


**Fig. 2.** Results for the studied MSC.

**Table 3.** Amount of each batch of animal used in each raw material batch at each processing plant.

| | | $Qa_{hafime}$(kg) | | | | | | | | | |
|---|---|---|---|---|---|---|---|---|---|---|---|
| | | f1 | | | | f3 | | | | | |
| | | a1 | | | | a1 | | | | a2 | |
| | | h1 | h2 | h3 | h4 | h1 | h2 | h3 | h4 | h1 | h2 |
| Processing plant 1 | | | | | | | | | | | |
| m1 | i1 | | | | 1135.9 | | 3528.4 | | | | |
| | i2 | | 1764.2 | | | | | | | | |
| m2 | i1 | | | | 820.4 | | | 2548.3 | 1728.0 | | |
| | i2 | | 1274.2 | 1274.2 | | | | | | | |
| m4 | i1 | 686.1 | | 686.1 | 441.7 | | 1372.2 | | | | |
| | i2 | | 686.1 | | | | | 1372.2 | | | |
| | i3 | | | | | 1372.2 | | | | | |
| | i4 | | | | | | | | 930.5 | | |
| m7 | i1 | | | 1274.2 | | 2548.3 | | | | | |
| | i2 | | | | | | 2548.3 | | | | |
| | i3 | 1274.2 | 1274.2 | | | | | | | | |
| | i4 | | | | | | | 2548.3 | | | |
| | i5 | | | | 820.4 | | | | 1728.0 | | |
| Processing plant 2 | | | | | | | | | | | |
| m1 | i1 | | | 1764.2 | | | | | 2392.6 | | |
| | i2 | | | | | | | 3528.5 | | | |
| | i3 | | | | | 3528.5 | | | | | |
| | i4 | 1764.2 | | | | | | | | | |
| m2 | i1 | | | | | | 2548.3 | | | | |
| | i2 | | | | | 2548.3 | | | | | |
| | i3 | 1274.2 | | | | | | | | | |
| m3 | i1 | | | | | | | | | | 2742.9 |
| | i2 | | | | | | | | | 2742.9 | |
| m5 | i1 | 1372.2 | | | 883.5 | | 2744.4 | | | | |
| | i2 | | | | | | | 2744.4 | 1860.9 | | |
| | i3 | | 1372.2 | | | 2744.4 | | 3332.4 | | | |
| | i4 | | | 1372.2 | | | | | | | |
| m6 | i1 | | | 1666.2 | 1072.8 | | | | 2259.7 | | |
| | i2 | | 1666.2 | | | | | | | | |
| | i3 | 1666.2 | | | | | 3332.4 | | | | |
| | i4 | | | | | 3332.4 | | | | | |

**Table 4.** Values for $Qp_{kper}$ for processing plant1.

| | $Qp_{kper}$ (kg) | | | | |
|---|---|---|---|---|---|
| | Processing plant1 | | | | |
| | p1 | p2 | p7 | p8 | p9 |
| | k1 | k1 | k1 | k1 | k1 |
| r1 | 100 | 200 | 1200 | | |
| r2 | | 100 | | | |
| r3 | | 100 | | 750 | 1000 |

**Table 5.** Values for $Qp_{kper}$ for processing plant 2.

| | $Qp_{kper}$ (kg) | | | | | | | |
|---|---|---|---|---|---|---|---|---|
| | Processing plant 2 | | | | | | | |
| | p3 | p4 | | p5 | | | p6 | |
| | k1 | k1 | k2 | k1 | k2 | k3 | k1 | k2 |
| r1 | 300 | 1900 | 100 | 2000 | 1000 | | 1950 | 50 |
| r2 | 100 | | 1000 | | 1000 | 700 | | 1000 |
| r3 | 200 | | 800 | | | | | 900 |
| | p12 | | | p11 | | | p10 | |
| | k1 | | | k1 | k2 | k3 | k1 | k2 |
| r1 | | | | | | | | |
| r2 | 1600 | | | 583.2 | 308.4 | 308.4 | 1750 | 1750 |
| r3 | | | | | | | | |

## 5 Conclusions

In this work, a MILP model has been proposed to solve the problem of batch dispersion in an MSC in order to optimize the batch traceability throughout the supply chain.

An MSC composed of 3 echelons was studied: slaughterhouses where different types of animals are slaughter, meat processing plants that produce different meat products, and retailers that demand these products. The posed formulation allows determining the type and amount of animal supplied by each slaughterhouse, the number and size of the batches, and the quantity of each of them that is sent to each processing plant. In turn, in this latter, the model also decides which raw materials and products to process, the number and size of their batches, the quantity of components to produce, and what product batches to use and in which amount to meet the retailers' demands. Furthermore, by minimizing the sum of the dispersions of all batches of animals, raw materials, and products in the MSC, the proposed model minimizes, in case of a problem with a product batch, the costs of its withdrawal from the market.

The presented case study shows the capability of the model to address the traceability in the MSC through an integrated approach, by determining the batch sizes in each echelon of the CS, while solving the problem of batch dispersion.

In future work, the MSC's total cost will be incorporated as an additional performance measure to obtain a multi-objective formulation of this problem.

## References


1. OMS Homepage. https://www.who.int/es/news-room/fact-sheets/detail/food-safety. Accessed 29 July 2020
2. Codex Alimentarius Commission: Principles of traceability/product tracing as a tool within food inspection and certification system. CAC/GL 60-2006 (2006)
3. Dabbene, F., Gay, P., Tortia, C.: Traceability issues in food supply chain management: a review. Biosyst. Eng. **120**, 65–80 (2014)
4. Dupuy, C., Botta-Genoulaz, V., Guinet, A.: Batch dispersion model to optimise traceability in food industry. J. Food Eng. **70**, 333–339 (2005)
5. Tamayo, S., Monteiro, T., Sauer, N.: Deliveries optimization by exploiting production traceability information. Eng. Appl. Artif. Intell. **22**, 557–568 (2009)
6. Hu, Z., Jian, Z., Ping, S., Xiaoshuan, Z., Weisong, M.: Modeling method of traceability system based on information flow in meat food supply chain. WSEAS Trans. Inf. Sci. Appl. **7**(6), 1094–1103 (2009)
7. Dabbene, F., Gay, P.: Food traceability systems: performance evaluation and optimization. Comput. Electron. Agric. **75**, 139–146 (2011)
8. Vélez Cervantes, P.: Estudio de optimización de una cadena de suministros pesquera usando un enfoque de ingeniería de sistemas de procesos. Tesis de Maestría, UNS (2016)
9. Mohammed, A., Wang, Q.: The fuzzy multi-objective distribution planner for a green meat supply chain. Int. J. Prod. Econ. **184**, 47–58 (2017)

# Energy Optimization for the Operation of a Sawmill

Nicolás Vanzetti[1(✉)], Néstor G. Steitzer[3], Gabriela Corsano[1,2], and Jorge M. Montagna[1]

[1] Instituto de Desarrollo y Diseño (INGAR, CONICET-UTN), Avellaneda 3657, S3002GJC Santa Fe, Argentina
{nvanzetti,gcorsano,mmontagna}@santafe-conicet.gov.ar

[2] Facultad de Ingeniería Química, Universidad Nacional del Litoral, Santiago del Estero 2829, Santa Fe, Argentina

[3] Suabia Maderas SRL, Av. Fundador Oeste 3380, Eldorado, Misiones, Argentina

**Abstract.** The wood drying is an important stage in the production process of a sawmill since it is where lumber acquires the level of humidity required by the market. Generally, it is a batch process that is carried out in dryers that operate in parallel out of phase. To cover the energy needs of the process, boilers fed with the by-products generated in the productive stages of the sawmill are used. To operate efficiently, the filling and scheduling of the dryers and the boiler feed and operation must be planned simultaneously. In this paper, these decisions are addressed through a mixed integer linear programming model (MILP) using state-task network concepts based on a discrete time formulation. Different work criteria related to sawmill operation are analysed, achieving very significant results for the performance of this industry.

**Keywords:** Drying kilns · Sawmills · State-Task Network · Planning · Scheduling · MILP model

## 1 Introduction

The forest industry plays an important economic, environmental and social role in the development of the northeast region of Argentina. In this region there are more than 85% of the forest plantations for industrial purposes in the country and it is made up of more than 1,000 factories, of which sawmills represent 95% and, therefore, lumber production becomes a fundamental pillar of this industry.

The process in the sawmills consists of several production stages, where one of the most important is the drying. In this stage, the boards acquire the characteristics required by the market, increasing the value and quality of the wood. Usually, this stage is a batch process that is carried out in dryers or drying kilns, which may have different characteristics. To heat these furnaces, boilers are employed that mainly use the by-products generated in the activities of the sawmill as fuel (bark, sawdust, wood shavings and chip). With the steam generated, the air is heated by means of a heat exchanger and is forced to circulate through the dryers.

© Springer Nature Switzerland AG 2021
D. A. Rossit et al. (Eds.): ICPR-Americas 2020, CCIS 1407, pp. 71–84, 2021.
https://doi.org/10.1007/978-3-030-76307-7_6

To carry out the planning of this stage, the filling and scheduling of the dryers must be simultaneously considered with the feeding and operation of the boilers. To achieve even drying of the products in each drying cycle, each kiln must be filled with boards of the same thickness. The duration of each cycle depends on the thickness of the boards and the dimension of the kiln. Likewise, for each board thickness there is a set of drying programs that vary the cycle time and, consequently, the energy demand.

With the rise of the plants generating electricity from forest biomass, the by-products of the sawmills have increased in value, ceasing to be just a residue of the production that was used to feed the boilers. Therefore, the use of sawmill by-products must be appropriately planned in order to cover the energy needs and increase the processes profitability, avoiding the tradeoffs between both productions: lumber and electric power.

In the literature, few papers study the drying operation and most focus on improving the energy efficiency of the process or addressing the implementation of new technological alternatives such as solar dryers [1, 2]. To the best of our knowledge, only the work of Gaudreault et al. [3] approaches the planning of this operation through the development of two models: a mixed integer linear programming (MILP) and a constrained programming. Both approaches minimize the delay of product delivery, determining the batch allocation to each kiln and the operating times, but they do not take into account energy requirements for drying.

The planning of the drying stage can be classified as a programming problem of non-identical parallel units, and has been the subject of several studies [4, 5]. For the approach presented in this work, a State-Task Network (STN) formulation based on discrete times was chosen, since it allows a better control of the resources involved in the process [6, 7]. STN formulation was introduced by Kondili et al. [8], and in Castro et al. [9] a description of its application in different scheduling problems is presented. This approach has proven to be very effective in tackling this problem because it has allowed a very complex problem to be represented in a simple way.

In this work a MILP model is presented for the optimal planning of the drying stage in a sawmill for a given time horizon, where all involved decisions are made simultaneously. In this way, an adequate drying plan is obtained according to the availability of boards to be dried, by-products to cover the energy demand and the production policies of the firm, assessing all the involved trade-offs.

## 2 Problem Description

The drying of the wood is an intermediate stage in the production process of the sawmills. Its correct planning avoids the accumulation of wet boards generated in the previous stage and delivers boards with the required humidity levels to the next stage (Fig. 1).

Boards enter this stage with the objective of reducing their humidity level to comply with market requirements. These boards are grouped according to their thickness and put together with the inventory of non-dry boards, which are waiting to be dried. The frequency in which the boards arrive at this stage depends on the production in the previous stage of sawing.

The sawmill has a set of drying kilns that do not necessarily have the same dimensions. When the dryer is loaded, it begins a drying cycle. To ensure even drying, each

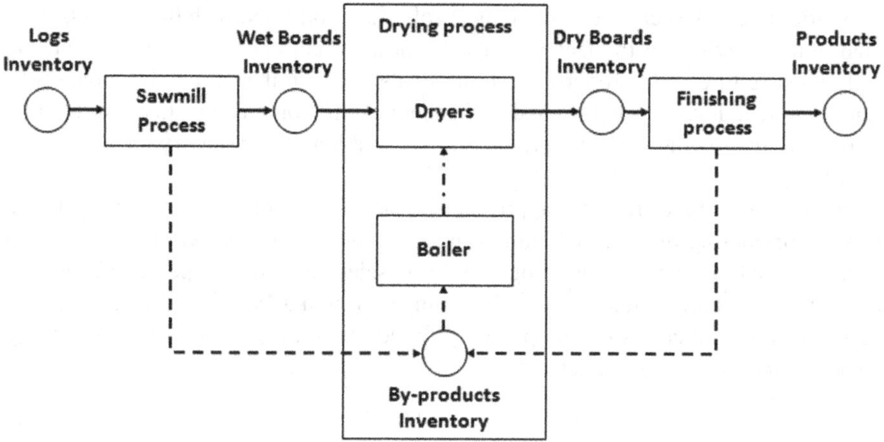


**Fig. 1.** Lumber production process

drying cycle must contain boards of a single thickness. On the other hand, to guarantee the efficient use of the dryers, a minimum load is established for each kiln. The duration of each cycle varies according to the size of the kiln and the thickness of the boards to be dried. At the same time, for each board thickness there is a set of drying programs with a different duration of the drying cycle and, consequently, a different energy demand. Each drying program considers three stages: kiln heating, drying, and cooling (Fig. 2). In addition, there is a setup time between each drying cycle, which depends on each dryer and corresponds to its filling.

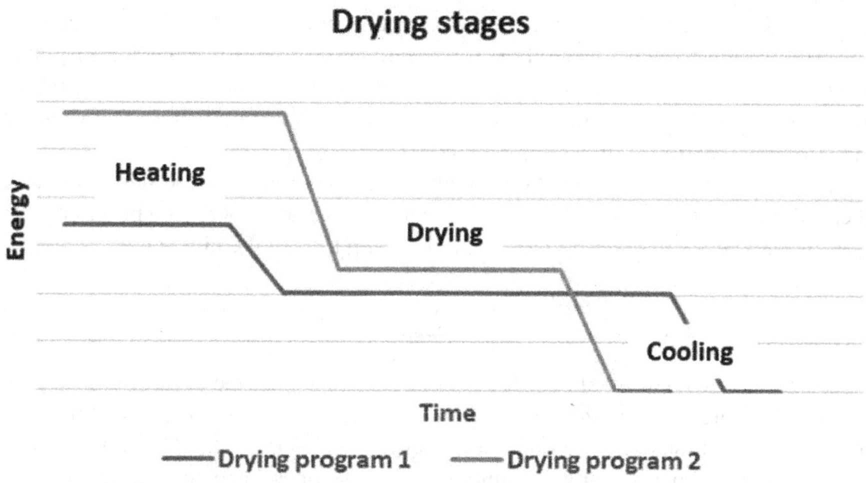


**Fig. 2.** Stages of a drying program

In order to cover energy requirements, a solid fuel boiler fed with by-products generated in the activities of the factory is used. For a correct operation of the boiler, the supply must fulfil certain proportion requirements between the various by-products to ensure humidity levels, volatility and ash generation in combustion. Just like boards, according to the sawing stage, by-products are available in the inventory for being used as fuel.

The objective of the presented approach is to plan the drying operation during a time horizon, considering different planning criteria. Explicitly, the thickness of boards to be dried in each drying kiln, the programs to be selected, and the operating times for the different kilns are simultaneously determined through a MILP model according to diverse objective functions for the planning. In addition, the resources used to satisfy the energy requirement are selected.

## 3 Model Formulation

The proposed formulation is based on an STN model that uses a discrete time grid. The time horizon is divided into discrete times $t$ of equal duration, which are determined according to the drying operation times and the regularity in the delivery of boards. In the STN, states are represented through index $e$ and represent the thickness of the boards to be dried. The drying task is implemented according to program $p$ in dryer $d$.

As resources for the drying of boards, the by-products generated in the productive stages of the sawmill are considered as types of fuel $b$, and are used to cover the energy requirements of the drying (Fig. 1).

Let $x_{dtep}$ be the binary variable that determines the beginning of a drying cycle in dryer $d$ in time $t$ using program $p$ for boards of thickness $e$. Each drying program $p$ is characterized by its duration ($Tp_{dep}$) and its energy requirement, taking into account kiln $d$ being used and board thickness $e$. Also, energy consumption is determined according to the different phases explained in Fig. 1. It is estimated through parameter $ES_{dt'ep}$ that indicates the energy required in the period between times $t'$ and $t' + 1$, again taking into account the selected program $p$, board thickness $e$ and dryer d being used when the drying cycle has started at time $t$, i.e. $x_{dtep}$ equal 1.

Equation (1) states that at most one single drying program $p$ can be started in dryer $d$ at each time $t$. This expression takes into account the duration of drying program $p$ for boards of thickness $e$ in dryer $d$ ($Tp_{dep}$) and the setup time of the kiln ($Ts_d$).

$$\sum_{t'}^{t=t+Tp_{dep}+Ts_d-1} \sum_e \sum_p x_{dt'ep} \leq 1 \qquad \forall\, d, t \tag{1}$$

Let $NE_{te}$ be the quantity of boards of thickness $e$ that enter the drying stage in time $t$. These boards are added to the inventory of non-dried boards with the same thickness from the previous time ($I_{t-1,e}$). Equation (2) states the material balance at time $t$, which establishes that the boards in stock at time $t - 1$ plus the ones that enter at time $t$ must be equal to the boards to be dried plus the remaining ones kept in inventory. $Qp_{dte}$ represents the quantity of boards of thickness $e$ that enter dryer $d$ at time $t$.

$$I_{t-1,e} + NE_{te} = I_{te} + \sum_d Qp_{dte} \qquad \forall t > 1, e \tag{2}$$

For a correct operation, the load of the dryers is between minimum and maximum values ($CMin_d$ and $CMax_d$, respectively) established by the company while taking into account the capacity and technology of dryers (Eq. (3)).

$$CMin_d \sum_p x_{dtep} \le Qp_{dte} \le CMax_d \sum_p x_{dtep} \quad \forall\, d, t, e \tag{3}$$

Equation (4) determines the energy required by each dryer at time $t$ ($ER_{dt}$). This expression considers the different stages if a drying cycle has started and takes into account consumption between times $t$ and $t + 1$.

$$ER_{dt} = \sum_e \sum_p \sum_{t'=0}^{Tp_{dep}-1} ES_{d(t')ep}\, x_{d(t-t')ep} \quad \forall\, d, t < t' \tag{4}$$

The total energy required by dryers at time $t$ cannot exceed the maximum energy that can be provided by the boiler ($EMax$) and must exceed a minimum value ($EMin$) for its proper operation (Eq. (5)).

$$EMin \le \sum_d ER_{dt} \le EMax \quad \forall\, t \tag{5}$$

Equation (6) determines the types of fuel $b$ used by the boiler to cover the required energy.

$$\sum_d ER_{dt} = f \sum_b Qb_{tb} cp_b \quad \forall\, t \tag{6}$$

where $Qb_{tb}$ is the amount of fuel $b$ consumed by the boiler in time $t$, $cp_b$ is the calorific value of fuel $b$, and $f$ is a factor of boiler efficiency.

Equation 7 determines the total amount of fuel ($Tb_t$) used at each time $t$.

$$Tb_t = \sum_b Qb_{tb} \quad \forall\, t \tag{7}$$

Fuel $b$ being used at time $t$ must not exceed a percentage ($fb_b$) of the total amount ($Tb_t$) being used in order to assure the proper functioning of the boiler (Eq. (8)).

$$Qb_{tb} \le fb_b\, Tb_t \quad \forall\, t, b \tag{8}$$

The availability of fuel $b$ is determined according to Eq. (9), where $IB_{tb}$ represents the inventory of fuel $b$ at time $t$ and $NB_{tb}$ is the amount of fuel $b$ that arrives at time $t$ from sawmill activities for the boiler.

$$IB_{t-1,b} + NB_{tb} = IB_{tb} + Qb_{tb} \quad \forall t, b \tag{9}$$

The operating criteria in the drying stage can vary according to both the level of production of sawmills and the policy adopted by the firm. These criteria can be contrasted among them, and this analysis is really interesting. Next, different objectives are posed.

### 3.1 Objective Function 1 (FO 1): Maximization of the Quantity of Dried Boards

This criterion tries to dry as many boards as possible over the time horizon. It is a widely used measure, but it loses sight of the costs generated in the process.

$$max\ FO\,1 = \sum_d \sum_t \sum_e Qp_{dte} \tag{10}$$

### 3.2 Objective Function 2 (FO 2): Costs Minimization

The second criterion (Eq. (11)) minimizes costs due to the use of the drying kilns ($Cd$) and the cost of fuel consumption ($Cf$).

$$min\ FO\ 2 = Cd + Cf \tag{11}$$

$Cd$ cost represents a fixed cost for each performed drying cycle, which depends on the dryer being used, the thickness of the boards to be dried, and the selected drying program ($Cdry_{dep}$). It is calculated as follows:

$$Cd = \sum\nolimits_d \sum\nolimits_t \sum\nolimits_e \sum\nolimits_p Cdry_{dep} x_{dtep} \tag{12}$$

$Cf$ cost is obtained from the Eq. (13), where $Cb_b$ is the unit cost of fuel $b$.

$$Cf = \sum\nolimits_t \sum\nolimits_b Cb_b Qb_{tb} \tag{13}$$

Furthermore, in order to assure a minimum quantity of dried boards, the inventory of boards by the end of the planning horizon must not exceed a certain percentage $fp$ of the total available boards through the planning horizon ($Te$) (Eq. (14)). $Te$ is defined as the sum of the tables that were in inventory at the beginning of the horizon and those that enter the drying stage (Eq. (15)). Another different and admissible approach is to set conditions for each board thickness instead of using a global condition. The following expression is used in this work:

$$I_{t=|T|,e} \leq fp\ Te \tag{14}$$

$$Te = fp \sum\nolimits_e \left( I_{t=0,e} + \sum\nolimits_t NE_{te} \right) \tag{15}$$

### 3.3 Objective Function 3 (FO 3): Makespan Minimization

The aim of this objective function is to dry the highest number of boards as soon as possible.

$$min\ FO\ 3 = MK \tag{16}$$

where $MK$ corresponds to the end time of the last planned drying cycle in the time horizon, which is calculated in Eq. (17).

$$MK \geq ord(t) x_{dtep} + \left( Tp_{dep} + Ts_d \right) x_{dtep} \tag{17}$$

where ord($t$) corresponds to the position in which time $t$ is in the set. In this case, a requirement is added on the minimum of boards to be dried such as Eqs. (14 and 15).

## 4 Examples

In order to evaluate the performance of the proposed model, an example corresponding to a sawmill with 4 dryers is presented and two cases with different availability of fuel will be analysed. In Case I, there is limited availability, while in Case II, availability is enough to meet any required demand. Table 1 shows the capacity of each dryer.

At the beginning of the planning, two kilns are busy with drying cycles from the previous planning. The drying cycles of this plan must begin before time $t = 60$. At the beginning of the planning horizon, the inventory of tables and fuel is null ($I_{t=0,e} = 0$ and $IB_{t=0,b} = 0$, respectively) and the boards are classified in 3 thicknesses which each one has 3 drying programs. Table 2 displays the duration of each drying cycle according to the dryer, thickness of boards and drying program, and the kilns setup time required for each dryer. Table 3 shows the fixed cost for each drying program, while Table 4 presents the costs for fuels and their calorific value.

Tables 5 and 6 indicate the quantity of boards and fuel that enter the drying stage throughout the studied horizon, respectively.

In this example, the boiler can deliver up to 150000 MJ/h, but, at the beginning of the time horizon, a fraction of this power is reserved to cover the energy required by the previous drying plan. The example is implemented and solved in GAMS using CPLEX solver in an Intel(R) Core(TM) i7-3770, 3.40 GHz. The computational time limit is fixed to 3600 CPU sec. The model consists of 10834 equations, 8061 continuous and 2043 binary variables.

The global results are shown in Table 7. The value obtained for each objective function is highlighted in each column. For FO 2, the maximum allowed level for boards inventory at the end of the planning horizon (*fp*) varies from 25% to 5% of the available boards for drying. This is done in order to analyse different planning scenarios and the impact of the final inventory level when cost is minimized. Also, in order to carry out appropriate comparisons, it is required for FO 3 to keep the same stock as the solution obtained with FO 1. The last 3 columns of the table correspond to Case II. For this case, only $fp = 25\%$ is analysed for FO 2. As it can be seen, optimal solutions are obtained with reasonable computing times in almost all cases, except when $fp = 15\%$, which results in an optimality gap equal to 0.7.

Analysing FO 1 in Case I, 1,380 tons of boards are dried in 19 drying cycles, which end at time t = 100, leaving 4.2% of the total boards without drying in inventory. By not considering the costs within the objective function, the shortest programs are selected, with a higher consumption of fuel, which allow drying a greater number of boards.

**Table 1.** Dryer capacity [*tn*]

| Dryer | *CMin* | *CMax* |
|---|---|---|
| $d_1$–$d_2$ | 45 | 60 |
| $d_3$–$d_4$ | 75 | 100 |

Note: tn, tons

**Table 2.** Processing times

| Dryer | Board thickness | Drying program | | | Setup time |
|---|---|---|---|---|---|
| | | *p1* | *p2* | *p3* | |
| $d_1$–$d_2$ | $e_1$ | 13 | 10 | 8 | 1 |
| | $e_2$ | 18 | 14 | 11 | |
| | $e_3$ | 23 | 21 | 17 | |
| $d_3$–$d_4$ | $e_1$ | 22 | 18 | 14 | 2 |
| | $e_2$ | 30 | 23 | 19 | |
| | $e_3$ | 39 | 30 | 24 | |

**Table 3.** Drying program cost [*$*]

| Dryer | Board thickness | Drying program | | |
|---|---|---|---|---|
| | | *p1* | *p2* | *p3* |
| $d_1$–$d_2$ | $e_1$ | 43.9 | 33.8 | 27.0 |
| | $e_2$ | 56.3 | 43.8 | 34.4 |
| | $e_3$ | 58.8 | 53.7 | 43.5 |
| $d_3$–$d_4$ | $e_1$ | 58.8 | 48.1 | 37.4 |
| | $e_2$ | 74.0 | 56.7 | 46.9 |
| | $e_3$ | 76.1 | 58.5 | 46.8 |

**Table 4.** Fuel parameters

| Fuel | Costs [*$/tn*] | Calorific value [*MJ/tn*] | Boiler feed [*%*] |
|---|---|---|---|
| Bark | 2.39 | 5998.22 | 15 |
| Chip | 5.33 | 7117.39 | 100 |
| Wood shavings | 4.02 | 16486.45 | 20 |
| Sawdust | 3.39 | 6363.77 | 15 |

Note: tn, tons; MJ, mega joules

With FO 2, as *fp* decreases, and with this the need to dry more boards at the same time, the number of drying cycles increases and simultaneously the use of the short drying program p3 tends to increase, generating higher fuel costs. Using FO 3, it is possible to dry the same as FO 1 but with lower cost and faster.

In Case II, the greater availability of fuel allows to improve the results of all objective functions. Using FO 1, the number of dried boards increases, but fails to dry the total, leaving 30 tons in inventory for the next planning. With FO 2 for $fp = 25\%$ the cost is

**Table 5.** Quantity of boards that enters drying stage in each time [*tn*]

| Board thickness | Time | | | | |
|---|---|---|---|---|---|
| | $t_0$ | $t_{12}$ | $t_{24}$ | $t_{36}$ | $t_{48}$ |
| $e_1$ | 30 | 130 | 190 | 60 | 10 |
| $e_2$ | 170 | 170 | 170 | 80 | 50 |
| $e_3$ | 150 | 50 | 160 | 0 | 20 |

Note: tn, tons

**Table 6.** Fuel entering each time in Case I [*tn*]

| Fuel | Time | | | | |
|---|---|---|---|---|---|
| | $t_0$ | $t_{12}$ | $t_{24}$ | $t_{36}$ | $t_{48}$ |
| Bark | 15 | 9 | 8 | 11 | 12 |
| Chip | 169 | 107 | 91 | 127 | 140 |
| Sawdust | 84 | 53 | 45 | 63 | 70 |
| Wood shavings | 25 | 16 | 13 | 19 | 21 |

Note: tn, tons

decreased by 1.1%, and FO 3 shows a reduction of 17 units of time in manufacturing, but with a 40.7% increase in costs, with respect to the same objective function of Case I.

Figure 3 shows the Gantt charts, depicting in red the drying cycles performed for boards of thickness $e_1$, in green those of thickness $e_2$ and in blue the corresponding to thickness $e_3$. The blocks with squares correspond to drying programs $p_1$; those with oblique lines are cycles with drying program $p_2$; and those in solid colour correspond to drying program $p_3$. The drying cycles from a previous drying plan are displayed in grey. For FO 2, only the planning for the case considering $fp = 25\%$ is shown.

It can be seen in Fig. 3 that in FO 2, the dryers have the amount of chip a greater number of idle times, due to the low number of drying cycles. In FO 3 of Case I, the lack of fuel forces to select slower drying programs, preventing a significant reduction in the time of completion of the drying plan if compared to FO 1. This can be better understood by looking at Fig. 4a, which represents the consumption and inventory of each fuel over the horizon. Although at the end of the horizon even sawdust is available (the only one in stock), it is not allowed to feed the boiler only with this fuel since the boiler does not work correctly (Eq. (7)).

Figure 4b depicts the percentage of the type of fuel used to feed the boiler. Depending to the limit value of each fuel ($fb_b$), the amount of chips used at any given time is greater than 50% and, in some cases, corresponds to 100% of the boiler supply. Figure 5 shows the energy generated by the boiler to cover the energy demand during the time horizon and the maximum limit it can generate. As it can be seen, at the beginning, the maximum

**Table 7.** Results of example

| | Case I | | | | | | | Case II | | |
|---|---|---|---|---|---|---|---|---|---|---|
| | FO 1 | FO 2 | | | | | FO 3 | FO 1 | FO 2 | FO 3 |
| | | Maximum inventory of board (*fp*) | | | | | | | | |
| | | 25% | 20% | 15% | 10% | 5% | | | | |
| Drying cost [$] | 814.9 | 805.7 | 794.1 | 826.1 | 821 | 820.1 | 820.1 | 759 | 805.7 | 747.2 |
| Fuel cost [$] | 4345.1 | 2012.2 | 2551.6 | 3049.6 | 3574.7 | 4288.5 | 4306 | 6445.5 | 1980.5 | 6466.5 |
| Total cost [$] | 5160 | **2817.9** | **3345.7** | **3875.8** | **4395.7** | **5108.6** | 5126.1 | 7204.5 | **2786.2** | 7213.7 |
| Dried boards[*tn*] | **1380** | 1080 | 1160 | 1240 | 1300 | 1380 | 1380 | **1410** | 1080 | 1410 |
| Final inventory [*tn*] | 60 | 360 | 280 | 200 | 140 | 60 | 60 | 30 | 360 | 30 |
| Inventory [%] | 4.2 | 25 | 19.4 | 13.9 | 9.7 | 4.2 | 4.2 | 2.1 | 25 | 2.1 |
| Performed programs | | | | | | | | | | |
| Drying program 1 | 3 | 12 | 7 | 7 | 5 | 4 | 4 | 1 | 12 | 1 |
| Drying program 2 | 5 | 2 | 9 | 7 | 7 | 4 | 4 | 1 | 2 | 1 |
| Drying program 3 | 11 | 0 | 0 | 2 | 5 | 11 | 11 | 17 | 0 | 17 |
| Total drying cycles | 19 | 14 | 16 | 16 | 17 | 19 | 19 | 19 | 14 | 19 |
| Total consumed fuel [*tn*] | | | | | | | | | | |
| Bark | 55 | 55 | 55 | 55 | 55 | 55 | 55 | 2.8 | 68.5 | 2.8 |
| Chip | 634 | 240.8 | 329.7 | 413.9 | 502.7 | 623.5 | 634 | 1204.5 | 228.4 | 1208.5 |
| Sawdust | 136.4 | 68.3 | 84.5 | 99.3 | 115 | 136.3 | 124.5 | 0 | 68.5 | 0 |
| Wood shavings | 94 | 91 | 94 | 94 | 94 | 94 | 94 | 3.8 | 91.4 | 3.8 |
| Makespan [*t*] | 100 | 91 | 90 | 91 | 100 | 100 | **98** | 88 | 90 | **81** |
| CPU Time [*sec*] | 142 | 2 | 383 | 3610 | 161 | 287 | 383 | 20 | 4 | 81 |
| Gap [%] | 0 | 0 | 0 | 0.7 | 0 | 0 | 0 | 0 | 0 | 0 |

Note: tn, tons

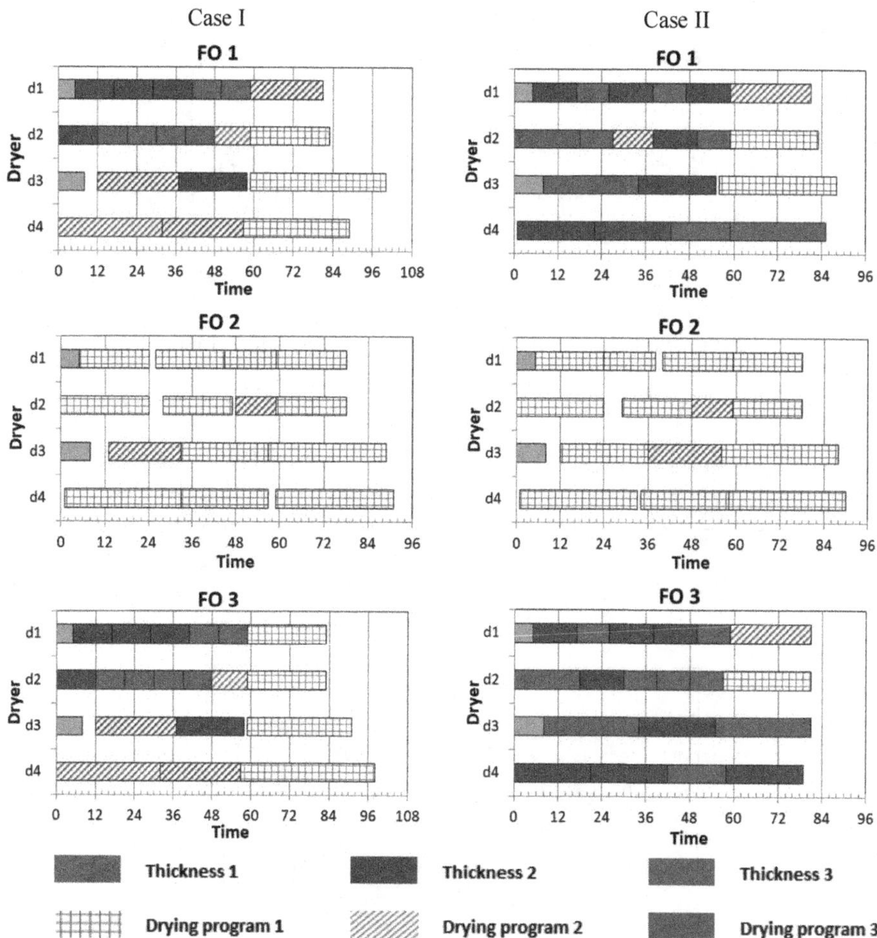


**Fig. 3.** Gantt chart of the drying plan (Color figure online)

limit of the boiler is reduced because the difference is reserved for the energy requirement of the previous drying plan.

## 4.1 Comparative Analysis and Discussion of Results

By comparing the obtained results, different drying plans are reached according to the considered objective function. Although FO 1 is usually more common in the forest industry when planning is addressed, many times with more heuristic tools or approaches, the presented approach allows evaluating different drying process scenarios. In this way, new perspectives are opened for companies in the sector in relation to the optimal management of the drying operation. This is reinforced if new options for the use of by-products, based on the use of biomass, are considered in the analysis. In this

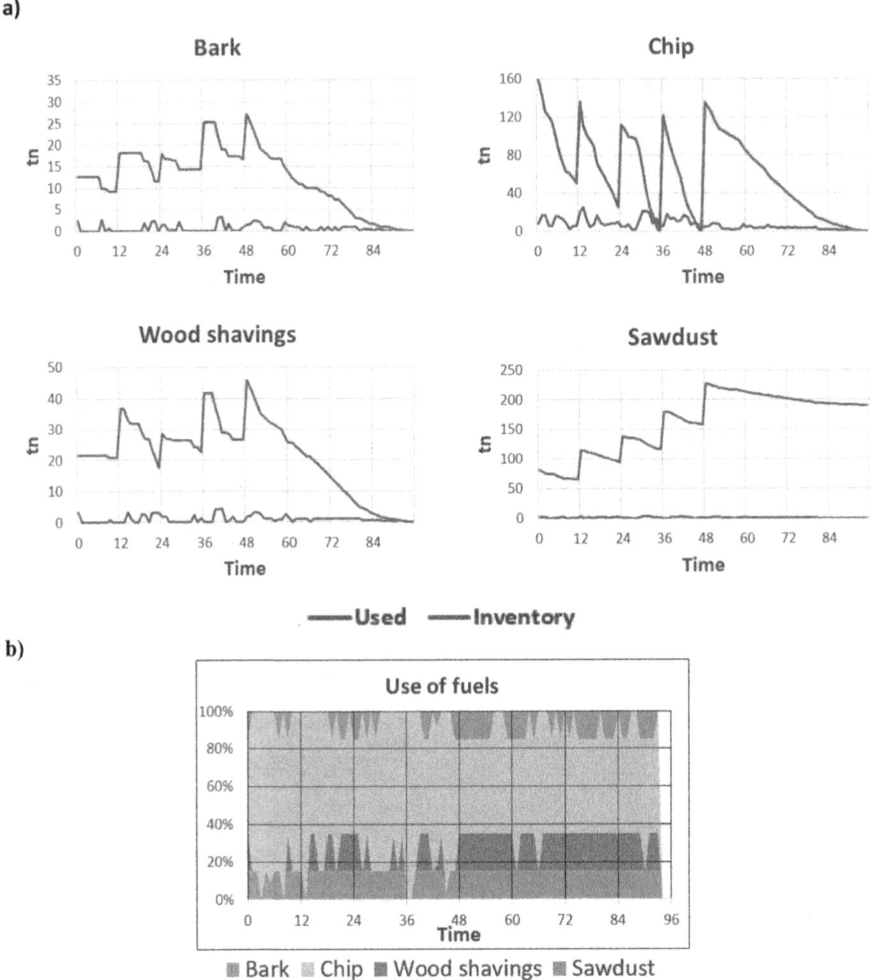


**Fig. 4.** Use of fuel for FO 3 of Case I

section, a discussion of the different criteria considered and the trade-offs between them is presented.

For FO 1, a greater number of boards are dried using a greater number of drying cycles. By not considering costs in this objective function, shorter drying programs are used, with higher energy consumption, generating the highest total costs. With FO 2, it is observed that the number of dry boards and drying cycles increases as *fp* decreases, using longer drying programs with less energy and fuel consumption, and consequently lower cost. As expected, when considering FO 3, the drying plan is carried out in less time, using a number of cycles similar to that of FO 1 but the downtimes between cycles decrease. By-product availability is an important factor in the planning, as can be glimpsed in Case I. All the objectives functions are improved when greater amount

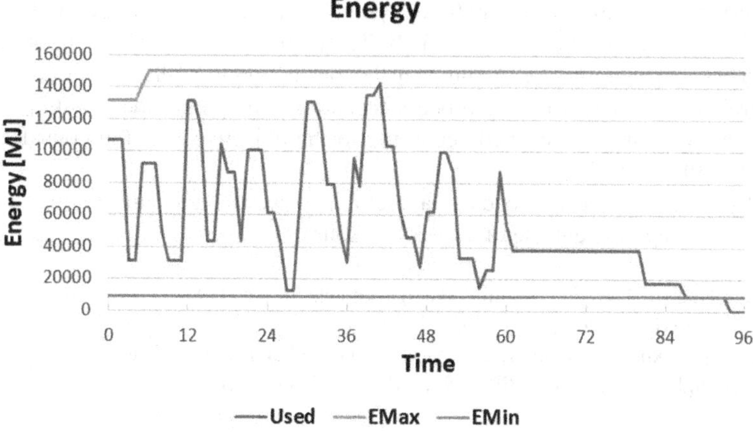


**Fig. 5.** Energy provided by the boiler for FO 3 of Case I

of fuel is available. However, the new possibilities to take advantage of these resources generate scenarios that this formulation can help to analyse.

Finally, the trade-offs between the different criteria considered are analysed. Obviously, generating an objective function that incorporates these 3 criteria would allow an adequate evaluation to be carried out. However, this is not possible as these criteria are expressed in different units (product, money and time). In the case of FO 1, the value contributed in the drying operation could be weighed when the wet boards are transformed into dry boards, but as in reality the wet boards lack commercial value, it is an abstraction that is very difficult to be specified. Therefore, a more appropriate strategy to plan the operation is to incorporate restrictions that involve the criteria considered. In the previous example, this strategy has been implemented establishing conditions on the production level through restrictions on the final inventory of undried boards.

In the case of FO 2, which minimizes costs, the drying plan has been analysed by varying the final inventory of undried boards. The conditions on the time horizon could have been added using Eq. 17 and incorporating a limit for makespan. Similarly, in FO 3, MK is minimized and production limitations were considered. Here you could add restrictions to limit the cost of operation. In short, the formulation admits the combination of the different criteria, which enhances its capacity for analysis and evaluation. Given the difficulty of including the criteria in a single expression as an objective function, the best alternative is to incorporate them through restrictions in the formulation.

Therefore, evaluating and adjusting these criteria can guide planners to make a good decision to analyze the wood drying process.

## 5 Conclusions

A MILP model with a State-Task Network formulation based on discrete times was presented for optimal planning of the drying stage of a sawmill over a time horizon. This allows to simultaneously determine the type and quantity of boards to dry, the selected drying program and the start time, together with the energy required and how this requirement is covered with the use of fuels. Three objective functions were presented. Although FO 1 is generally more common in the forestry industry, the presented approach allows evaluating different drying process scenarios. Comparing the results obtained, different drying plans are reached according to the objective function considered, in reasonable CPU times.

Finally, the proposed approach is a useful tool for planning the drying stage of sawmills, allowing efficient use of available resources.

**Acknowledgement.** The authors would like to acknowledge Nestor Steitzer and Suabia Maderas S.R.L for their predisposition and collaboration.

The authors acknowledge financial support from CONICET, FONCyT and UTN for research activities through their projects PIP 0352, PICT 2017-4004, and PID SIUTIFE0005246TC, respectively.

## References

1. Simo-Tagne, M., Zoulalian, A., Rémond, R., Rogaume, Y.: Mathematical modelling and numerical simulation of a simple solar dryer for tropical wood using a collector. Appl. Therm. Eng. **131**, 356–369 (2018)
2. Khouya, A., Draoui, A.: Computational drying model for solar kiln with latent heat energy storage: case studies of thermal application. Renew. Energy **130**, 796–813 (2019)
3. Gaudreault, J., Frayret, J., Rousseau, A., D'Amours, S.: Combined planning and scheduling in a divergent production system with co-production: a case study in the lumber industry. Comput. Oper. Res. **38**, 1238–1250 (2011)
4. Fumero, Y., Corsano, G., Montagna, J.M.: Simultaneous batching and scheduling of batch plants that operate in campaign-mode, considering nonidentical parallel units and sequence-dependent changeovers. Ind. Eng. Chem. Res. **53**, 17059–17074 (2014)
5. Lee, H., Maravelias, C.T.: Mixed-integer programming models for simultaneous batching and scheduling in multipurpose batch plants. Comput. Chem. Eng. **106**, 621–644 (2017)
6. Méndez, C.A., Cerdá, J., Grossmann, I.E., Harjunkoskic, I., Fahlc, M.: State-of-the-art review of optimization methods for short-term scheduling of batch processes. Comput. Chem. Eng. **30**(6–7), 913–946 (2006)
7. Wu, Y., Maravelias, C.T.: A general model for periodic chemical production scheduling. Ind. Eng. Chem. Res. **59**(6), 2505–2515 (2020)
8. Kondili, E., Pantelides, C.C., Sargent, R.W.H.: A general algorithm for short-term scheduling of batch operations – 1. MILP formulation. Comput. Chem. Eng. **17**, 211–227 (1993)
9. Castro, P.M., Grossmann, I.E., Zhang, Q.: Expanding scope and computational challenges in process scheduling. Comput. Chem. Eng. **114**, 14–42 (2018)

# A Production Planning MILP Optimization Model for a Manufacturing Company

Juan Antonio Cedillo-Robles[1](✉), Neale R. Smith[1](✉), Rosa G. González-Ramirez[2](✉), Julio Alonso-Stocker[2](✉), Joaquín Alonso-Stocker[2](✉), and Ronald G. Askin[3](✉)

[1] Tecnologico de Monterrey, 64849 Monterrey, NL, Mexico
{a00997352,nsmith}@tec.mx
[2] Universidad de los Andes, Chile, 12455 Santiago, RM, Chile
rgonzalez@uandes.cl, {jjalonso,jalonso1}@miuandes.cl
[3] Arizona State University, Tempe, AZ 85281, USA
Ron.Askin@asu.edu

**Abstract.** This paper proposes a mixed-integer linear programming (MILP) model that is implemented based on a rolling horizon scheme to solve an aggregate production planning decision problem of a manufacturing company that produces snacks in Monterrey, Mexico. The demand of the company is characterized by trends and seasonality. The proposed solution is evaluated by means of computational experiments to determine the relation between demand uncertainty and flexibility of a production system. A $2^k$ factorial experimental design and a multivariate regression were performed. Results show forecast bias and length of frozen period in the rolling horizon have a strong effect on total profit. The safety stock level was also found to be a significant factor, depending on the level of bias.

**Keywords:** Production planning · Rolling horizon · Forecast error · Flexibility factors

## 1 Introduction

This study considers a production planning problem for a snack factory located in the metropolitan area of Monterrey, Mexico. The company has presented significant costs due to high inventories, overproduction, and stockouts. Executives consider that the company experiences a bullwhip effect due to the lack of a precise forecast and an assertive communication between departments in the company. This can be also a problem related to the flexibility of the production system of the company, as even in the cases in which forecasts have a good accuracy, the company does not achieve the desired results. The company operates like this. The production manager receives a demand forecast estimated by the sales department. Then, the production manager defines an aggregate production plan every month, and that plan is executed on a weekly basis by sales and production departments. The level of uncertainty and variability of the demand of

© Springer Nature Switzerland AG 2021
D. A. Rossit et al. (Eds.): ICPR-Americas 2020, CCIS 1407, pp. 85–96, 2021.
https://doi.org/10.1007/978-3-030-76307-7_7

the company are important factors to be considered for their planning operations. This uncertainty can be related to sales, prices, technology development, etc.

One strategy that can be employed to modeling the decision-making process of the production manager is by means of a rolling horizon scheme [14]. Unlike a fixed-horizon production planning process, in a rolling horizon strategy demand forecasts are updated every iteration of the planning horizon [10]. Hence, under a rolling horizon planning scheme, several different optimization sub-problems are solved considering a forecast of demand. After the realization of the demand, the resulting plans are updated, and the iterations continue until the production plan for the entire planning horizon has been determined [1].

Based on the characteristics of the company considered as a case study, a mixed integer linear programming model is proposed, to support the production planning decisions of the company. The proposed model is implemented considering a rolling horizon scheme. A set of experiments is conducted to evaluate the mathematical relation between demand forecast errors and the flexibility of the production system.

The remainder of this paper is organized as follows. Section 2 presents a literature review regarding districting problems. Section 3 presents the proposed mathematical formulation and solution approach. Section 4 describes the experimental design and the analysis of results. Conclusions and recommendations for future research are given in Sect. 5.

## 2 Literature Review

Production planning is an area of great interest for both practitioners and academics. Since the pioneering studies that proposed linear decision rules and suggested some basic applications methods, innovative models and solution methodologies to handle these problems have been proposed in the literature. Jalmania et al. [5] present a comprehensive literature survey and future research directions for aggregate production planning problems under uncertainty. They analyze and classify the literature and identify future research paths, highlighting that few studies on rolling horizon implementations for aggregate production planning can be found in the literature, in both the deterministic case and the case under uncertain conditions.

Among the contributions that consider a rolling horizon for Aggregate Production Planning (APP) models, we can mention the study conducted by Demirel et al. [1]. They proposed a mixed integer linear programming model under a rolling horizon scheme with flexibility requirements where demand is regarded as an uncertain variable. Jain and Palekar [3] and Hahn and Brandenburg [2] consider the APP in process industries. The former contribution develops a stochastic linear programming method to study APP with resource limitation considerations in a continuous food producing company. The latter contribution develops a sustainable APP decision model for a chemical process industry by applying stochastic queuing networks. Only Jain and Palekar [3] consider a rolling horizon implementation.

Lin and Uzsoy [7] use two different chance-constraint formulations to reduce planned release changes while maintaining a desired service level under stochastic demand in a rolling horizon environment and compare them with an LP formulation that does not consider uncertainty. Vogel et al. [11] develop a hierarchical and an integrated model considering two levels of hierarchical production planning, the APP and the Master Production Scheduling (MPS), in order to show that it is possible to integrate them into a single model and, at the same time, fulfill their requirements. The interaction of both levels was implemented on a rolling horizon approach. Li and Ierapetritou [6] propose a methodology to derive the production capacity constraints based on short-term scheduling model through parametric programming technique and use this capacity information in the rolling horizon framework to improve the final solution's quality.

Zhang et al. [14] develop a two-stage integer stochastic planning model for an international enclosure manufacturing company with seasonal demand and market growth uncertainty. The model was implemented considering different solution methodologies, including two rolling horizon algorithms, comparing their performance. Sitompul and Aghezzaf [8] propose a preventive maintenance model that considers a non-cyclical planning, which means that the preventive maintenance periods do not necessarily fall at equally distant times. They solve the problem at the detailed level using a rolling horizon approach and calculate the average of the total costs at the detailed level. Stephan et al. [9] propose a multi-stage stochastic dynamic programming method to support real world capacity provision and adjustment decisions in automotive manufacturing networks. They compared their proposed methodology with respect to a rolling horizon two-stage stochastic planning approach. Wu and Ierapetritou [12] present a hierarchical approach for production planning and scheduling under uncertainty, using a multi-stage formulation. The optimal schedule is determined using a rolling horizon strategy. Yao et al. [13] address a medium-term capacity and production planning model to plan and support production operations for meeting multiple product demands considering a multi-facility manufacturing system. The system has limited product flexibility, and, continuous and discrete capacity change options with varying implementation time and cost. For testing, the MILP problem approach was implemented with a rolling horizon schedule as solution method in practice.

Table 1 summarizes in chronological way the most related contributions to this work, including those previously described. The table classifies the contribution with respect to the type of model proposed, the solution methodology and the type of planning horizon (fixed or rolling one). The last row corresponds to the contribution presented by this manuscript.

**Table 1.** Related literature on production planning models under uncertainty

| Publication | Modeling approach | | | Solution methodology | | | Planning horizon | |
|---|---|---|---|---|---|---|---|---|
| | I | II | III | A | B | C | 1 | 2 |
| Jain and Palekar [3] | | X | | X | | | | X |
| Wu and Ierapetritou [12] | | | X | X | | | | X |
| Li and Ierapetritou [6] | | X | | | | X | | X |
| Stephan et al. [9] | | | X | X | | | | X |
| Sitompul and Aghezzaf [8] | | | X | X | | | | X |
| Zhang et al. [14] | | | X | X | | | | X |
| Lin and Uzsoy [7] | X | | | X | | | | X |
| Vogel et al. [11] | | X | | X | | | | X |
| Hanh and Brandenburg [2] | | | X | | | X | X | |
| Jamalnia et al. [4] | | | X | X | | | X | |
| Jalmania et al. [5] | | X | X | X | X | | X | X |
| Demirel et al. [1] | | X | | X | | | | X |
| Yao et al. [13] | | X | | | | X | | X |
| Cedillo et al. (This work) | | | X | X | | | | X |

[I] LP [II] MILP [III] Stochastic; [A] Optimization [B] Simulation [C] Heuristics; [1] Fixed Horizon [2] Rolling Horizon

## 3 Mathematical Formulation and Solution Approach

In this section we present the mathematical formulation for the aggregate production planning problem (Sect. 3.1) and the solution approach based on a rolling horizon scheme (Sect. 3.2).

### 3.1 Mathematical Formulation

In this paper we consider a manufacturing company that produces set of $I$ products, manufactured in a set of $L$ product lines during a planning horizon of $T$ periods. The aim of the company is to maximize the global net profit balancing the sales revenue and operational costs subject to the plant capacity over the planning horizon $T$. The indexes, sets, constants, and variables are listed below, followed by the model formulation.

**Objective**
$z$: total profit [$].

**Sets**

$I$: set of products $i$.
$L$: set of product lines $l$.
$T$: set of time periods $t$.

**Variables**

$Inv_{it}$: inventory units of product $i$ in period $t$ [units].
$B_{it}$: unfulfilled units of product $i$ in period $t$ [units].
$Z_{ilt}$: binary variable that takes value of 1 if a set-up is made for product $i$ on production line $l$ in period $t$ [0, 1].
$Mp_{ilt}$: manufactured units of product $i$ on production line $l$ in period $t$ [units].
$S_{it}$: sold units of product $i$ in period $t$ [units].

**Constants**

$M$: A big number.
$Cs_{il}$: set-up cost of product $i$ manufactured on production line $l$ [$].
$Cp_{il}$: production cost of product $i$ manufactured on production line $l$ [$].
$Ch_i$: holding cost of product $i$ on inventory [$].
$Cb_i$: cost of unfulfilled product $i$ [$].
$Ts_{il}$: set-up time of product $i$ manufactured on production line $l$ [time units].
$Tp_{il}$: time required to produce product $i$ on production line $l$ [time units].
$Q_l$: maximum capacity of the production line $l$ [time units].
$\beta_{it}$: availability of the production line $l$ in period $t$ [0-1].
$d_{it}$: demand forecast for product $i$ in period $t$ [units].
$P_{it}$: sale price of product $i$ in period $t$ [$].
$Inv0_i$: initial inventory units of product $i$ [units].
$SS_{it}$: safety stock units of product $i$ in period $t$ [units].

**Objective Function**

$$\begin{aligned} Max\ z = & \sum_{i=1}^{I}\sum_{t=1}^{T} P_{it}S_{it} \\ & -\sum_{i=1}^{I}\sum_{l=1}^{L}\sum_{t=1}^{T} Cs_{il}Z_{ilt} - \sum_{i=1}^{I}\sum_{l=1}^{L}\sum_{t=1}^{T} Cp_{il}Mp_{ilt} - \sum_{i=1}^{I}\sum_{t=1}^{T} Ch_i Inv_{it} \\ & -\sum_{i=1}^{I}\sum_{t=1}^{T} Cb_i B_{it} \end{aligned} \tag{1}$$

**Subject to**

$$\textstyle\sum_{i=1}^{I} (Ts_{il}Z_{ilt} + Tp_{il}Mp_{ilt}) \le \beta_{lt}Q_l \qquad \forall l, t \tag{2}$$

$$Mp_{ilt} \le MZ_{ilt} \qquad \forall l, t \tag{3}$$

$$Inv_{it} = Inv_{it-1} + \textstyle\sum_{l=1}^{L} Mp_{ilt} - S_{it} \qquad \forall i, t = 2, \ldots, T \tag{4}$$

$$Inv_{i1} = Inv0_i + \textstyle\sum_{l=1}^{L} Mp_{i1} - S_{i1} \qquad \forall i, t \tag{5}$$

$$S_{it} = d_{it} - B_{it} \qquad \forall i, t \tag{6}$$

$$Inv_{it} \geq SS_{it} \quad \forall i, t \tag{7}$$

$$S_{it}, Mp_{ilt}, Inv_{it}, B_{it} \geq 0; \quad \forall i, l, t \tag{8}$$

$$Z_{ilt} = \{0, 1\} \quad \forall i, l, t \tag{9}$$

In the objective function (1), the first expression represents the total net profit. The second and third term describe the setup and production costs. The last terms represents the inventory and stockout costs. Constraint (2) is the capacity constraint in time units for each production line. Constraint (3) guarantees that there can be production in a period and production line, only if there is a setup. Constraints (4) and (5) enforce inventory balance. Constraint (6) defines the amount of products $i$ to be sold according to the demand forecast and the unfulfilled units for each period of time $t$ and demand scenario k. Constraint (7) sets the minimum safety stock for each product $i$, each period $t$. Constraints (8) and (9) are non-negativity constraint for all variables, specifying $Z_{ilt}$ as binary.

### 3.2 The Rolling Horizon Method

For this study, we define $H$ as the number of periods for which we wish to create production plans. The planning horizon each time the model is run is of $T$ periods, $T < H$. Parameter $R$ is introduced to help describe the general rolling horizon strategy. $R$ is a horizon ($R < T < H$) in which production plan is fixed and the real demand can be observed and used to update forecast, adjusting the plan over a planning horizon of $T$ periods. The steps of the algorithm are further described:

1. Make the forecast for the demand in $t = 1 \ldots T$ with its respective measures of error and deviation. Run the model to get the production plan $Mp_{ilt}$. Set the current period as $c = 1$.
2. For $t = c \ldots R$, fix the resulting production plan and take the values of actual demand in this horizon. Save these values as parameters $Mp_{ilt}$ and $d_{it}$ respectively in the MILP model. Run the model with this update and get the values of $Inv_{it}$ and $B_{it}$.
3. For $t = R + 1, \ldots, R + T$, update the forecast and run the model in the same way as step 1 recognizing the current period is updated by $R$ periods to the current time, and considering the value of $Inv_{iR} = Inv0_i$ in this horizon.
4. If $c < H$, update $H = H\text{-}R$ and return to step 2; else, stop.

## 4 The Experimental Design

The experimental design was performed to test the impact of specific factors, both forecast error and production flexibility on total profits of the company. We propose a factorial design $2^k$. For the statistical analysis, *Minitab 18* was used in all combinations generated of three factors that were varied in two levels, high and low. In total, we will consider $2^3$ instances* 5 replicates = 40 aleatory experiments.

### 4.1 Instance Generation

The bias of the forecast demand error is considered in order to evaluate the inaccuracy of the method employed by the company to forecast the demand. With respect to the flexibility of the production system, we consider the level of safety stocks. The length of the frozen horizon is also included in the experiment. The value of $Tp_{il}$ is equal to 1 s per piece and $Ts_{il}$ is 25 min. Given that the company does not have capacity limitations, we assume that capacity is high enough to produce all the units required by the company. Table 2 presents the rest of the parameters and their values for the high and low level of the experiments. The values were defined according to the characteristics of the case study under consideration. The bias factor is interpreted as the percentage by which the real demand is underestimated. Thus, a value of 60% indicates that the forecast is on average only 40% of the real demand.

**Table 2.** Parameters to evaluate in the factorial experimental design.

| Parameter | Nomenclature | High Level | Low Level |
|---|---|---|---|
| SS | Safety stock service level | 95% | 85% |
| bias | Absolute value of bias *(underestimated)* | 60% | 10% |
| R | Number of periods in the frozen horizon | 6 | 1 |

To generate the instances, a data set of real demand faced by the company corresponding to the periods from year 2016 was used for experimental purpose. The forecasts were prepared with the specified bias and Root-Mean-Square Error (RMSE) as representative for the company's historical forecast error.

### 4.2 Numerical Results and Discussion

The proposed MILP was implemented in AMPL TeamEAT Version 20120911 (MS VC++ 6.0, 32-bit) with Gurobi 5.0.1 as the solver. The rolling horizon procedure was automated with Python 3.6, using the *amplpy* library for interacting with AMPL. Likewise, *pandas* and *xlrd* libraries were used to read instances files and write files with the results of the experiments. To solve the test instances, an Intel Core (TM) i3-6006U CPU @ 2.00 GHz 1.99 GHz computer, with 8gb RAM was used. The runtime for 40 instances was 3 h, 1 min and 20 s. The overall results of the experiments are summarized in Table 3. The minimum, mean and maximum values obtained for each of the variables is shown, as well as the minimum, mean and maximum computational times for the solved instances. Table 4 through 6 present the results by parameter level.

As can be observed in previous tables, the parameter related to the bias has an important effect on profits. For the rest of the parameters, the profits do not present a significant variation in the results obtained by each level. However, the frozen horizon $R$ has an impact on the values of inventory, with a higher mean and maximum value when we consider the low level of $R = 1$.

**Table 3.** Global results after implementing the solution method with Python.

| Value | *Mp* | *S* | *B* | *Inv* | Profit (*z*) | Computational time |
|---|---|---|---|---|---|---|
| Min | 558 | 558 | 155 | 0 | \$48376.00 | 01:12 min |
| Mean | 939 | 905 | 533 | 44 | \$99909.25 | 04:32 min |
| Max | 1443 | 1284 | 880 | 207 | \$160806.00 | 05:59 min |

**Table 4.** Performance of safety stock level on variables and total profit

| SS (1-α) | Value | *Mp* | *S* | *B* | *Inv* | Profit (*z*) |
|---|---|---|---|---|---|---|
| High level = 95% | Min | 559 | 559 | 155 | 0 | \$48415.00 |
| | Mean | 965 | 913 | 526 | 65 | \$99780.25 |
| | Max | 1443 | 1284 | 880 | 207 | \$160806.00 |
| Low level = 85% | Min | 558 | 558 | 155 | 0 | \$48376.00 |
| | Mean | 917 | 903 | 535 | 31 | \$100972.36 |
| | Max | 1251 | 1241 | 880 | 56 | \$153985.00 |

**Table 5.** Performance of bias on variables and total profit

| Bias | Value | *Mp* | *S* | *B* | *Inv* | Profit (*z*) |
|---|---|---|---|---|---|---|
| High level = 60% | Min | 558 | 558 | 843 | 0 | \$48376.00 |
| | Mean | 571 | 571 | 867 | 0 | \$50423.50 |
| | Max | 595 | 595 | 880 | 0 | \$54109.00 |
| Low level = 10% | Min | 1250 | 1195 | 155 | 38 | \$139106.00 |
| | Mean | 1307 | 1239 | 200 | 89 | \$149395.00 |
| | Max | 1443 | 1284 | 880 | 207 | \$160806.00 |

In addition to the results shown in the above tables, to verify the significance of the factors studied on the variables of interest, a multivariate regression analysis was performed. For this, Minitab 18 version was used. The analysis considered the step-by-step method to eliminate non-significant terms. The only coefficient that resulted significant for total profit as an individual factor was the bias. The rest of the parameters only have an impact interacting directly with bias, that is, their behavior depends on the level of bias: the safety stock level $SS\ (1\text{-}\alpha)$ and the number of periods of frozen horizon $R$. $Tp_{il}$ and $Ts_{il}$ depend strictly on previous interaction to be significant. An R-square adjusted value of 84.8% was obtained. Table 7 summarizes these results.

**Table 6.** Performance number of periods in frozen horizon on variables and total profit

| R | Value | *Mp* | *S* | *B* | *Inv* | Profit (*z*) |
|---|---|---|---|---|---|---|
| High level = 6 | Min | 558 | 558 | 155 | 0 | $48376.00 |
| | Mean | 913 | 910 | 528 | 23 | $102895.50 |
| | Max | 1284 | 1284 | 880 | 54 | $160806.00 |
| Low level = 1 | Min | 574 | 574 | 203 | 0 | $50794.00 |
| | Mean | 966 | 900 | 539 | 66 | $96923.00 |
| | Max | 1443 | 1236 | 865 | 207 | $143683.00 |

**Table 7.** Significance of the coefficients.

| Term | Coefficient | Standard error coefficient | T-value | P-value | Variance Inflation Factor (VIF) |
|---|---|---|---|---|---|
| Constant | 126505 | 2098 | 60.31 | 0 | |
| bias | −103504 | 5552 | −18.64 | 0 | 2.19 |
| SS*bias | −179520 | 74970 | −2.39 | 0.018 | 1 |
| R*bias | −21675 | 2374 | −9.13 | 0 | 1.19 |

Considering the factors and interactions that resulted more significant, we present a comparative analysis of them with respect to profits: (i) bias; (ii) Interaction of bias and *R*; and (iii) Interaction of bias and safety stocks.

(i) **Comparative analysis of bias with respect to profits**
Figure 1 presents the behavior of high and low level of bias on profits during experimentation. We can observe a decrease on profits when bias is higher, so the lower the value of this, the higher the benefits, which lead us to conclude that profits are highly influenced by the bias.

(ii) **Comparative analysis of the interaction of bias and R with respect to profits**
Figure 2 illustrates the behavior of bias and *R* with respect profits. As we can observe, when the bias has the biggest value (60%), the profits are low, even for a shorter length of the frozen period *R* ($R = 1$), but these short-frozen period tends to raise the profits a little by fast reaction. On the other hand, when the bias is small (10%), more stable and better profits are achieved when the length of the frozen period *R* is higher ($R = 6$).

(iii) **Interaction bias and Safety Stocks on profits comparative analysis**
Figure 3 presents the corresponding graphs to compare the bias and safety stock with respect to profits. As we can observe in the figure, for the same safety stock level, the level of bias has an important effect on total profits, being significantly higher when the bias is small (10%). The same conclusions can be derived when

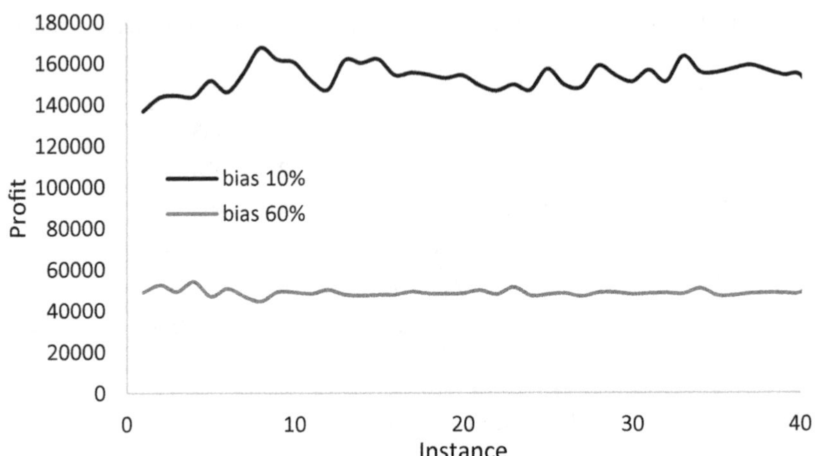


**Fig. 1.** Comparative analysis of the levels of bias with respect to total profit

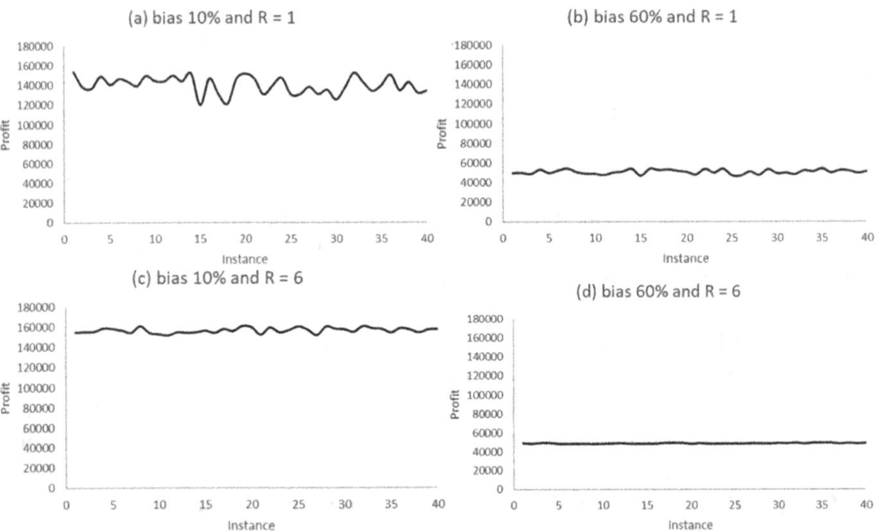


**Fig. 2.** Analysis of the interaction of bias and the frozen period *R* with respect to total profits.

considering a higher safety stock level (95%), significant differences on the profits are achieved depending on the level of bias.

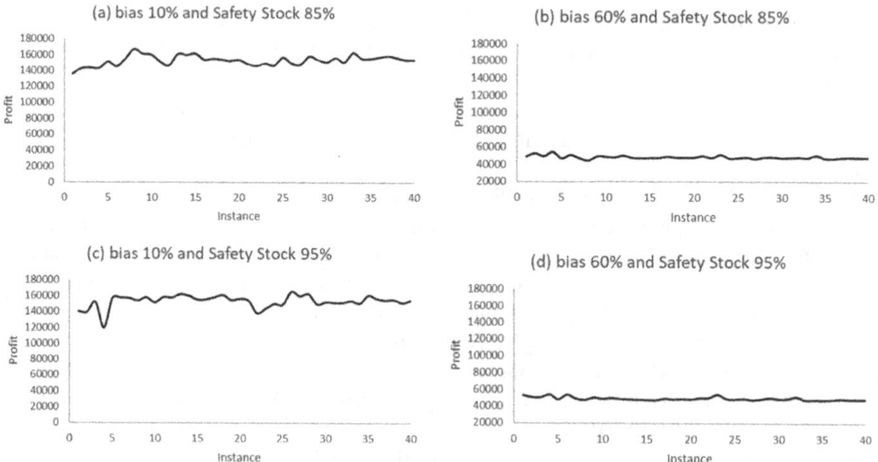


**Fig. 3.** Analysis of the interaction of bias and safety stocks with respect to total profits

## 5 Conclusions and Recommendations for Future Research

In this paper, we have proposed a Mixed-Integer Linear Programming (MILP) optimization planning model, motivated by a case study company that produces snacks in Monterrey, Mexico. To solve the model, a rolling horizon scheme is considered in order to resemble current practices of the production managers of the company, that determine their aggregated plans for a medium term planning horizon, and adjust them according to the demand faced. The proposed framework has been implemented in Python and AMPL.

Numerical experiments were performed, for which a set of instances were generated. Values of the parameters were defined to resemble the operations of the company. The experiments incorporated the forecast bias as a factor to be considered, to evaluate the impact that this has on profits and the other variables of the model (production amount, inventories, etc.). We also performed a statistical analysis to determine the factors that are most significant on the profits. Results show that the bias is the most significant factor affecting profits, and the other variables. The rest of the factors are not significant. Only the interaction of the bias and the frozen period *R*, and the interaction of bias and safety stock.

For future research we propose to extend this work in two ways. First, consider a stochastic version of the problem that can consider different demand scenarios. And the second aspect is to incorporate a recourse action during the frozen period *R*, considering the situation in which the company is more flexible and can adjust their production plans according to demand changes or changes in the customer's orders.

## References

1. Demirel, E., Özelkan, E.C., Lim, C.: Aggregate planning with Flexibility Requirements Profile. Int. J. Prod. Econ. **202**(May), 45–58 (2018). https://doi.org/10.1016/j.ijpe.2018.05.001

2. Hahn, G.J., Brandenburg, M.: A sustainable aggregate production planning model for the chemical process industry. Comput. Oper. Res. **94**, 154–168 (2018). https://doi.org/10.1016/j.cor.2017.12.011
3. Jain, A., Palekar, U.S.: Aggregate production planning for a continuous reconfigurable manufacturing process. Comput. Oper. Res. **32**(5), 1213–1236 (2005). https://doi.org/10.1016/j.cor.2003.11.001
4. Jamalnia, A., Yang, J.B., Xu, D.L., Feili, A.: Novel decision model based on mixed chase and level strategy for aggregate production planning under uncertainty: case study in beverage industry. Comput. Ind. Eng. **114**, 54–68 (2017). https://doi.org/10.1016/j.cie.2017.09.044
5. Jamalnia, A., Yang, J., Feili, A., et al.: Aggregate production planning under uncertainty: a comprehensive literature survey and future research directions. Int. J. Adv. Manuf. Technol. **102**, 159–181 (2019). https://doi.org/10.1007/s00170-018-3151-y
6. Li, Z., Ierapetritou, M.G.: Rolling horizon-based planning and scheduling integration with production capacity consideration. Chem. Eng. Sci. **65**(22), 5887–5900 (2010). https://doi.org/10.1016/j.ces.2010.08.010
7. Lin, P.C., Uzsoy, R.: Chance-constrained formulations in rolling horizon production planning: an experimental study. Int. J. Prod. Res. **54**(13), 3927–3942 (2016). https://doi.org/10.1080/00207543.2016.1165356
8. Sitompul, C., Aghezzaf, E.H.: An integrated hierarchical production and maintenance-planning model. J. Qual. Maintenan. Eng. **17**(3), 299–314 (2011). https://doi.org/10.1108/13552511111157407
9. Stephan, H.A., Gschwind, T., Minner, S.: Manufacturing capacity planning and the value of multi-stage stochastic programming under Markovian demand. Flex. Serv. Manuf. J. **22**(3–4), 143–162 (2010). https://doi.org/10.1007/s10696-010-9071-2
10. Venkataraman, R., D'Itri, M.P.: Rolling horizon master production schedule performance: a policy analysis. Prod. Plann. Control **12**(7), 669–679 (2001). https://doi.org/10.1080/09537280010016774
11. Vogel, T., Almada-Lobo, B., Almeder, C.: Integrated versus hierarchical approach to aggregate production planning and master production scheduling. OR Spectrum **39**(1), 193–229 (2017). https://doi.org/10.1007/s00291-016-0450-2
12. Wu, D., Ierapetritou, M.: Hierarchical approach for production planning and scheduling under uncertainty. Chem. Eng. Process. **46**(11), 1129–1140 (2007). https://doi.org/10.1016/j.cep.2007.02.021
13. Yao, X., Almatooq, N., Askin, R., Gruber, G.: Capacity and production planning for automobile assembly with limited model-specific tooling flexibility. Working Paper, Arizona State University (2020)
14. Zhang, X., Prajapati, M., Peden, E.: A stochastic production planning model under uncertain seasonal demand and market growth. Int. J. Prod. Res. **49**(7), 1957–1975 (2011). https://doi.org/10.1080/00207541003690074

# 2D Nesting and Scheduling in Metal Additive Manufacturing

Ibrahim Kucukkoc[1(✉)], Zixiang Li[2], and Qiang Li[3]

[1] Department of Industrial Engineering, Balikesir University, Balikesir, Turkey
ikucukkoc@balikesir.edu.tr
[2] Industrial Engineering Department, Wuhan University of Science and Technology, Wuhan, China
[3] College of Engineering, Mathematics and Physical Sciences, University of Exeter, Exeter, UK

**Abstract.** Additive manufacturing (or 3D printing) is considered to be the future of manufacturing. Flexibility, accuracy, rapidness, cost efficiency and lightness are among the advantages those additive manufacturing provides. Astronauts can now produce their own tools and parts in the space with no waiting for another launch. Planning of additive manufacturing machines is a recent hot topic. Efficient use of such resources plays an important role to reduce the costs of additively manufactured parts as well as disseminate this technology making its advantages more apparent. This research addresses the metal additive manufacturing machine scheduling problem with multiple unidentical selective laser melting machines. Machines may have different specifications (dimension, speed, and cost parameters) and parts may have different characteristics (width, length, height, volume, release date and due date). The objective is to obtain a schedule such that total tardiness is minimized and parts are allocated on platforms (or building trays) with no overlap. A genetic algorithm approach is proposed to solve the problem within reasonable times. Results of the computational tests show the promising performance of the proposed method.

**Keywords:** Metal additive manufacturing · 3D printing · Scheduling · 2D nesting · Production planning

## 1 Introduction

The manufacturing industry has faced an enormous advancement with the utilization of additive manufacturing (or 3D printing) technologies. *Selective laser melting (SLM), selective laser sintering, fused deposition modelling, electronic beam melting* and *stereolithography* are some of the most popular ones belonging to this technology. Different from traditional manufacturing techniques which subtract material to produce a final product, 3D printing technologies produce 3D objects through a layer-by-layer production process [1].

Additive manufacturing provides several advantages as opposed to conventional subtractive techniques. Design flexibility, high accuracy, resource efficiency, lightness,

© Springer Nature Switzerland AG 2021
D. A. Rossit et al. (Eds.): ICPR-Americas 2020, CCIS 1407, pp. 97–112, 2021.
https://doi.org/10.1007/978-3-030-76307-7_8

strength-to-weight ratio and material efficiency are some of those advantages. However, design flexibility is probably the most important one among them. This is because, additive manufacturing allows very complex design structures with the flexibility of layer-by-layer production. With the shift from rapid prototyping to direct digital manufacturing, this feature helps to position additive manufacturing in a higher technology level.

SLM is used to melt and fuse metallic powders together via a high-power density laser. The powder is very fine and deposited on a substrate using a roller with a thickness of usually between 20 μm and 100 μm. The high-power density laser beam melts the 2D cross section of the first layer and so fuses those areas. Afterwards, a new layer of powder is deposited on the substrate and corresponding areas are melted and fused via laser. This layer-by-layer process is repeated until a final product is achieved [2].

The application areas of SLM technique, especially for titanium alloys, are expanding even more recently. SLM has become a widely used additive manufacturing technique to fabricate critical parts (such as turbopumps) used in the aerospace industry. NASA has decided to additively manufacture a rocket engine fuel pump in one piece instead of mounting hundreds of components (including a turbine that spins at over 90,000 rpm) [3]. So that the additively manufactured pump has 45% fewer parts in comparison with traditionally manufactured pumps [1].

The increase in the wide application of AM technologies and demand on 3D printed parts force companies use their resources more efficiently. In this context, considerations for planning and scheduling of additive manufacturing resources are gaining more importance. This research proposes a genetic algorithm (GA) approach to efficiently schedule multiple AM machines for fabricating part orders coming from several customers with the aim of minimizing total tardiness. 2D nesting of parts on the machine platforms (building bed) are also considered as a constraint. Machines are unidentical; their speed, capacity and cost parameters are different. Parts have different dimensions, volume, due dates and release dates. As a unique contribution to literature, this is the first study, which minimizes total tardiness considering 2D nesting of parts, in AM machine scheduling domain.

The remainder of this paper is organized as follows. The following section reviews the literature. Section 3 describes the problem characteristics. Section 4 provides the basic steps of the proposed GA approach. Section 5 reports the results of computational tests and finally Sect. 6 concludes the paper with future research directions.

## 2 Literature Review

While the literature on AM and 3D printing is rather extensive, the number of studies on planning and scheduling of AM machines is quite limited. This section reviews the literature on planning and scheduling of AM machines and shows the gap in the literature.

The first study on AM machine scheduling problem relates to Kucukkoc et al. [4]. They addressed the characteristics of planning production on AM machines and proposed a mathematical formulation to calculate job production times. The model proposed by Kucukkoc et al. [4] aims at increasing the utilization of additive manufacturing machines considering order delivery times. Another preliminary work on production planning of AM machines belongs to Li et al. [5]. They proposed a mathematical model and first heuristics (i.e. best-fit and adapted best-fit algorithms) to minimize average production cost per volume of material. Following this, the topic attracted many researches. Chergui et al. [6] addressed nesting and scheduling in additive manufacturing. They proposed a mathematical formulation and a heuristic to satisfy demands by customers considering due dates. Kucukkoc et al. [7] proposed a GA approach to minimize maximum lateness in a multiple AM machine environment. Fera et al. [8] developed a modified GA approach for time and cost optimization of single AM machine scheduling problem. Dvorak et al. [9] proposed heuristics for scheduling AM machines considering due dates. Kucukkoc [1] proposed mixed-integer linear programming models to minimize makespan in single, parallel identical and parallel non-identical AM machine scheduling problems. Li et al. [10] introduced the dynamic order acceptance and scheduling problem and proposed a strategy-based metaheuristic decision-making approach. The proposed approach helps companies decide whether to accept an order and when to produce it on which machine.

This survey shows that there is no research in metal additive manufacturing field that minimizes total tardiness while considering 2D nesting constraints. Thus, this paper contributes to literature proposing a GA approach to minimize total tardiness while ensuring 2D nesting feasibility.

## 3 Problem Definition

As clearly described by Kucukkoc [1], the problem of production planning and scheduling of metal AM machines is NP-Hard to solve. A group of AM machines in different specifications (i.e. width, length, speed, etc.) are available to produce parts received from individual customers. The parts also have different characteristics (i.e. width, length, height, volume, and due date) which need to be taken into account to obtain a feasible and efficient solution. The problem is to group parts into batches (called jobs) and allocate them to machines such that a performance criterion is optimized. The performance criterion aimed in this research is total tardiness minimization.

Although the problem can be supposed to be a classical bin packing problem, actually it is not. This is a multi-faced problem which involves many different problems inside: *bin packing*, *scheduling* and *nesting* (see Fig. 1). Furthermore, grouping parts and assigning them to machines can be considered a *batch scheduling* problem. However, Kucukkoc [1] shows why this problem cannot be considered as a batch scheduling problem.

Figure 2 depicts a schematic representation of the addressed problem in an environment with multiple AM machines of SLM type. The two questions which need to be answered are as follows:

- Which parts will be grouped together into which job batch?
- Which job batch will be processed when and on which machine?

One can suppose that there is a two-stage decision making process. However, the parts are grouped as a job batch considering not only other parts in the same job, but also the specifications of the machine that job will be fabricated on. Therefore, these two questions are tightly interrelated to each other and the solution process cannot be divided into two consecutive stages. The problem must be handled through a holistic view.

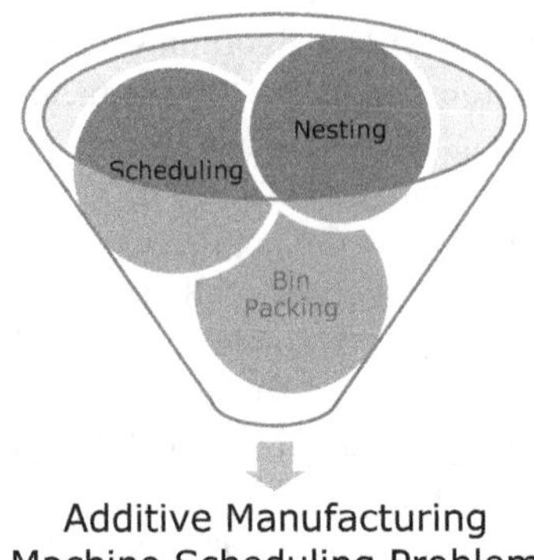


**Fig. 1.** Three basic components of the AM machine scheduling problem.

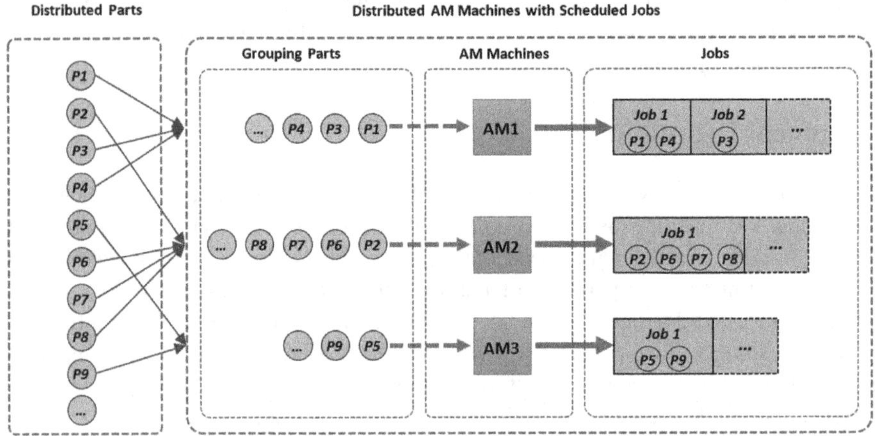


**Fig. 2.** A schematic representation of the multiple AM machine scheduling problem [1].

Another complicated feature of this problem is that the processing times of jobs are dynamically calculated via a formula. That means the processing times are not known before the content of the job is decided. For example, in a classical batch scheduling problem, the maximum processing time of a part in a batch determines the processing time of that batch. However, in a metal AM scheduling problem, the processing time of a job (or batch) is calculated via a formula (see Eq. (1)) considering the maximum height and total volume of the parts assigned to that job.

Each part ($i \in I$) has a width ($w_i$), length ($l_i$), height ($h_i$), area ($a_i$), volume ($v_i$), release date ($r_i$) and due date ($d_i$). Each machine ($m \in M$) has a maximum width ($MW_m$), length ($ML_m$), height ($MH_m$) and area ($MA_m$) supported by its tray or building platform. Machines also have other parameters, i.e. $VT_m$, $HT_m$, and $SET_m$. These parameters are used for calculating job processing times and costs. $VT_m$ is the time spent to form per unit volume of material. $HT_m$ is the time spent for powder-layering (it is repeated for each layer based on the highest part produced in the job). $SET_m$ is the set-up time needed for initialising and cleaning before/after each job. Equation (1) is used to calculate the production time of a job ($j \in J$) on a machine ($m$).

$$PT_{mj} = SET_m \cdot Z_{mj} + VT_m \cdot \sum_{i \in I} v_i \cdot X_{mji} + HT_m \cdot \underset{i \in I}{max}\{h_i \cdot X_{mji}\} \tag{1}$$

where $X_{mji}$ and $Z_{mj}$ are binary variables. $X_{mji}$ is equal to 1 if part $i$ is assigned to job $j$ on machine $m$, otherwise it is 0. $Z_{mj}$ is equal to 1 if job $j$ on machine $m$ is utilized; 0, otherwise. A job is considered to be utilized if at least one part is assigned to that job.

As seen in Eq. (1), $PT_{mj}$ is the sum of three terms: (i) set-up time, (ii) material releasing time (based on total volume) and (iii) powder layering time (based on maximum height). This equation indicates that different combinations and allocations of same parts on to different machines yield different processing times and so completion times of parts. Therefore, grouping parts with similar heights may provide an advantage in reducing the processing time of a job.

The completion time of a job on a machine ($CT_{mj}$) can be stated as in Eq. (2).

$$CT_{mj} \geq ST_{mj} + PT_{mj} \quad \forall m \in M,\ j \in J. \tag{2}$$

where $ST_{mj}$ is the start time of job $j$ on machine $m$. A job can start when all parts in this job are ready to be built and no part can be removed until *all* parts in the job are completed. Therefore, the start time of a job ($ST_{mj}$) must be greater than or equal to the maximum of release dates in that job, i.e. $ST_{mj} \geq \max_{i \in I_{mj}}\{r_i\}$ where $I_{mj}$ denotes the set of parts assigned to job $j$ on machine $m$. Considering the basic assumption of processing one job at a time on a machine, it is necessary to start a job after the completion of a previous job on that machine ($ST_{mj} \geq CT_{m,j-1}$). Therefore, the start time of a job can be expressed as in Eq. (3), where $CT_{m0} = 0$.

$$ST_{mj} \geq \max\left\{CT_{m,j-1}, \max_{i \in I_{mj}}\{r_i\}\right\} \quad \forall m \in M,\ j > 1. \tag{3}$$

The completion time of a part ($C_i$) is equal to the completion time of job in which this part is assigned; $C_{i \in I_{mj}} = CT_{mj}$. Thus, the tardiness of a part ($T_i$) can simply be calculated as $T_i = \max\{0, C_i - d_i\}$. Thus, the objective function is to minimize the total tardiness $T = \sum_i T_i$.

A 2D nesting problem occurs to ensure a feasible allocation of parts into job batches. One can assume that it is enough to satisfy that the sum of part areas does not exceed the building area capacity of the machine's platform. However, there is a chance that this assumption can yield infeasible solutions when the parts are physically located on the machine platform. Therefore, a 2D nesting problem has also been combined with the machine scheduling problem in this research.

Basic assumptions of the problem are given as follows:

- The maximum capacity (width, length and height) of at least one machine is large enough to fabricate the largest part in terms of width, length and height.
- Part specifications are known and deterministic.
- Machine specifications are known and deterministic.

- Machines may have different capacity and specifications.
- No breakdowns or failures.
- Only one job can be processed on a machine at any time.
- All parts need to be assigned to a job exactly once.
- No parts can be added to a job after the job starts.
- A part can be removed from a job after all parts have been completed in the same job.
- All machines are available at the start of planning period.
- Building orientations of parts are known and constant.
- 2D nesting problem is considered to ensure that the allocations do not exceed machines' width and length.

Under these assumptions, the problem is to find the optimum allocation of parts to jobs on machines such that the total tardiness is minimized. Following section presents the proposed GA based approach with numerical examples.

## 4 Proposed Method

The problem studied here is neither a classical machine scheduling problem nor a nesting problem itself. Therefore, a sophisticated solution method is necessary to obtain efficient as well as high-quality solutions.

In this research, the GA approach proposed by Kucukkoc et al. [4] is modified to minimize total tardiness considering the 2D nesting constraints. The pseudocode of the proposed GA approach is given in Fig. 3.

```
GA
Generate initial population
Evaluate chromosomes in the population (considering 2D
nesting constraints)
WHILE (no improvement in the last MaxItNoImp iterations)
    Tournament selection
    Genetic operators (Crossover & Mutation)
    Evaluate newly generated chromosomes
    Form new generation (replace the worst in the popula-
    tion)
ENDWHILE
Stop and report the best solution
```

**Fig. 3.** Pseudocode of the proposed GA approach.

The algorithm starts with generating random chromosomes to assure diversity. Figure 3 presents the pseudocode of the proposed GA approach. The number of chromosomes in the population is constant. After generating the population, each chromosome in the population is evaluated using the procedure given in Fig. 4 considering the nesting constraints. Parents are selected using tournament selection (given in Fig. 5) and genetic operators are applied. The procedures used for crossover and mutation are given in Figs. 6 and 7, respectively. After crossover and mutation, the newly obtained individuals (children) are evaluated using the same decoding procedure and the new population is formed replacing the worst chromosomes with a better child (if any). Each chromosome is made up of part numbers (or IDs) generated with permutation coding. So that, each gene represents a part number and each part is represented on the chromosome only once. Therefore, the length of a chromosome equals the number of parts to be produced. If there are 8 parts, a sample chromosome could be `chrom = {3, 8, 4, 1, 7, 2, 6, 5}`. The order of parts on the chromosomes are used in the decoding procedure. Parts are selected one by one based on their sequence on the chromosome and allocated to jobs on machines based on the decoding procedure provided in Fig. 4 (where `AP` denotes the list of parts not assigned and `nm` is the number of machines).

**Decoding Procedure**

```
Add all parts into set AP in the order that they appear on the chromosome
Set j ← 1 //job index
WHILE (AP ≠ ∅)
  FOR (m = 1 TO nm STEP 1) //start from the first machine
    totArea ← 0, I_mj ← ∅
    FOREACH (i in AP) //try each part
      IF (totArea + a_i ≤ MA_m and h_i ≤ MH_m and w_i ≤ MW_m and l_i ≤ ML_m)
        IF (totArea = 0 or Nesting(i, I_mj) = true)
          Add part i to I_mj
          totArea ← totArea + a_i
          Remove part i from AP
        ENDIF
      ENDIF
    ENDFOREACH
    j ← j + 1
  ENDFOR
ENDWHILE
return
```

**Fig. 4.** Pseudocode of the decoding procedure.

In the decoding procedure, nesting constraints are satisfied via an external function, called `Nesting()`, for which the source code can be retrieved from GitHub [11]. This function gets a machine ID and a list of parts as parameter and returns true if they fit on the machine tray with no exceedance. The algorithm starts with the first machine and

```
Selection (Tournament)
Set min ← MAX_Value, winner ← ∅, tourSize ← Ceil(popSize/6)
FOR (t = 1 TO tourSize STEP 1)
  Choose a random chromosome (p) from the population
  IF (Fitness(p) < min)
    Set winner ← p
  ENDIF
ENDFOR
return
```

**Fig. 5.** The procedure used for tournament selection.

```
Crossover
cp ← [1,chromSize] //Determine a random cut point (cp)
Initialise child1 and child2; child1 ← ∅, child2 ← ∅
head1 ← parent1[1,cp]; head2 ← parent2[1,cp]
temp1 ← parent1[cp + 1,chromSize] + parent1[1,cp]
temp2 ← parent2[cp + 1,chromSize] + parent2[1,cp]
Set tail1 ← ∅, tail2 ← ∅
FOREACH (i in temp2) //Build tail1
  IF (i ∉ head1)
    Add part i to tail1
  ENDIF
ENDFOREACH
FOREACH (i in temp1) //Build tail2
  IF (i ∉ head2)
    Add part i to tail2
  ENDIF
ENDFOREACH
child1 ← head1 + tail1, child2 ← head2 + tail2
return
```

**Fig. 6.** The procedure used for crossover.

first job. The first part on the chromosome is selected and checked if it fits on the tray. If the part's width, length, height and area are smaller than or equal to the machine's width, length, height and area, the first phase of the control process is completed. In

```
Mutation
rand ← Random[0,1] //to determine swap or insert
r1 ← Random[1,chromSize] //determine a random location r1
r2 ← Random[1,chromSize] //determine a random location r2
WHILE (r1 = r2)
   r2 ← Random[1,chromSize]
ENDWILE
IF (rand < 0.5) //swap operation
   temp ← parent[r1]
   parent[r1] ← parent[r2]
   parent[r2]] ← temp
ENDIF
IF (rand ≥ 0.5) //insert operation
   Move part at r1 to its new location r2 on parent
ENDIF
return
```

**Fig. 7.** The procedure used for mutation.

the second phase of the control process, the machine ID and part's data are sent to the `Nesting()` function together with existing parts list in the same job on the same machine. If the returned result is true, the part is assigned to the current job and the `totArea` is increased by $a_i$. If the result is false, the next part is checked following the same procedure. When no more part can be allocated to a job, the machine ID is altered and the unallocated parts are tried to be assigned after a two-stage feasibility control process. When all the machines are full, new jobs are opened and the assignment process continues starting from the first machine again. This cycle continues until all parts have been allocated. As seen from this procedure, a two-stage control process is utilized to check the assignment feasibility of a part to a job on a machine and the nesting condition is checked if the first stage is completed successfully. This is because `Nesting()` function has a time consuming procedure and it is called only when the first phase of the control is completed successfully. So that any unnecessary nesting efforts are avoided.

The parents for genetic operators are selected using a tournament selection procedure given in Fig. 5. To determine two parents, the procedure is applied twice and the winners are designated as parents. The tournament size (`tourSize`) is determined as `Ceil(popSize/6)`, where `Ceil(X)` returns the smallest integral value that is greater than or equal to `X`. So that the diversity is ensured and chromosomes with better fitness values have more chance for reproduction. However, all individuals have a chance to be selected.

Winners of the tournament selection (`parent1` and `parent2`) undergo crossover and mutation, respectively. Crossover helps mostly better chromosomes generate new individuals and form next generations. It is applied via determining a random cut point (`cp`) and systematically combining the genes of the two parents referencing that cut point (see Fig. 6). The head of `child1` is retrieved from the first part of `parent1` (from the first gene to `cp`). The remaining genes of `child1` is completed based on the sequence of missing genes after `cp` on `parent2`. Similarly, the head of `child2` is the head of `parent2` (from the first gene to `cp`) and the remaining part (`tail2`) of `child2` is completed based on the order of missing genes after `cp` on `parent1`. So that, one crossover is completed. This is repeated several times based on a parameter called `crossover rate` whose value is determined by the user.

Mutation is another important genetic operator. It helps algorithm scan different areas of the search space and prevents algorithm stuck in local optima. Figure 7 shows the procedure used for mutation. A parent is selected from the population via tournament selection and either `swap` or `insert` is applied based on the value (between [0, 1]) of a random variable, i.e. `rand`. The two randomly determined genes are swapped and a mutant is constituted if the random variable is lower than 0.5. If not, a randomly determined gene is moved to another randomly determined location to obtain a `mutant`. Mutation rate controls how many times mutation will be applied on a newly selected parent.

New generation is formed replacing the worst chromosomes in the population with the best individuals (children) obtained from genetic operators. In this process, it is very important to avoid duplication of chromosomes. Otherwise, a couple of iterations later the population may be just a multiple copy of the same chromosome.

The algorithm runs until there is no improvement in a predefined number of iterations and the best solution is reported.

## 5 Tests and Results

The proposed GA approach has been coded in JAVA and run on Intel® Core™ i7-6700HQ CPU @2.60 GHz with 16 GB of RAM to solve the test problems given in Table 1. Number of parts (*nbParts*), number of machines (*nbMachines*), range of parts, and machines used to fabricate parts are provided for each test instance. For example, I6 is constituted of 14 parts (i.e. P15–P28 in part dataset) and three SLM machines (two of which are M1 and the other one is of M2). Test problems have been organized in ascending order of their complexity. The problem complexity increases when the number of parts and the number of machines get larger. Machine park has been differentiated to obtain a more diversified problem set. Details of SLM machine specifications are obtained from Kucukkoc [1] and provided in Table 2. Part dataset is also provided in Table 4 (see Appendix).

The population size, crossover rate and mutation rate are considered to be 30, 0.6 and 0.1, respectively. The algorithm was terminated when there is no improvement within 2500 iterations for test instances I1–I6 and 5000 iterations for test instances I7–I20. The values of these parameters have been determined through some preliminary tests considering the problem size [7].

The results of the experimental tests have been reported in Table 3. The column *nbJobs* indicates the total number of jobs utilized on machines. While the ultimate goal of the GA is to minimize *total tardiness*; *maximum tardiness* and *the number of late parts* are also reported as additional performance indicators. Moreover, the time consumed to solve each problem has been reported in seconds (s) for the comparison purposes in future researches.

**Table 1.** Data on test instances.

| Test instance | nbParts | nbMachines | Range of parts | Machines |
|---|---|---|---|---|
| I1 | 12 | 2 | P1–P12 | M1[1], M2[1] |
| I2 | 12 | 2 | P13–P24 | M1[1], M2[1] |
| I3 | 14 | 2 | P1–P14 | M1[1], M2[1] |
| I4 | 14 | 2 | P15–P28 | M1[1], M2[1] |
| I5 | 14 | 3 | P1–P14 | M1[1], M2[2] |
| I6 | 14 | 3 | P15–P28 | M1[2], M2[1] |
| I7 | 18 | 2 | P1–P18 | M1[1], M2[1] |
| I8 | 18 | 2 | P11–P28 | M1[1], M2[1] |
| I9 | 18 | 2 | P15–P32 | M1[1], M2[1] |
| I10 | 18 | 3 | P19–P36 | M1[2], M2[1] |
| I11 | 24 | 3 | P1–P24 | M1[2], M2[1] |
| I12 | 24 | 3 | P1–P24 | M1[1], M2[2] |
| I13 | 24 | 4 | P17–P40 | M1[2], M2[2] |
| I14 | 24 | 5 | P17–P40 | M1[3], M2[2] |
| I15 | 30 | 5 | P1–P30 | M1[3], M2[2] |
| I16 | 30 | 6 | P1–P30 | M1[3], M2[3] |
| I17 | 30 | 7 | P11–P40 | M1[4], M2[3] |
| I18 | 30 | 8 | P11–P40 | M1[4], M2[4] |
| I19 | 40 | 11 | P1–P40 | M1[6], M2[5] |
| I20 | 40 | 12 | P1–P40 | M1[6], M2[6] |

**Table 2.** Machine specifications.

| Machine ID | Height (cm) | Width (cm) | Length (cm) | Area ($cm^2$) | Set-up time (h) | HT (hour/cm) | VT (hour/$cm^3$) |
|---|---|---|---|---|---|---|---|
| M1 | 32.5 | 25 | 25 | 625 | 2.00 | 1 | 0.030864 |
| M2 | 32.5 | 25 | 25 | 625 | 1.00 | 1 | 0.030864 |

The GA algorithm finds reasonable solutions within short CPU times. The CPU time requirement increases with the increase in the number of parts and so the problem

**Table 3.** Results of the experimental tests.

| Test instance | nbJobs | Maximum tardiness (h) | Number of late parts | Total tardiness (h) | CPU (s) |
|---|---|---|---|---|---|
| I1 | 5 | 0 | 0 | 0 | 31 |
| I2 | 6 | 57.89 | 2 | 66.43 | 39 |
| I3 | 6 | 10.35 | 2 | 18.57 | 62 |
| I4 | 6 | 56.58 | 3 | 98.55 | 46 |
| I5 | 5 | 0 | 0 | 0 | 66 |
| I6 | 6 | 0 | 0 | 0 | 52 |
| I7 | 8 | 0 | 0 | 0 | 142 |
| I8 | 7 | 135.44 | 5 | 335.08 | 118 |
| I9 | 8 | 97.12 | 7 | 308.61 | 89 |
| I10 | 9 | 0 | 0 | 0 | 100 |
| I11 | 9 | 10.22 | 1 | 10.22 | 183 |
| I12 | 9 | 7.54 | 1 | 7.54 | 172 |
| I13 | 9 | 114.30 | 3 | 213.50 | 175 |
| I14 | 10 | 34.84 | 2 | 36.92 | 231 |
| I15 | 12 | 28.19 | 2 | 38.36 | 432 |
| I16 | 11 | 26.21 | 3 | 58.16 | 321 |
| I17 | 13 | 110.64 | 5 | 174.66 | 225 |
| I18 | 12 | 94.06 | 3 | 139.80 | 339 |
| I19 | 15 | 62.46 | 6 | 148.56 | 386 |
| I20 | 15 | 76.52 | 3 | 152.44 | 586 |

complexity. This is because of the exponential increase in the search space. Each problem has been solved three consecutive times and the best results are reported.

When the results are investigated, some problems have no tardiness. See, for example, I1, I5–I7 and I10. That means no part is late in these cases. In I3 and I5, the same parts (i.e. P1–P14) are scheduled with a different machine park. For I3, there is only one machine in each type (M1 and M2). So, this problem is solved with a total of two machines. However, for I5, there are three machines in total (two machines of type M2 and one M1) which yields a better fitness value (0 instead of 18.57) as expected. The maximum of objective value was observed for I8 with a total tardiness of 335.08 h.

A part of the convergence graph of the proposed GA approach is given in Fig. 8 for I12 where total tardiness is '7.54'. As seen from the graph, the best fitness value in the initial population was over 700 h of total tardiness. GA produces efficient schedules which gradually decreases total tardiness to '7.54' in the first 500 iterations.

The detailed nesting result of the best solution for I12 is also provided in Fig. 9. It is worthy to note here that the nesting function is used just to make sure that the assigned

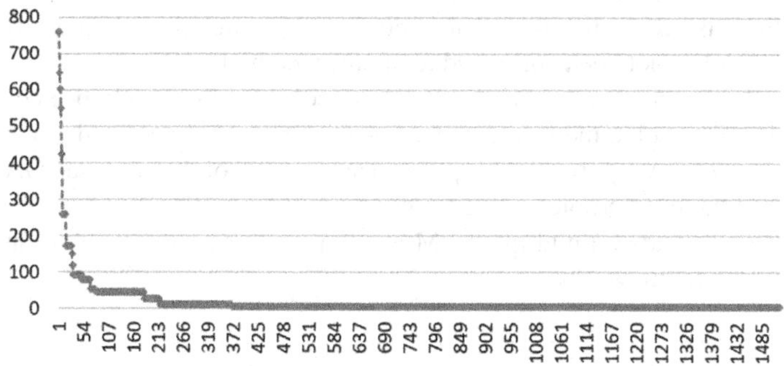


**Fig. 8.** Convergence of the proposed algorithm (the first section)

parts to a job do not exceed the dimensions of the tray, not to maximize the utilization of the trays. This is because minimization of the *nbJobs* does not mean the minimization of the total tardiness as contextualized in previous sections.

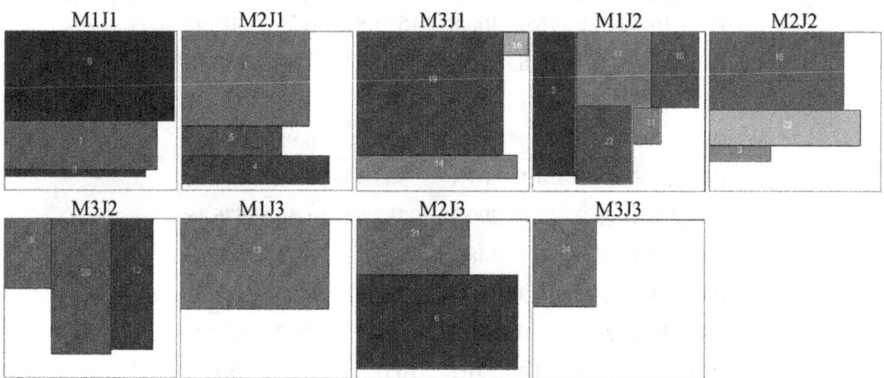


**Fig. 9.** Allocation of parts to jobs.

## 6 Conclusions and Future Research

This paper proposed a GA approach for scheduling metal additive manufacturing machines to minimize total tardiness considering 2D nesting constraints in a multiple unrelated parallel machine environment. The proposed method employs a nesting function which ensures that the parts assigned to a job on a machine do not exceed tray capacity (in terms of width and length) as well as height. The production time of a job is calculated via a function based on the content of the job, i.e. the maximum height and total volume of parts assigned to that job. Pseudocodes of GA mechanisms have been provided and test problems have been solved. Test data is given in details to provide comparable data in future researches. The results reported exhibits the promising solution building capacity and convergence of the proposed approach. However, as the problem is new and there is no comparable result reported in the literature, the performance of the proposed GA has not been compared to another method.

In future studies, new metaheuristics (e.g. ant colony, tabu search, bee colony etc.) can be developed to solve the test instances provided in this paper and results can be compared to that of GA reported in this paper. The capacity of the proposed GA can also be improved with some problem specific genetic operators. Furthermore, fabrication of irregular shaped parts on multiple SLM machines can be considered as an original research topic for future works.

## Appendix

(See Table 4).

**Table 4.** Part dataset used for test problems.

| Part | $h_i$ | $w_i$ | $l_i$ | $a_i$ | $v_i$ | $r_i$ | $d_i$ |
|---|---|---|---|---|---|---|---|
| P1 | 16.7 | 18.8 | 16.0 | 300.8 | 1573.8 | 6.25 | 305.90 |
| P2 | 8.8 | 6.7 | 22.8 | 152.8 | 421.3 | 7.35 | 282.18 |
| P3 | 20.3 | 9.3 | 2.1 | 19.5 | 147.8 | 9.83 | 378.25 |
| P4 | 7.4 | 21.6 | 3.9 | 84.2 | 285.2 | 20.95 | 214.67 |
| P5 | 27.3 | 14.9 | 4.1 | 61.1 | 583.3 | 36.58 | 148.98 |
| P6 | 25.8 | 23.2 | 12.9 | 299.3 | 3282.5 | 51.50 | 576.18 |
| P7 | 14.5 | 22.2 | 6.7 | 148.7 | 1265.5 | 56.35 | 240.42 |
| P8 | 3.5 | 24.6 | 15.3 | 376.4 | 723.3 | 69.90 | 211.63 |
| P9 | 20.4 | 20.5 | 1.0 | 20.5 | 278.5 | 75.07 | 330.87 |
| P10 | 23.3 | 6.8 | 13.4 | 91.1 | 1051.8 | 86.00 | 387.98 |
| P11 | 26.3 | 4.0 | 5.0 | 20.0 | 201.9 | 93.03 | 447.13 |
| P12 | 12.8 | 5.9 | 21.0 | 123.9 | 866.1 | 93.10 | 445.10 |
| P13 | 28.4 | 21.8 | 15.3 | 333.5 | 7347.6 | 97.15 | 634.57 |
| P14 | 10.9 | 23.2 | 3.2 | 74.2 | 333.6 | 97.25 | 177.55 |
| P15 | 14.1 | 19.6 | 13.7 | 268.5 | 2956.0 | 101.75 | 355.30 |
| P16 | 3.9 | 3.6 | 3.1 | 11.2 | 32.5 | 106.95 | 258.75 |
| P17 | 3.2 | 10.6 | 13.1 | 138.9 | 265.6 | 107.58 | 295.55 |
| P18 | 24.6 | 7.3 | 12.6 | 92.0 | 1387.4 | 115.55 | 509.62 |
| P19 | 7.1 | 21.2 | 20.0 | 424.0 | 1086.0 | 128.37 | 294.25 |
| P20 | 25.2 | 8.4 | 21.6 | 181.4 | 3559.2 | 132.32 | 575.72 |
| P21 | 29.6 | 16.4 | 10.6 | 173.8 | 2902.9 | 134.15 | 569.97 |

*(continued)*

**Table 4.** (*continued*)

| Part | $h_i$ | $w_i$ | $l_i$ | $a_i$ | $v_i$ | $r_i$ | $d_i$ |
|---|---|---|---|---|---|---|---|
| P22 | 21.1 | 10.8 | 8.0 | 86.4 | 854.6 | 136.65 | 353.23 |
| P23 | 25.2 | 22.0 | 4.9 | 107.8 | 1986.0 | 138.30 | 326.28 |
| P24 | 28.2 | 9.5 | 14.9 | 141.6 | 2974.3 | 149.48 | 515.72 |
| P25 | 22.9 | 6.3 | 4.6 | 29.0 | 408.2 | 164.08 | 525.87 |
| P26 | 24.2 | 17.7 | 9.4 | 166.4 | 1271.4 | 167.45 | 362.63 |
| P27 | 19.0 | 1.0 | 20.9 | 20.9 | 252.4 | 177.23 | 358.03 |
| P28 | 26.6 | 11.2 | 19.7 | 220.6 | 3740.9 | 206.40 | 589.43 |
| P29 | 18.5 | 12.5 | 8.2 | 102.5 | 991.9 | 219.62 | 560.93 |
| P30 | 10.1 | 15.4 | 11.9 | 183.3 | 1377.6 | 228.55 | 653.97 |
| P31 | 24.5 | 24.5 | 24.2 | 592.9 | 11122.9 | 242.90 | 1177.72 |
| P32 | 30.6 | 23.0 | 11.9 | 273.7 | 5234.1 | 244.87 | 739.32 |
| P33 | 3.1 | 14.7 | 9.9 | 145.5 | 274.6 | 245.08 | 553.43 |
| P34 | 18.9 | 17.4 | 14.8 | 257.5 | 2628.4 | 248.18 | 536.87 |
| P35 | 11.2 | 1.7 | 23.9 | 40.6 | 206.9 | 258.98 | 524.58 |
| P36 | 4.6 | 19.5 | 10.4 | 202.8 | 639.5 | 268.42 | 646.72 |
| P37 | 23.4 | 2.6 | 15.9 | 41.3 | 431.5 | 279.77 | 691.40 |
| P38 | 7.5 | 13.7 | 18.5 | 253.5 | 968.2 | 286.25 | 566.03 |
| P39 | 12.7 | 8.2 | 1.4 | 11.5 | 95.0 | 290.73 | 335.78 |
| P40 | 22.7 | 14.3 | 12.5 | 178.8 | 2306.2 | 294.53 | 473.53 |

## References


1. Kucukkoc, I.: MILP models to minimise makespan in additive manufacturing machine scheduling problems. Comput. Oper. Res. **105**, 58–67 (2019)
2. Nematollahi, M., et al.: Additive manufacturing (AM), Chap. 12. In: Niinomi, M. (ed.) Metals for Biomedical Devices, 2nd edn, pp. 331–353. Woodhead Publishing (2019)
3. NASA. Additive Manufacturing-Pioneering Affordable Aerospace Manufacturing. George C. Marshall Space Flight Center (2016). https://www.nasa.gov/sites/default/files/atoms/files/additive_mfg.pdf
4. Kucukkoc, I., Li, Q., Zhang, D.Z.: Increasing the utilisation of additive manufacturing and 3D printing machines considering order delivery times. In: 19th International Working Seminar on Production Economics, Innsbruck, Austria, 22–26 February 2016 (2016)
5. Li, Q., Kucukkoc, I., Zhang, D.Z.: Production planning in additive manufacturing and 3D printing. Comput. Oper. Res. **83**, 157–172 (2017)
6. Chergui, A., Hadj-Hamou, K., Vignat, F.: Production scheduling and nesting in additive manufacturing. Comput. Ind. Eng. **126**, 292–301 (2018)
7. Kucukkoc, I., et al.: Scheduling of multiple additive manufacturing and 3D printing machines to minimise maximum lateness. In: 20th International Working Seminar on Production Economics, Innsbruck, Austria, 19–23 February 2018 (2018)

8. Fera, M., et al.: A modified genetic algorithm for time and cost optimization of an additive manufacturing single-machine scheduling. Int. J. Ind. Eng. Comput. **9**(4), 423–438 (2018)
9. Dvorak, F., Micali, M., Mathieu, M.: Planning and scheduling in additive manufacturing. Inteligencia Artif. **21**, 40–52 (2018)
10. Li, Q., et al.: A dynamic order acceptance and scheduling approach for additive manufacturing on-demand production. Int. J. Adv. Manuf. Technol. **105**(9), 3711–3729 (2019)
11. GitHub: 2DPackingAlgorithm [Software] (2017). https://github.com/shubhampuranik/2DPackingAlgorithm

# Metaheuristics and Algorithms

# Proposal of a Heuristic for Cluster Analysis with Application in Allocation of Anaerobic Co-digesters for Biogas Production

Monique Schneider Simão(✉), José Eduardo Pécora, and Gustavo Valentim Loch

Federal University of Paraná, Curitiba, Paraná, Brazil
monique.schneider@ufpr.br

**Abstract.** Cluster analysis refers to optimally segmenting a set of entities, and there are several types of algorithms to reach this purpose. It can be applied in decision making, machine learning, data mining, and pattern recognition. This work aims to develop a simple and fast partition heuristic to solve the cluster analysis problem, where the optimality criteria are the smallest distances between the vertices of the same cluster. To verify the proposed heuristic, tests were made in instances ranging from 20 to 5000 nodes and different degrees of complexity. The developed algorithm can be applied to large size datasets were time could be a limitation factor. As an example of application, the heuristic developed was used to choose the location for installing anaerobic co-digesters for the production of biogas in the state of Paraná, Brazil.

**Keywords:** Cluster analysis · Heuristics · Biogas

## 1 Introduction

Cluster analysis involves optimally partitioning a given set of entities into a predetermined number of mutually exclusive clusters, where the criterion of optimality depends on the application in which the grouping is being employed [1]. Clustering is useful in several exploratory analyses, such as decision making, machine learning, data mining and pattern recognition [2].

Clustering algorithms can be divided into two main approaches: hierarchical and partition. The hierarchical approach produces a series of nested segments, agglomerating or dividing the previous clusters to achieve the desired segmentation. The partition system performs segmentations that optimize a given objective criterion or function. For each approach, there are numerous types of algorithms (genetic algorithm, ant colony, k-means, etc.) that have characteristics that make them more appropriate for solving a particular kind of problem [3].

Within the ways to solve cluster analysis problems is the use of heuristics, which are methods for solving specific problems that use shortcuts to produce solutions that, although not always optimal, are sufficiently adequate in a timely manner. This work aims to develop a simple and fast heuristic to partition a given dataset into a predefined number

© Springer Nature Switzerland AG 2021
D. A. Rossit et al. (Eds.): ICPR-Americas 2020, CCIS 1407, pp. 115–125, 2021.
https://doi.org/10.1007/978-3-030-76307-7_9

of clusters in a way that the sum of distances within all clusters is the smallest possible. We tested the proposed heuristic in instances ranging from 20 to 5000 nodes and different degrees of complexity. This paper's main contribution is present a heuristic that could be applied to large size datasets where time may be a limitation factor and high accuracy isn't crucial. Examples of this work application include multiple facility allocation where an extensive area is considered and the installation of a facility where it can be utilized by a number of persons large as possible (e.g., medical center, supermarket, etc.). We applied the heuristic to find the best location for biogas production in the state of Paraná, Brazil. This application was chosen due to the current relevance of renewable energy sources as an alternative to meet the global energy demand without significant environmental impacts [4].

## 2 Problem Definition

The optimization model used was proposed in [3], which was based on the work [1] and is described by (1) to (7).

Parameters:
$k \in K = \{1, \ldots, M\}$ defines the sets of cluster;
$i,j \in V = \{1, \ldots, N\}$ defines the sets of points;
$M$ is the total number of clusters;
$N$ is the number of vertices.
Decision variables:
$x_{ik}$ is a binary variable whose value is equal to 1 if the vertex $i$ is in cluster $k$ and 0 otherwise;
$y_{i,j}$ is the binary variable whose value is equal to 1 if the vertices $i$ e $j$ are in the same cluster and 0 otherwise;
$d_{i,j}$ is the distance between vertices $i$ e $j$.
Objective Function:

$$Min \sum_{i=1}^{N-1} \sum_{j=i+1}^{N} d_{ij} y_{ij} \tag{1}$$

Subjected to:

$$\sum_{k=1}^{M} x_{ik} = 1, \forall k \in K \tag{2}$$

$$\sum_{i=1}^{N} x_{ik} \geq 1, \forall i \in V \tag{3}$$

$$y_{ij} \geq x_{ik} + x_{jk} - 1, \forall i \in V, j \in \{i+1, \ldots, N\}, k \in K \tag{4}$$

$$y_{ij} \geq 0, \forall i \in V, j \in \{i+1, \ldots, N\} \tag{5}$$

$$x_{ik} \in \{0, 1\}, \forall i \in V, k \in K \tag{6}$$

$$y_{i,j} \in \{0, 1\} \forall i, j \in V \tag{7}$$

The objective function (1) aims to minimize the total distance among vertices of the same cluster. The set of constraints (2) assure that each vertex belongs to only one cluster. The set of constraints (3) assure that each cluster has at least one vertex. The constraints (4) and (5) assure that the variable $y_{ij}$ is equal to 1 if both $x_{ik}$ and $x_{jk}$ are equal to 1. The domain of the variables is defined by Eqs. (6) and (7).

## 3 Proposed Heuristics

The methodology developed in this paper was grounded on the work [5] where the authors proposed an algorithm that search for spikes of density, based on the assumption that clusters are density spikes distant from each other and surrounded by a less dense neighborhood. The heuristic proposed in this work and in [5] works only for data defined by a set of coordinates.

First, to identify which regions had the highest density of attributes, the search space is divided into a number of squares, defined as follows: (i) to form the squares, the $X$ and $Y$ axes are first subdivided into $n*z$ segments, were $z$ is arbitrarily defined and $n$ is the number of clusters desired; (ii) the number of subdivisions of each axis is increased until the sum of attempts that did not improve the objective function is equal to a predetermined value.

Then, to determine the density, sum the quantity of a particular attribute (e.g., points, cities, production potential, number of consumers, etc.) in each region. In sequence, order the regions in relation to density in a decreasing way, after that, select the first $n$ regions.

Finally, cluster's centroid is determined by searching, in each of the $n$ regions selected previously, the point that has the smallest accumulated distance from the other points belonging to the same square. The heuristic's pseudocode is presented below.

### 3.1 Pseudocode

While the maximum number of attempts without improvements isn't reached do:

Divide the search space by $n * z$, where $z \geq 0$ and $n$ is the number of clusters;

Sum the points or attributes within each region to obtain the density;
Select the $n$ densest regions;
For each area selected above, define as cluster's centroid the point with the smallest sum of distances from other points in the same region;
Assign the remaining points to the cluster of the nearest centroid;
Verify if the restrictions were satisfied;
Evaluate the objective function;
If the actual objective function value is better than the previous, then set the best cluster's configuration as the actual solution;
If the value is not better, then add one to the number of attempts without improvement;
Add one to z's value;

Return the best cluster configuration.

## 4 Results

The method was tested on instances generated by the authors. To obtain the exact solution for the clustering problem on these instances, we developed a program in C++ and solved it using CPLEX 12. The results obtained are in Table 1.

**Table 1.** Results for the exact solution.

| Instance | Nodes | Cluters | Solution | Computing time (s) | Status | GAP | Lower bound |
|---|---|---|---|---|---|---|---|
| A20 | 20 | 2 | 8348.6836 | ≈505 | Optimal | | |
| B20 | 20 | 2 | 8009.9048 | ≈602 | Optimal | | |
| C20 | 20 | 2 | 8201.6025 | ≈444 | Optimal | | |
| D20 | 20 | 2 | 6891.7427 | ≈218 | Optimal | | |
| E20 | 20 | 2 | 8410.6553 | ≈428 | Optimal | | |
| F20 | 20 | 2 | 9365.5303 | ≈430 | Optimal | | |
| G20 | 20 | 2 | 7845.5913 | ≈487 | Optimal | | |
| H20 | 20 | 2 | 6390.4751 | ≈550 | Optimal | | |
| I20 | 20 | 2 | 9667.8535 | ≈691 | Optimal | | |
| J20 | 20 | 2 | 8991.2773 | ≈446 | Optimal | | |
| K20 | 20 | 2 | 12523.0234 | ≈594 | Optimal | | |
| L20 | 20 | 2 | 12014.8584 | ≈517 | Optimal | | |
| M20 | 20 | 2 | 10866.3984 | ≈454 | Optimal | | |
| N20 | 20 | 2 | 10337.6152 | ≈468 | Optimal | | |
| A50 | 50 | 4 | 32359.9023 | ≈31208 | Optimal | | |
| B50 | 50 | 4 | 30562.3027 | ≈30529 | Optimal | | |
| C50 | 50 | 4 | 30824.7422 | ≈1369 | Optimal | | |
| D50 | 50 | 4 | 27175.5898 | ≈27493 | Optimal | | |
| E50 | 50 | 4 | 27483.0547 | ≈11000 | Optimal | | |
| F50 | 50 | 4 | 29114.4160 | ≈1925 | Optimal | | |

(*continued*)

**Table 1.** (*continued*)

| Instance | Nodes | Cluters | Solution | Computing time (s) | Status | GAP | Lower bound |
|---|---|---|---|---|---|---|---|
| G50 | 50 | 4 | 32634.8594 | ≈35417 | Optimal | | |
| A200 | 200 | 4 | 1216165.875 | ≈35885 | Feasible | 0.9978 | 2694.0615 |
| B200 | 200 | 4 | 1075722.125 | ≈35958 | Feasible | 0.9779 | 23771.430 |
| C200 | 200 | 4 | 1605665.375 | ≈35737 | Feasible | 0.9745 | 41002.098 |
| D200 | 200 | 4 | 1545418.375 | ≈35933 | Feasible | 0.9971 | 4543.4970 |
| E200 | 200 | 4 | 1418379.125 | ≈35909 | Feasible | 0.9847 | 21750.123 |
| F200 | 200 | 10 | 248272.7813 | ≈35713 | Feasible | 1 | 0 |
| G200 | 200 | 10 | 210943.4219 | ≈35656 | Feasible | 1 | 0 |
| H200 | 200 | 10 | 282760.9375 | ≈35826 | Feasible | 1 | 0 |
| I200 | 200 | 10 | 282364.25 | ≈35613 | Feasible | 1 | 0 |
| J200 | 200 | 10 | 311180.5 | ≈35769 | Feasible | 0.9999 | 28.8968 |
| K200 | 200 | 10 | 318806.1875 | ≈35774 | Feasible | 1 0 | 0 |
| L200 | 200 | 10 | 354271.5 | ≈35726 | Feasible | 1 | 0 |
| M200 | 200 | 10 | 294376.71875 | ≈35819 | Feasible | 1 | 0 |
| N200 | 200 | 10 | 237300.84375 | ≈35835 | Feasible | 1 | 0 |
| 0200 | 200 | 10 | 306800 | ≈35863 | Feasible | 1 | 0 |

The algorithm for the heuristic solution was implemented in Python and executed on an Intel Core i5 1.80 GHz RAM notebook. We set the initial value of $z$ as 1 and increased it by one until reach 5 attempts without improvement. The results obtained by the proposed heuristic, along with the comparison with the exact results, are shown in Table 2.

Table 1 show that the heuristic obtained values near the optimal solution obtained by the exact method in a much smaller time. It is also possible to verify that in the cases where the exact method couldn't reach the optimal solution, the value achieved by the heuristic was better. The heuristic proposed was also tested in larger instances, obtained in [6]. The values of the objective functions and the test execution time for these instances are shown in Table 3, and the distribution of clusters for instances "S1" and "S2" is illustrated in Fig. 1.

For "S" instances the algorithm´s performance varied significantly as the size of the data set increased and several points were assigned to the wrong clusters, as can be seen in Fig. 1, which shows the distribution of the points. This problem could be solved by considering the distance between the density spikes, selecting as clusters the densest locations that are the most far from the others. However, the attempts made with this methodology had the execution time as a limiting factor, both because Python is an interpreted language that needs special compilation libraries to run faster, and the configuration of the computer used to run the tests.

## 5 Application and Final Considerations

The developed heuristic was applied to choosing locations for installing anaerobic co-digesters for the production of biogas in the state of Paraná, Brazil. The solution aimed

**Table 2.** Results for the heuristics solution.

| Instance | Solution | Difference from the exact solution | Computing time (s) |
|---|---|---|---|
| A20 | 8348.6832 | ≈0 | ≈2 |
| B20 | 8009.9048 | ≈0 | ≈2 |
| C20 | 8201.6027 | ≈0 | ≈2 |
| D20 | 6891.7428 | ≈0 | ≈2 |
| E20 | 8410.6554 | ≈0 | ≈2 |
| F20 | 9365.5298 | ≈0 | ≈2 |
| G20 | 7845.5912 | ≈0 | ≈2 |
| H20 | 6390.4749 | ≈0 | ≈2 |
| I20 | 9667.8537 | ≈0 | ≈2 |
| J20 | 8991.2770 | ≈0 | ≈2 |
| K20 | 12523.0237 | ≈0 | ≈2 |
| L20 | 12014.8582 | ≈0 | ≈2 |
| M20 | 11122.5866 | +256.188 | ≈2 |
| N20 | 10337.6151 | ≈0 | ≈2 |
| A50 | 47627.3856 | +15267.5 | ≈2 |
| C50 | 30824.7420 | ≈0 | ≈2 |
| D50 | 27303.4368 | +127.847 | ≈2 |
| E50 | 28732.1027 | +1249.05 | ≈2 |
| F50 | 29188.0602 | +73 | ≈2 |
| G50 | 34364.1975 | +1729.34 | ≈2 |
| A200 | 713661.6293 | −502504.2 | ≈3 |
| B200 | 597733.9354 | −477988.2 | ≈3 |
| C200 | 969365.1094 | −636300.3 | ≈3 |
| D200 | 848980.4534 | −696437.9 | ≈3 |
| E200 | 829846.7959 | −588532.3 | ≈3 |
| F200 | 202176.5698 | −46096.21 | ≈4 |
| G200 | 179009.6952 | −31933.73 | ≈4 |
| H200 | 203059.6423 | −79701,3 | ≈4 |
| I200 | 166949.95 | −115414.3 | ≈4 |
| J200 | 195687.4640 | −115493 | ≈4 |
| K200 | 170802.1563 | −31709817 | ≈4 |
| L200 | 297823.4845 | −35129327 | ≈4 |
| M200 | 240254.331 | −29197418 | ≈4 |
| N200 | 185526.2464 | −23544558 | ≈4 |
| 0200 | 194504.4108 | −30485496 | ≈4 |

to minimize the transport distances, the total cost of the supply chain, and the associated carbon emissions. The main logistical activities in the biogas supply chain are substrate harvesting or collection, substrate storage, substrate transport, biogas production, biogas transport, and consumption of biogas [7].

Geography and distance can be essential factors because the transformation of biomass into energy is highly dependent on geographical factors since the supply of substrates and the points of demand for biogas are often widely dispersed. The use of cluster analysis to solve this problem is supported by the fact that deploying regional or

**Table 3.** Results for the "S" instances.

| Instance | Nodes | Clusters | Exact solution[a] | Proposed heuristic results | Computing Time |
|---|---|---|---|---|---|
| S1 | 5000 | 15 | 41304133758.6787 | 42278068222.2418 | ≈8 min |
| S2 | 5000 | 15 | 50100392668.3150 | 51514320860.6473 | ≈8 min |
| S3 | 5000 | 15 | 56666466350.1669 | 61896790663.16 | ≈8 min |
| S4 | 5000 | 15 | 53703664418.8502 | 64399923015.9137 | ≈8 min |

[a]Value of the objective function calculated using the ground truth centroids found in [6]

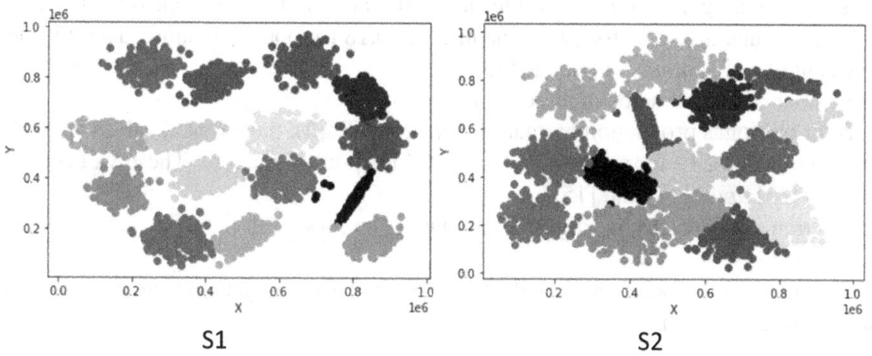


**Fig. 1.** Results for "S1" and "S2" instances.

centralized anaerobic digesters is an effective way to serve various rural establishments and consumers [8].

Biogas is formed in nature by the biological degradation of organic matter under anaerobic conditions. The main components of biogas are methane and carbon dioxide, which can be captured to generate energy in heat and electricity. The generation of biogas can be maximized by using techniques such as co-digestion, which is the simultaneous digestion of two or more organic substrates. Co-digestion can increase biogas production by 25%-400%, compared to isolated digestion of the same substrates [9]. In addition to increased biogas production, co-digestion also provides the following benefits: (i) improves the process stability; (ii) decreases the concentration of inhibitory substances; (iii) better balance of nutrients and moisture; (iv) increase the organic load; and (v) economic advantages due to equipment and cost-sharing [10].

In this case study, the choice of raw materials that will compose the mixture fed in the co-digester was made based on the article [11], where the tailings produced in Brazil were quantitatively and qualitatively analyzed to list which ones had the greatest potential for biogas generation. According to [11], more than 60% of the total biogas generation potential in Brazil comes from cattle manure, followed by sugarcane vinasse, swine manure, and poultry manure.

We obtained data on the number of heads of each of the types of farming considered and of tons of sugarcane produced in each of the 399 Paraná's cities from [12]. From these data, we calculated the biogas production potential for each city in the state. To calculate the potential for biogas production from the manure of different animals we used Eq. (8) obtained from [11].

$$Q_{CCH4} = Q_{SCH4} = Q_{PCH4} = \frac{I_M.NA.f_m.\varepsilon}{1000} \tag{8}$$

Where:

$Q_{CCH4}$ is the potential for biogas production from cattle manure ($m^3$/year);
$Q_{SCH4}$ is the potential for biogas production from swine manure ($m^3$/year);
$Q_{PCH4}$ is the potential for biogas production from poultry manure ($m^3$/year).
$I_M$ is the manure generation rate of the animal herd (kg/year), the adopted values were: 10 for cattle manure, 2,25 for swine manure and 0,18 for poultry manure. These values were found in [13] *apud* [14] and [15];
*NA* number of animal herds in all territory;
$f_m$ factor of biogas production by manure type ($m^3$ $CH_4$/t), the adopted values were: 12 for cattle manure, 12 for swine manure and 27 for poultry manure. These values were found in [13] *apud* [14] and [15];
$\varepsilon$ collection efficiency in bio-digester (adopted to be 90% as in [11]).

To calculate the potential for biogas production from sugarcane vinasse we used Eq. (9), obtained from [11].

$$Q_{VCH4} = P_V.T_S.COD_V.E_{fv}.f_V.E_{cV} \tag{9}$$

Where:

$Q_{VCH4}$ is the potential for biogas production from sugarcane vinasse ($m^3$/year);
$P_V$ is the vinasse production by mass unity of sugarcane ($m^3$/year);
$T_S$ is the sugarcane produced amount (tons);
$COD_V$ is the typical COD for vinasse, whose value was assumed as equal to 35,5 kg/$m^3$ [16];
$E_{fv}$ is the removing efficiency of vinasse organic load in an anaerobic reactor, whose value was assumed as equal to 62% based on [17];
$f_V$ is the specific factor of biogas production per mass of COD removed in the anaerobic reactor, whose value was assumed as equal to 0,29 $m^3$/kg $COD_{removed}$ and obtained from [18];
$E_{cV}$ is the collecting efficiency of biogas in the anaerobic reactor, assumed as 90%.

The total methane production potential of each city was calculated using Eq. (10). The map of production potential's obtained is shown in Fig. 2.

$$Pot_{ch4} = Q_{VCH4} + Q_{CCH4} + Q_{SCH4} + Q_{PCH4} \tag{10}$$

Where:

$Pot_{ch4}$ is the total biogas production potential ($m^3$/year);

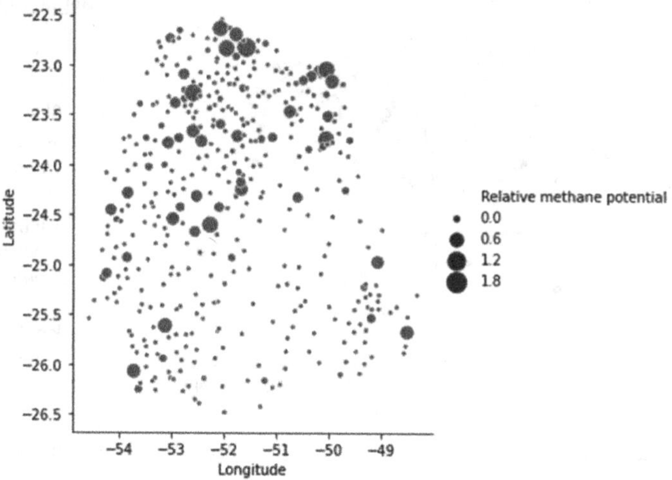


**Fig. 2.** Relative methane production potential for each city in the state of Paraná, Brazil.

The heuristic developed was used to segment the state of Paraná into seven clusters to minimize transport distances and the total cost of the biogas generation supply chain. The number of cluster was defined arbitrarily.

First, to identify which regions had the highest potential density for biogas production, the geographical space of Paraná was subdivided into a number of segments, the initial value of $z$ was set as 0.5. The number of subdivisions of each axis was increased by one until reach five attempts that failed to improve the objective function.

Then, to determine the density of the regions previously obtained, each city's total biogas production potential in the region was added. The regions were then ranked according to the density of the biogas production potential in a decreasing way. After that, the first $n$ regions were selected.

To choose the centroid we considered not only the distance from other points but also the production differences, to do so, we multiplied the distance by the methane potential, making the centroids closer to high production points. Finally, each city was assigned to the clusters of the nearest centroid. The results obtained are shown in Fig. 3.

Comparing Figs. 2 and 3, it is possible to observe that the algorithm was able to identify as centroid the spots with higher potential for methane production. We can conclude that despite the simplicity of the proposed heuristic, it can obtain satisfactory results for different sizes of spaces and numbers of clusters. Also, the technique is flexible and can be easily adapted to solve problems with different scopes.

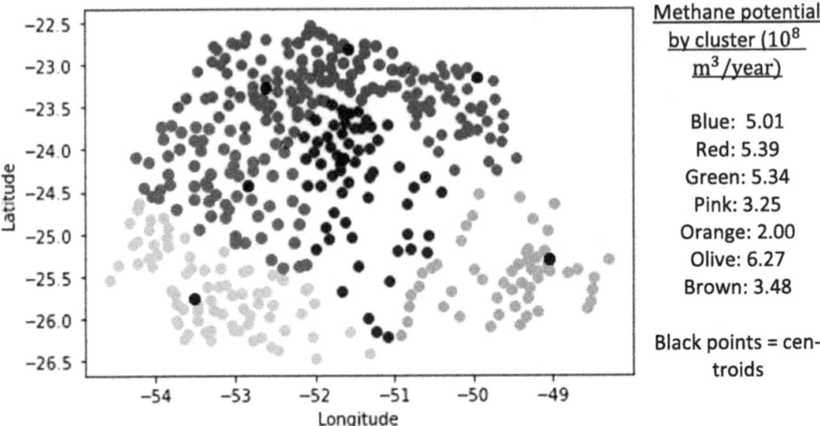


**Fig. 3.** Application of the cluster analysis heuristic for an anaerobic co-digester allocation in the state of Paraná, Brazil.

## References


1. Rao, M.R.: Cluster analysis and mathematical programming. J. Am. Stat. Assoc. **66**(335), 622–626 (1971)
2. Jain, A.K., Murty, M.N., Flynn, P.J.: Data clustering: a review. ACM Comput. Surv. **31**(3), 264–323 (1999)
3. Nascimento, M.C.V., Toledo, F.M.B., DE Carvalho, A.C.P.L.F.: Investigation of a new GRASP based clustering algorithm applied to biological data. Comput. Oper. Res. **37**(8), 1381–1388 (2010)
4. Esposito, G., et al.: Anaerobic co-digestion of organic wastes. Rev. Environ. Sci. Bio/Technol. **11**(4), 325–341 (2012)
5. Rodriguez, A., Laio, A.: Clustering by fast search and find of density peaks. Science **344**(6191), 1492–1496 (2014)
6. Fränti, P., Sieranoja, S.: K-means properties on six clustering benchmark datasets. Appl. Intell. **48**(12), 4743–4759 (2018). http://cs.uef.fi/sipu/datasets/
7. Zhang, J., et al.: An integrated optimization model for switchgrass-based bioethanol supply chain. Appl. Energy **102**, 1205–1217 (2013)
8. Park, Y.S., Szmerekovsky, J., Dybing, A.: Optimal location of biogas plants in supply chains under carbon effects: insight from a case study on animal manure in north dakota. J. Adv. Transp. (2019)
9. Ali, S.F., et al.: Co-digestion, pretreatment and digester design for enhanced methanogenesis. Renew. Sustain. Energy Rev. **42**, 627–642 (2015)
10. Hagos, K., et al.: Anaerobic co-digestion process for biogas production: progress, challenges and perspectives. Renew. Sustain. Energy Rev. **76**, 1485–1496 (2017)
11. Dos Santos, I.F.S., et al.: Assessment of potential biogas production from multiple organic wastes in Brazil: impact on energy generation, use, and emissions abatement. Resour. Conserv. Recycl. **131**, 54–63 (2018)
12. IBGE – INSTITUTO BRASILEIRO DE GEORAFIA E ESTATÍSTICA - Homepage, https://www.ibge.gov.br/estatisticas/economicas/agricultura-e-pecuaria/21814-2017-censo-agropecuario.html?=&t=resultados. Accessed on 08 May 2020
13. Sganzerla, E.: Biodigestors: A Solution. Agropecuária, Porto Alegre, Brazil (In Portuguese) (1983)

14. Colatto, L., Langer, M.: Biodigestor - solid waste for energy production. Unoesc & Ciência – ACET, Joaçaba, Brazil **2**(2), 119–128 (In Portuguese) (2011)
15. Brazilian Ministry of Cities. Anaerobic digestion technologies with relevance toBrazil. Substrates, Digesters, and Biogas Use. PROBIOGAS, Brasília (Accessed 27 March 2017) (In Portuguese) (2015). http://www.cidades.gov.br/images/stories ArquivosSNSA/probiogas/probiogas-tecnologias-biogas.pdf
16. Oliveira, B.G., Carvalho, J.L.N., Chagas, M.F., Cerri, C.E.P., Cerri, C.C., Feigl, B.J.: Methane emissions from sugarcane vinasse storage and transportation systems: comparison between open channels and tanks. Atmos. Environ. **159**, 135–146 (2017)
17. Fuess, L.T., Kiyuna, L.S.M., Ferraz Júnior, A.D.N., Persinoti, G.F., Squina, F.M., Garcia, M.L., Zaiat, M. Thermophilic two-phase anaerobic digestion using an innovative fixed-bed reactor for enhanced organic matter removal and bioenergy recovery from sugarcane vinasse. Appl. Energy **189**, 480–491 (2017)
18. Harada, H., Uemura, S., Chen, A.C., Jayadevan, J.: Anaerobic treatment of recalcitrant distillery wastewater by a thermophilic UASB reactor. Bioresour. Technol. **55**, 215–221 (1996)

# Benefits of Multiskilling in the Retail Industry: k-Chaining Approach with Uncertain Demand

Yessica Andrea Mercado and César Augusto Henao(✉)

Universidad del Norte, Barranquilla, Colombia
{myessica,cahenao}@uninorte.edu.co

**Abstract.** In the retail industry, demand seasonality, uncertainty, and unscheduled personnel absenteeism create an unbalance between staffing and staff demand, which can significantly increases over/understaffing costs. The use of a multi-skilled workforce is an attractive flexibility source to minimize such mismatch. This paper presents a mixed integer linear programming model for structuring the multiskilling skills of a set of employees. To explicitly incorporate demand uncertainty, we develop a two-stage stochastic optimization approach. Then, using a Monte Carlo simulation we analyze the performance of the solutions for different levels of demand variability. To test the performance of the proposed stochastic approach, it is compared with the solutions of two myopic approaches: zero multiskilling and total multiskilling. The proposed methodology is applied to a case study in a Chilean retail store, using real and simulated data. Our results showed that the proposed approach reported the maximum possible benefit for all levels of variability. Finally, we provide information to decision makers to address a key aspect of multiskilling: how much to add.

**Palabras claves:** Multiskilling · Chaining · Personnel scheduling · Stochastic programming · Retail

## 1 Introduction

The service sector, and in particular the retail industry, must respond not only to predictable phenomena such as seasonal demand, but also to unpredictable phenomena such as demand uncertainty and unscheduled staff absenteeism. Both phenomena complicate the efficient planning of human resources (HR), as they make it difficult to match the staffing and personnel demand. Thus, even companies with an effective shift scheduling system may face problems of over and understaffing at various times during the course of a week [1–4]. The literature in this area suggests that several authors have effectively addressed the problems of under and overstaffing by using a multiskilled workforce.

In a retail store, a multiskilled workforce is one that has the ability to work in $k$ departments (with $k \geq 2$), which allows the operational manager to transfer multiskilled employees from overstaffed departments to understaffed departments. Many researchers have found the $k$-chaining approach to be the most cost-effective policy for structuring a multiskilled workforce [2, 3, 5–13]. Specifically, the $k$-chaining approach implies

© Springer Nature Switzerland AG 2021
D. A. Rossit et al. (Eds.): ICPR-Americas 2020, CCIS 1407, pp. 126–141, 2021.
https://doi.org/10.1007/978-3-030-76307-7_10

that certain employees are trained to work in $k$ departments with $k \geq 2$ and that these employees and departments assignment decisions are configured through a bipartite network involving closed long chains (CLC) and/or closed short chains (CSC). In a CLC there is one multiskilled employee per department forming a single chain connecting all departments in the store, while in a CSC there is a subset of multiskilled employees connecting a subset of departments. Particularly, the 2-chaining policy has been reported as the most popular and recommended policy in the literature [2, 3, 7, 8, 14]. Under this policy all multiskilled employees are trained in at most two departments, i.e., $k = 2$ (see Fig. 1).

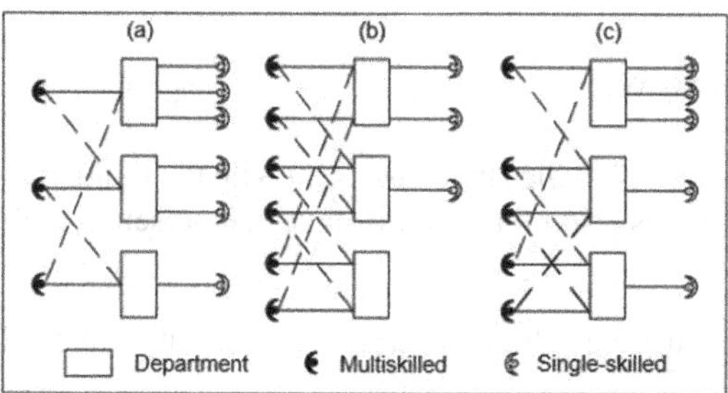


**Fig. 1.** 2-chaining configurations: (a) One CLC; (b) Two identical CLC; y (c) One CLC and one CSC. Source: Henao et al. [2].

Few studies have explored the benefits of implementing $k$-chaining policies with $k \geq 2$. Figure 2 shows examples of configurations of this type. Wang and Zhang [15] evaluated and compared the performance of CLCs using a 2-chaining policy versus a $k$-chaining policy with $k > 2$. They found that, at higher levels of demand variation, $k$-chaining policies with $k > 2$ retained high levels of performance as opposed to 2-chaining policies. This suggests that at high levels of demand uncertainty it is more beneficial to have a higher degree of flexibility in chaining. In turn, their results also showed that there are diminishing returns to scale as $k$ increases, i.e., the returns when $k = 5$ and $k = 6$ are almost identical. Therefore, extremely high levels of multiskilling may be impractical and unnecessary, and, consequently, clear policies are also required to define the most cost-effective levels of multiskilling.

In this paper we propose a novel methodology to evaluate the benefits of chaining with uncertain demand. First, the problem is formulated as a deterministic Mixed Integer Linear Programming (MILP) model. Our formulation differs from previous studies in the literature since multiskilling structures are defined through $k$-chaining policies with $k \geq 2$ guaranteeing the formation of closed chains (CLC and/or CSC). Second the MILP model is reformulated in a two-stage stochastic optimization (TSSO) model to incorporate demand uncertainty. Finally, the proposed methodology is applied to a case study in a Chilean retail store, considering different levels of demand variability.

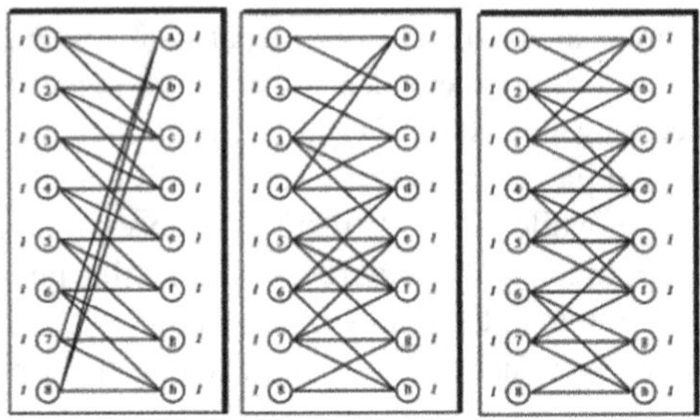

**Fig. 2.** $k$-chaining configurations with $k \geq 2$. Source: Iravani et al. [6].

## 2 Literature Review

Table 1 presents a review of studies that have dealt with the multiskilling problem in HR planning. The following is a summary of the elements present in Table 1:

**Table 1.** Characteristics of published studies with multiskilling in the HR management.

| References | HR-DL | MS | DU | SU | Chaining | NS | M | UM | AP |
|---|---|---|---|---|---|---|---|---|---|
| Cai & Li [16] | S+TS | Par | No | No | No | R | H | – | PS |
| Agnihothri et al. [17] | A | Par | No | No | No | R | S | – | MA |
| Eitzen & Panton [18] | TS | Par | No | No | No | NR | A | – | EPP |
| Bokhorst et al. [19] | A | Var | No | No | 2-chaining | R | OPT+S | – | MA |
| Hopp et al. [14] | A | Var | Yes | No | 2-chaining | R | S+MP | SA | MA |
| Wallace & Whitt [5] | S | Var | No | No | $k$-chaining | NR | S+A | – | CC |
| Iravani et al. [6] | A | Par | Yes | No | $k$-chaining | NR | S | SA | CC |
| Yang [20] | A | Par | No | Yes | $k$-chaining | NR | S | SA | MA |
| Sayin & Karabati [21] | A | Par | Yes | Yes | No | NR | OPT+S | SA | – |
| Heimerl & Kolisch [22] | S+TS | Par | No | No | No | NR | OPT+H | – | TE |
| Campbell [23] | A+DOS | Par | Yes | No | No | NR | OPT+H | SA | HE |
| Simchi-Levi & Wei [7] | A | Par | Yes | – | 2-chaining | R | A | CF | MA |
| Parvin et al. [24] | A | Par | No | No | $k$-chaining | NR | H | – | MA |
| Deng & Shen [25] | A | Par | Yes | – | $k$-chaining | NR | OPT+S+A | MSE | MA |
| Paul & McDonald [26] | S+A | Var | Yes | No | 2-chaining | R | OPT+H+A | MSE | HE |
| Gnanlet & Gilland [27] | S+A | Var | Yes | No | 2-chaining | R | OPT | SP, CF | HE |
| Wang & Zhang [15] | A | Par | Yes | – | $k$-chaining | NR | OPT+A | DRO, SA | MA |
| Henao et al. [1] | TS | Var | No | No | No | NR | OPT | – | RE |

(*continued*)

**Table 1.** (*continued*)

| References | HR-DL | MS | DU | SU | Chaining | NS | M | UM | AP |
|---|---|---|---|---|---|---|---|---|---|
| Cuevas et al. [28] | TS | Par | No | No | No | NR | OPT | – | RE |
| Henao et al. [2] | A | Var | Yes | No | 2-chaining | R | OPT+H | RO | RE |
| Liu [29] | S+A | Par | Yes | No | 2-chaining | R | OPT+S | SP | MA |
| Agrali et al. [30] | A | Par | No | No | No | NR | OPT+A | – | HE |
| Taskiran & Zhang [31] | S+TS | Var | No | No | No | NR | OPT | – | CC |
| Mac-Vicar et al. [32] | TS+A | Par | No | Yes | No | NR | OPT+H | SA | RE |
| Henao et al. [3] | A | Var | Yes | No | 2-chaining | R | OPT+H+A | CF | RE |
| Porto et al. [8] | S+TS | Var | Yes | No | 2-chaining | R | OPT | SA | RE |
| ***This paper*** | ***A*** | ***Var*** | ***Yes*** | ***No*** | ***k-chaining*** | ***NR*** | ***OPT*** | ***SP*** | ***RE*** |

1. *HR decision level* (HR-DL): Indicates the type of addressed staff scheduling problem in each study. It can be: (a) staffing (S); (b) daily shift scheduling (SS); (c) days-off scheduling (DOS), (c) tour scheduling (TS), where rest days and work shifts are assigned simultaneously; and (d) assignment (A). This last problem refers to the personnel assignment problem, which is addressed in our study.
2. *Multiskilling* (MS): Indicates whether multiskilling was modeled as a parameter (Par), or a decision variable (Var).
3. *Demand uncertainty* (DU): Indicates if the problem considered the demand uncertainty.
4. *Supply uncertainty* (SU): Indicates if the problem considered the supply uncertainty (i.e., unscheduled personnel absenteeism).
5. *Chaining:* Indicates if the study used a 2-chaining policy or a $k$-chaining policy with $k \geq 2$. Also indicates if the study did not consider any of those policies.
6. *Number of skills* (NS): Indicates if in the study there is a constraint to limit the maximum number of skills in the multiskilled employees. The study can be: (a) *restricted* (R), where each employee can have only two skills and (b) *not restricted* (NR), where the number of skills is not restricted, i.e., each multiskilled employee can have more than two skills in total.
7. *Method* (M): Indicates the used solution method, can be: (a) heuristics (H); (b) simulation (S); (c) optimization (OPT); y (d) analytics (A); y (e) Markov processes (MP).
8. *Uncertainty modeling* (UM): Indicates the approach used to evaluate the uncertainty associated with the parameters of a model. (1) *In-optimization*: approaches that explicitly incorporates the uncertainty in the math formulation, can be: (a) closed-form equation (CF); (b) stochastic programming (SP); (c) robust optimization (RO); y (d) distributionally robust optimization (DRO). (2) *Post-optimization*: approaches that evaluates the impact of uncertainty once the deterministic solution is obtained, can be: (a) sensibility analysis (SA); and (b) multiple scenarios evaluation (MSE).
9. *Application* (AP): Indicates the industry or sector that was studied, can be: (a) Postal services (PS); (b) Manufacture (MA); (c) Electric power plant (EPP); (d) Call-centers (CC); (e) Technology (TE); (f) Health (HE); and (g) Retail (RE).

### 2.1 Main Gaps

In the extensive literature on personnel multiskilling, it is possible to identify the following gaps.

First, Table 1 shows that although there are multiple articles studying multiskilling, most of them incorporate multiskilling as an input and not as a variable indicating whether an employee will be trained in a certain skill set or not. This has the negative effect of restricting design considerations and limiting the possibilities of finding better solutions. Similarly, it is also shown that there are studies evaluating the benefits of multiskilling using the chaining approach ($k = 2$ or $k \geq 2$). However, only a few of these studies incorporate multiskilling as a decision variable for the purpose of developing $k$-chaining policies [2, 3, 5, 8, 14, 19, 26, 27].

Second, Table 1 also shows that there are several studies evaluating the benefits of multiskilling in the face of uncertainty in either demand or supply. However, only [2, 3, 7, 15, 27, 29] explicitly incorporated this uncertainty within the mathematical formulation (i.e., *in-optimization* approach). Therefore, the solutions obtained in such works are considered good and robust in the face of different levels of uncertainty. This is not the case with the studies that evaluated uncertainty using a *post-optimization* approach, since such works cannot guarantee robustness or feasibility of the deterministic solution found in the presence of uncertainty.

Third, it is observed in Table 1 that most studies considering the chaining approach restrict the number of skills of multiskilled employees to a maximum of two. Moreover, with the exception of Wallace and Whitt [5] and Iravani et al. [6] whose studies are applied in call-centers, the studies that implement $k$-chaining with $k \geq 2$ focus mainly on the manufacturing sector. It should be noted that, unlike the manufacturing sector where the resource allocation problem is typically formulated as a balanced system (equal number of supply and demand nodes), in the retail industry the allocation problem is unbalanced in nature (greater number of supply nodes (employees) than demand nodes (departments)). Therefore, the $k$-chaining approach in retail requires new methodological challenges.

In summary, the exhaustive literature review shows that for the retail industry there are no studies that simultaneously evaluate the potential benefits of using $k$-chaining policies with $k \geq 2$, that model multiskilling as a decision variable, and that also explicitly incorporate the uncertainty of demand within the mathematical formulation. This paper will simultaneously address all of the listed gaps.

## 3 Problem Description

The problem proposed to be studied in this paper is to plan a multiskilling workforce at a strategic level for the retail industry. For a weekly planning horizon, the problem consists in designing a suitable training plan for a known workforce that is initially trained to work in a single department (i.e., single-skilled), through $k$-chaining structures where multiskilled employees can be trained in two or more departments and where each employee must work a given number of hours per week as specified in his/her contract. The solution to this model should provide an efficient solution to the assignment problem that will tell us how many hours per week each employee should work in the departments where he/she received training. The solution to the problem will decide how many employees should be single-skilled and how many multiskilled, how many additional skills each multiskilled employee should acquire, and under what levels of demand uncertainty a 2-chaining or $k$-chaining policy with $k \geq 2$ is more beneficial.

Our formulation considers the following assumptions: (1) Weekly demand per department has a full known probability distribution, and zero correlation between demands per department is assumed. (2) The staff demand for each department can be partially satisfied. This implies that the understaffing costs are included in the total cost function. (3) Overstaffing and training costs are also incorporated into the total cost function. (4) It is assumed that the over/understaffing and training costs are the same for all departments. (5) There is no employee absenteeism. (6) All employees have the same contract and therefore work the same number of working hours per week. (7) The deterministic assignment problem is fully balanced, i.e., the available hours for assignment in any department are equal to the value in hours of the average demand in that department. (8) Initially all employees are single-skilled (9) Multiskilled employees may work in two or more departments. (10) Employees are homogeneous, which means that the individual productivity of each employee is the same.

# 4 Methodology

## 4.1 Deterministic Optimization Model Using $K$-Chaining with $K \geq 2$

In this subsection we present the MILP model to solve the personnel assignment problem using the k-chaining approach with $k \geq 2$, as explained before. The mathematical notation of the problem is presented below:

**Sets**

$I$ Store employees indexed by $i$
$L$ Store departments indexed by $l$
$S$ Demand scenarios indexed by $s$
$I_l$ Employees under contract in department $l$ indexed by $i$, $\forall l \in L$

**Parameters**

$r_l(s)$ Random weekly demand in hours for department $l$ and scenario $s$, $\forall l \in L, s \in S$
$\overline{r}_l$ Average weekly demand in hours for department $l$, $\forall l \in L$
$c$ Training cost of an employee to work in any department
$u$ Understaffing cost per hour in any department
$b$ Overstaffing cost per hour in any department
$h$ Number of weekly hours an employee must work according to his/her contract
$m_i$ Department for which employee $i$ was initially trained, $\forall i \in I$

**Variables**

$x_{il}$ Equal to 1 if employee $i$ is trained for department $l$, otherwise 0, $\forall i \in I, l \in L$
$v_i$ Number of additional skills of employee $i$, $\forall i \in I$

$\omega_{il}$ Number of weekly working hours assigned to employee $i$ in department $l$, $\forall i \in I, l \in L$
$\kappa_l$ Understaffing in weekly hours in department $l$, $\forall l \in L$
$\delta_l$ Overstaffing in weekly hours in department $l$, $\forall l \in L$

The deterministic MILP model can be formulated as follow:

$$Min \underbrace{\sum_{i \in I} \sum_{l \in L: l \neq m_i} c x_{il}}_{(a)} + \underbrace{\sum_{l \in L} u \kappa_l}_{(b)} + \underbrace{\sum_{l \in L} b \delta_l}_{(c)} \tag{1}$$

s.t.

$$\sum\nolimits_{i \in I} \omega_{il} + k_l - \delta_l = \overline{r_l} \qquad \forall l \in L \tag{2}$$

$$\sum\nolimits_{l \in L} \omega_{il} = h \qquad \forall i \in I \tag{3}$$

$$\omega_{il} \leq h x_{il} \qquad \forall i \in I, l \in L \tag{4}$$

$$x_{il} = 1 \qquad \forall i \in I, l \in L : l = m_i \tag{5}$$

$$v_i = \sum\nolimits_{i \in L: l \neq m_i} x_{il} \qquad \forall i \in I \tag{6}$$

$$\sum\nolimits_{i \in I_l} v_i = \sum\nolimits_{i \in \{I - I_l\}} x_{il} \qquad \forall l \in L \tag{7}$$

$$x_{il} \in \{0, 1\} \qquad \forall i \in I, l \in L \tag{8}$$

$$v_i \geq 0 \qquad \forall i \in I \tag{9}$$

$$\omega_{il} \geq 0 \qquad \forall i \in I, l \in L \tag{10}$$

$$\kappa_l, \delta_l \geq 0 \qquad \forall l \in L \tag{11}$$

The objective function (1) minimizes the following weekly costs: (a) training of employees in additional departments; (b) understaffing; and (c) overstaffing. Constraints (2) account for demand fulfillment and allow us to calculate the level of over/understaffing associated with each department. Constraints (3) ensure that employees work exactly the number of hours required by their contracts. Constraints (4) guarantee that each employee can be assigned to work in, and only in, departments for which they have been trained. Constraints (5) determine the department for which each employee was initially trained. Constraints (6) indicate, for each employee, how many additional skills he or she has in total. Constraints (7) ensure a k-chaining policy with $k \geq 2$; these constraints ensure the formation of closed chains (CLC and/or CSC). Finally, constraints (8)–(11) define the domain of each variable.

### 4.2 Two Stage Stochastic Formulation

This subsection presents a reformulation of the deterministic model presented in Sect. 4.1 to two-stage stochastic optimization (TSSO) model considering stochastic demand. In this formulation there are first and second stage variables. The first-stage variables $x_{il}$ and $v_i$ are related to decisions about training employees in different departments. The second-stage variables, $\omega_{il}(s)$, $\kappa_l(s)$, and $\delta_l(s)$, are the operational adjustments that are made once the uncertain demand in each department is observed. That is, once the realizations of demand are known, over/understaffing may occur.

Following, we present the TSSO version of the formulation (1)–(11) using a method called *Sample Average Approximation* (SAA) [33].

$$Min \underbrace{\sum_{i\in I} \sum_{l\in L: l\neq m_i} cx_{il}}_{first-stage} + \underbrace{\frac{1}{|S|}\left[\sum_{s\in S}\sum_{l\in L} u\kappa_l(s) + \sum_{s\in S}\sum_{l\in L} b\delta_l(s)\right]}_{second-stage} \quad (12)$$

*s.t.*

$$\sum\nolimits_{i\in I} \omega_{il}(s) + k_l(s) - \delta_l(s) = r_l(s) \qquad \forall l \in L, s \in S \quad (13)$$

$$\sum\nolimits_{l\in L} \omega_{il}(s) = h \qquad \forall i \in I, s \in S \quad (14)$$

$$\omega_{il}(s) \leq hx_{il} \qquad \forall i \in I, l \in L, s \in S \quad (15)$$

$$x_{il} = 1 \qquad \forall i \in I, l \in L : l = m_i \quad (16)$$

$$v_i = \sum\nolimits_{i\in L: l\neq m_i} x_{il} \qquad \forall i \in I \quad (17)$$

$$\sum\nolimits_{i\in I_l} v_i = \sum\nolimits_{i\in\{I-I_l\}} x_{il} \qquad \forall l \in L \quad (18)$$

$$x_{il} \in \{0, 1\} \qquad \forall i \in I, l \in L \quad (19)$$

$$v_i \geq 0 \qquad \forall i \in I \quad (20)$$

$$\omega_{il}(s) \geq 0 \qquad \forall i \in I, l \in L, s \in S \quad (21)$$

$$\kappa_l(s), \delta_l(s) \geq 0 \qquad \forall l \in L, s \in S \quad (22)$$

The objective function (12) seeks to minimize the weekly training cost (first-stage decisions) and the expected weekly cost of over/understaffing (second-stage decisions). In the TSSO version, decisions about which employees will be trained in two or more departments are associated with the first stage of the stochastic problem. These decisions will affect the levels of over/understaffing, which are decisions associated with the second stage of the problem. Therefore, the objective in the TSSO approach is to choose the first-stage variables that minimize the sum of the first-stage costs and the expected value of the second-stage costs.

With respect to constraints (2)–(11) presented in the deterministic formulation, constraints (2)–(4) and (10)–(11) were modified to explicitly incorporate demand uncertainty. Note that all these constraints involve only second-stage variables. Therefore, constraints (5)–(9) did not receive any modification because they are associated with first-stage variables. Also, note that the interpretation of constraints (13)–(22) and constraints (2)–(11) is the same.

Unlike formulation (4.1), formulation (4.2) explicitly incorporates demand uncertainty. This allows multiskilling decisions to be feasible and robust to perturbations in forecasted demand. Finally, this implies that the deterministic formulation (4.1) may lead to inadequate staff planning and higher labor costs.

## 5 Case Study

### 5.1 Data Requirements

Here, we describe the characteristics and configuration of the sets and parameters used in this study. The database used contains real and simulated data, which is derived from a retail store located in Santiago, Chile. In the following, we present a description of both data sets.

**Real Data.** It is considered a retail store with 6 departments and 30 employees, i.e., $|L| = 6$ and $|I| = 30$. All employees have the same contract, such that all employees must work 45 h per week, i.e., $h = 45$. In relation to average demands by department, for department 1 (D1), $\overline{r}_1 = 315$ (i.e., 7 employees) while for the other 5 departments: (D2), $\overline{r}_2 = 225$; (D3), $\overline{r}_3 = 135$; (D4), $\overline{r}_4 = 135$; (D5), $\overline{r}_5 = 180$ and (D6), $\overline{r}_6 = 360$. With respect to cost parameters, we assume a minimum training cost per employee $\left(c = US\$1 - wk/employee\right)$ [34]. The over/understaffing costs are equal to $u = US\$60/h$ and $b = US\$15/h$, respectively.

**Simulated Data.** Two types of simulated data associated with uncertain demand are considered, these are: (i) in-sample data and (ii) out-of-sample data. In-sample data refer to the initial data that will be used to find the solutions to the first-stage variables using the TSSO approach. While out-of-sample data refer to the data that will be reserved to compare the performance of the solutions generated by the TSSO approach and the myopic approaches: Zero Multiskilling (ZM) and Total Multiskilling (TM). The myopic ZM approach represents an experiment where all employees are kept as single-skilled and, therefore, there are no multiskilled employees. The myopic TM approach represents an experiment where all employees will be multiskilled and can work in any department of the store.

In both types of simulated data, and in order to evaluate how multiskilling structures work with the $k$-chaining approach with $k \geq 2$ and different levels of uncertainty, we consider nine levels of variability for demand in each store department: $CV = 5, 10, 20, 30, 40, 50, 60, 70, 80\%$.

For both data sets, the non-negative realizations of stochastic demand for each store department are considered to follow a normal distribution, and to avoid generating outliers, the probability distribution functions are truncated at the 5th and 95th percentiles (i.e., we represent 90% of the data distribution). For the random parameters of the weekly hours of demand $r_l(s)$, we use a Monte Carlo simulation to randomly generate $|S| = 1000$ and $|S| = 10000$ scenarios for the in-sample and out-of-sample data respectively.

## 6 Results and Discussion

### 6.1 In-Sample Analysis

The results and discussions of this analysis are divided into the following two subsections: Computational times and model sizes, and How much multiskilling should be added?

**Computational Times and Model Sizes.** The TSSO model (12)–(22) was written in AMPL and the problem was solved using CPLEX commercial software running through NEOS server for 8 h (restrictive). For each coefficient of variation (CV), the TSSO approach was run considering 1000 demand realizations in each of the 6 store departments.

Table 2, for each CV, reports the computational times obtained by the TSSO approach (in hours) and the deterministic model (DT) (in seconds). In turn, the table shows the size of the models. In general, the solution times of the TSSO approach are much higher as a result of the difference in the size of the models. It should be clarified that since the decisions addressed in this study are at the strategic level with a weekly planning horizon, the times obtained by the TSSO model are considered reasonable. Additionally, considering all the CVs, we report that the optimality gaps with the TSSO approach were less than 0.05%.

**Table 2.** Computational times and model sizes.

| Model | Constraints | | | Binary variables | | | Continuous variables | | |
|---|---|---|---|---|---|---|---|---|---|
| TSSO | 216066 | | | 150 | | | 192060 | | |
| DT | 282 | | | 150 | | | 252 | | |
| Model | CPU time | | | | | | | | |
| | 5% | 10% | 20% | 30% | 50% | 60% | 70% | 80% | Average |
| TSSO | 1.2 h | 8 h | 8 h | 8 h | 6.3 h | 8 h | 8 h | 8 h | 6.9 h |
| DT | 1.5 s | | | | | | | | 1.5 s |

**How Much Multiskilling Should Be Added?** In this subsection we answer the following strategic question: How many employees should be multiskilled and in how many additional departments? Thus, we determine the optimal number of multiskilled employees required for each department and level of demand variability (CV). Figure 3 shows (in color) the optimal number of multiskilled employees per department and CV and includes the following four metrics. First, the percent of multiskilled employees relative to the total number of employees, %ME. Second, the percentage of multiskilled employees trained in an additional department and, therefore, trained in a total of two departments, %ME(2). Third, the percentage of multiskilled employees trained in two additional departments and, therefore, trained in a total of three departments, %ME(3). Fourth, the percentage of total multiskilling (%TM), which refers to the number of additional skills trained, relative to the theoretical maximum feasible number.

| | | **Average weekly demand (h)** | | | | | | **%ME** | **%ME(2)** | **%ME(3)** | **%TM** |
|---|---|---|---|---|---|---|---|---|---|---|---|
| | | 315 | 225 | 135 | 135 | 180 | 360 | | | | |
| | 5 | 1 | 1 | 1 | 1 | 1 | 1 | 20 | 20 | 0 | 4 |
| | 10 | 2 | 1 | 1 | 1 | 1 | 2 | 27 | 27 | 0 | 5 |
| | 20 | 3 | 2 | 1 | 1 | 2 | 3 | 40 | 40 | 0 | 8 |
| | 30 | 4 | 3 | 2 | 2 | 2 | 5 | 60 | 60 | 0 | 12 |
| %CV | 40 | 5 | 4 | 2 | 2 | 3 | 6 | 73 | 73 | 0 | 15 |
| | 50 | 6 | 4 | 3 | 3 | 4 | 7 | 90 | 90 | 0 | 18 |
| | 60 | 7 | 5 | 3 | 3 | 4 | 8 | 100 | 90 | 10 | 22 |
| | 70 | 7 | 5 | 3 | 3 | 4 | 8 | 100 | 70 | 30 | 26 |
| | 80 | 7 | 5 | 3 | 3 | 4 | 8 | 100 | 63 | 37 | 27 |
| | | D1 | D2 | D3 | D4 | D5 | D6 | | | | |
| | | **Store Department** | | | | | | | | | |

**Fig. 3.** Multiskilling metrics associated with each demand variability level.

The figure shows how the number of multiskilled employees in the store increases as the average demand value of a department and the coefficient of variation (CV) increases. It is also observed that when the variability is the lowest (CV = 5%), only one multiskilled employee per department is required despite the assumption of minimum training cost. This result can be interpreted as the minimum necessary investment in multiskilling (i.e., ME = 20% and TM = 4%). In turn, when demand variability is the highest (CV = 80%), the optimal number of multiskilled employees per department is equal to its total staff. In fact, for CVs equal to or greater than 60%, 100% of employees are required to be multiskilled.

In addition, it can be noted that for a CV of 5% to 50% the total number of multiskilled employees is trained in a total of two departments (i.e., one additional department). This indicates that for these scenarios of variability in demand a 2-chaining policy delivers the maximum benefit. However, for high levels of demand variability, CV = 60,70 and 80%, multiskilled employees trained in one or even two additional departments are required. Therefore, for no CV, multiskilled employees trained in three or more additional departments (i.e., 4 or more departments in total) are required. This indicates that values of $k = 2$ and $k = 3$ in $k$-chaining policies offer the most cost-effective results.

Additionally, for each CV, Fig. 4 presents in detail the number of multiskilled employees in the store who are trained in a total of two departments or three departments.

## 6.2 Out-of-Sample Analysis

The objective in this subsection is to make fair comparisons between the TSSO approach solutions and the myopic TM and ZM approaches. Table 3 presents for each combination

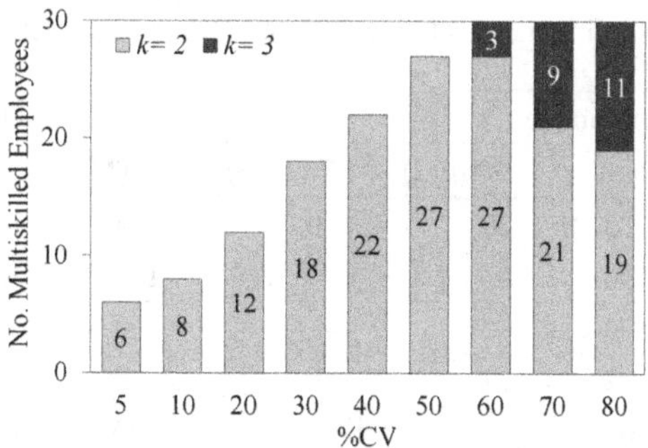


**Fig. 4.** Number of multiskilled employees in two or three departments for each CV.

of approach and CV, the values of the metrics associated with the in-sample and out-of-sample analyses. Columns 3, 4, 5 and 6 show the multiskilling requirements reported in the in-sample analysis (i.e., %ME, %ME(2), %ME(3), %TM) and showed in Fig. 3. Columns 7 and 8 are associated with the out-of-sample analysis, which are based on the 10000 demand scenarios generated through Monte Carlo simulation. Column 7 shows the percent savings in the average weekly cost of over/understaffing ($\%\overline{SSA}$) obtained for each approach. Column 8 reports the total average weekly cost ($\overline{\varphi}$) for each approach.

Note that, the metric $\%\overline{SSA}$ allows to evaluate the reliability and robustness of the solutions delivered by each approach. On the one hand, the myopic TM approach will always exhibit the highest possible reliability and robustness, i.e., it achieves the maximum possible reduction in the weekly cost of over/understaffing ($\%\overline{SSA} = 100$). However, it is the most conservative approach, as it requires the maximum multiskilling (%ME = %TM = 100%). In contrast, the myopic ZM approach always carries the highest average total cost, because it does not provide any level of protection against demand variability (i.e., $\%ME = \%TM = \%\overline{SSA} = 0\%$.). Regarding the TSSO approach, like the TM approach, also achieves the maximum savings in weekly over/understaffing cost ($\%\overline{SSA} = 100$) for each CV, but with a much lower investment in multiskilling. Therefore, the TSSO approach being less conservative than the TM approach, achieves for each CV the lowest average total cost. This result is interesting, since the TSSO approach with $k$-chaining obtains 100% of the potential benefits attainable, but with a much lower investment, which makes it profitable and attractive to decision makers.

**Table 3.** Multiskilling metrics for each approach and coefficient of variation.

| CV | Approach | In-sample | | | | Out-of-sample | |
|---|---|---|---|---|---|---|---|
| | | %ME | %ME(2) | %ME(3) | %TM | $\%\overline{SSA}$ | $\overline{\varphi}$ $(US\$)$ |
| 5% | TSSO | 20 | 20 | 0 | 4 | 100 | 709 |
| | TM | 100 | – | – | 100 | 100 | 853 |
| | ZM | 0 | – | – | 0 | 0 | 1670 |
| 10% | TSSO | 27 | 27 | 0 | 5 | 100 | 1426 |
| | TM | 100 | – | – | 100 | 100 | 1565 |
| | ZM | 0 | – | – | 0 | 0 | 3337 |
| 20% | TSSO | 40 | 40 | 0 | 8 | 100 | 2811 |
| | TM | 100 | – | – | 100 | 100 | 2948 |
| | ZM | 0 | – | – | 0 | 0 | 6640 |
| 30% | TSSO | 60 | 60 | 0 | 12 | 100 | 4213 |
| | TM | 100 | – | – | 100 | 100 | 4344 |
| | ZM | 0 | – | – | 0 | 0 | 9936 |
| 40% | TSSO | 73 | 73 | 0 | 15 | 100 | 5574 |
| | TM | 100 | – | – | 100 | 100 | 5698 |
| | ZM | 0 | – | – | 0 | 0 | 13246 |
| 50% | TSSO | 90 | 90 | 0 | 18 | 100 | 7084 |
| | TM | 100 | – | – | 100 | 100 | 7199 |
| | ZM | 0 | – | – | 0 | 0 | 16654 |
| 60% | TSSO | 100 | 90 | 10 | 22 | 100 | 8356 |
| | TM | 100 | – | – | 100 | 100 | 8460 |
| | ZM | 0 | – | – | 0 | 0 | 19859 |
| 70% | TSSO | 100 | 70 | 30 | 26 | 100 | 10623 |
| | TM | 100 | – | – | 100 | 100 | 10720 |
| | ZM | 0 | – | – | 0 | 0 | 23342 |
| 80% | TSSO | 100 | 63 | 37 | 27 | 100 | 12945 |
| | TM | 100 | – | – | 100 | 100 | 13028 |
| | ZM | 0 | – | – | 0 | 0 | 26636 |

## 7 Conclusions and Future Research

This paper addresses a multiskilled staff assignment problem with k-chaining and demand uncertainty. Our results allow answer the following fundamental question: how much multiskilling to add. The results showed that even when the training cost is minimal, total multiskilling is not needed to ensure the maximum benefits of multiskilling. Furthermore, that the number of required multiskilled employees increases as the average demand of a department and the coefficient of variation of demand are higher. The results also showed that the best way to add multiskilling is to generate staffing structures that combine single-skilled and multiskilled employees.

In addition, the results help us conclude that the TSSO approach exhibits excellent returns by achieving 100% of the potential benefits of total multiskilling, but with a lower investment. Also, the results show that, for demand variability scenarios less than or equal to 50%, a 2-chaining policy yields the maximum possible benefits. But with variability levels greater than 50%, higher levels of multiskilling are required, so in these scenarios it is more beneficial to use $k$-chaining policies with $k \geq 2$. However, in our case study, it was found that multiskilled employees who are skilled in more than three departments are not needed to achieve maximum benefits.

In relation to future research, the proposed methodology could be modified to: (1) Evaluate a flexibility policy that simultaneously considers multiskilling and hiring decisions. (2) Consider that employees are heterogeneous, i.e., that the individual productivity of each employee may vary depending on the number of departments in which he/she is trained. (3) Allowing the training and over/understaffing costs to vary from one department to another.

**Acknowledgements.** This research was supported by "*Fundación para la Promoción de la Investigación y la Tecnología (FPIT)*" under Grant 4.523.

## References

1. Henao, C.A., Muñoz, J.C., Ferrer, J.C.: The impact of multi-skilling on personnel scheduling in the service sector: a retail industry case. J. Oper. Res. Soc. **66**(12), 1949–1959 (2015). https://doi.org/10.1057/jors.2015.9
2. Henao, C.A., Ferrer, J.C., Muñoz, J.C., Vera, J.A.: Multiskilling with closed chains in the service sector: a robust optimization approach. Int. J. Prod. Econ. **179**, 166–178 (2016). https://doi.org/10.1016/j.ijpe.2016.06.013
3. Henao, C.A., Muñoz, J.C., Ferrer, J.C.: Multiskilled workforce management by utilizing closed chains under uncertain demand: a retail industry case. Comput. Ind. Eng. **127**, 74–88 (2019). https://doi.org/10.1016/j.cie.2018.11.061
4. Álvarez, E., Ferrer, J.C., Muñoz, J.C., Henao, C.A.: Efficient shift scheduling with multiple breaks for full-time employees: a retail industry case. Comput. Ind. Eng. **150**, 106884 (2020). https://doi.org/10.1016/j.cie.2020.106884
5. Wallace, R., Whitt, W.: A staffing algorithm for call centers with skill-based routing. Manuf. Serv. Oper. Manage. **7**(2), 276–294 (2005). https://doi.org/10.1287/msom.1050.0086
6. Iravani, S., Van Oyen, M., Sims, K.: Structural flexibility: a new perspective on the design of manufacturing and service operations. Manage. Sci. **50**(2), 151–166 (2005). https://doi.org/10.1287/mnsc.1040.0333
7. Simchi-Levi, D., Wei, Y.: Understanding the performance of the long chain and sparse designs in process flexibility. Oper. Res. **60**(5), 1125–1141 (2012). https://doi.org/10.1287/opre.1120.1081
8. Porto, A.F., Henao, C.A., Lopez, H., González, E.R.: Hybrid flexibility strategic on personnel scheduling: retail case study. Comput. Ind. Eng. **133**, 220–230 (2019). https://doi.org/10.1016/j.cie.2019.04.049
9. Henao, C.A., Batista, A., Pozo, D., Porto, A.F. y González, V.I.: Multiskilled personnel assignment problem under uncertain demand: a benchmarking analysis. Submitted to Computers & Operations Research, Under 1st review (2021)

10. Fontalvo-Echavez, O., Fuentes-Quintero, L., Henao, C.A., González, V.I.: Two-stage stochastic optimization model for personnel days-off scheduling using closed-chained multiskilling structures. In: Tohmé, F., Mejía, G., Rossit, D. (eds.) Production Research. ICPR-Americas 2020. Communications in Computer and Information Science, vol 1407, Springer Nature Switzerland AG (2021). https://doi.org/10.1007/978-3-030-76307-7_2
11. Abello, M.A., Ospina, N.M., De la Ossa, J.M., Henao, C.A., González, V.I.: Using the k-chaining approach to solve a stochastic days-off-scheduling problem in a retail store. In: Rossit, D.A., Tohmé, F., Mejía, G. (eds.) Production Research. ICPR-Americas 2020. Communications in Computer and Information Science, vol 1407, Springer Nature Switzerland AG (2021). https://doi.org/10.1007/978-3-030-76307-7_12
12. Vergara, S., Del Villar, J., Masson, J., Pérez, N., Henao, C.A., González, V.I.: Impact of labor productivity and multiskilling on staff management: A retail industry case. In: Rossit, D.A., Tohmé, F., Mejía, G. (eds.) Production Research. ICPR-Americas 2020. Communications in Computer and Information Science, vol 1408, Springer Nature Switzerland AG (2021). https://doi.org/10.1007/978-3-030-76310-7_18
13. Henao C.A.: Diseño de una fuerza laboral polifuncional para el sector servicios: caso aplicado a la industria del retail (Doctoral Thesis, Pontificia Universidad Católica de Chile, Santiago Chile) (2015). https://repositorio.uc.cl/handle/11534/11764
14. Hoop, W., Tekin, E., Van Oyen, M.: Benefits of skill chaining in serial production lines with cross-trained workers. Manuf. Serv. Oper. Manage. **50**(1), 83–98 (2004). https://doi.org/10.1287/mnsc.1030.0166
15. Wang, X., Zhang, J.: Process flexibility: a distribution-free bound on the performance of k-chain. Oper. Res. **63**(3), 555–571 (2015). https://doi.org/10.1287/opre.2015.1370
16. Cai, X., Li, K.: A genetic algorithm for scheduling staff of mixed skills under multicriteria. Eur. J. Oper. Res. **125**, 359–369 (2000). https://doi.org/10.1016/S0377-2217(99)00391-4
17. Agnihothri, S., Mishra, A., Simmons, D.: Workforce cross-training decisions in field service systems with two job types. J. Oper. Res. Soc. **54**, 410–418 (2003). https://doi.org/10.1057/palgrave.jors.2601535
18. Eitzen, G., Panton, D.: Multi-skilled workforce optimisation. J. Ann. Oper. Res. **127**, 359–372 (2004). https://doi.org/10.1023/B:ANOR.0000019096.58882.54
19. Bokhorst, J., Slomp, J., Gaalman, G.: Assignment Flexibility in a Cellular Manufacturing System - Machine Pooling versus Labor Chaining-. Flexible Automation and Intelligent Manufacturing, FAIM2004 (2004)
20. Yang, K.K.: A comparison of cross-training policies in different job shops. Int. J. Prod. Res. **45**(6), 1279–1295 (2007). https://doi.org/10.1080/00207540600658039
21. Sayin, S., Karavati, S.: Assigning cross-trained workers to departments: A two-stage optimization model to maximize utility and skill improvement. Eur. J. Oper. Res. **176**, 1643–1658 (2007). https://doi.org/10.1016/j.ejor.2005.10.045
22. Heimerl, C., Kolisch, R.: Scheduling and staffing multiple projects with a multi-skilled workforce. J. OR Spectr. **32**, 343–368 (2010). https://doi.org/10.1007/s00291-009-0169-4
23. Campbell, G.M.: A two-stage stochastic program for scheduling and allocating cross-trained workers. J. Oper. Res. Soc. **62**(6), 1038–1047 (2011). https://doi.org/10.1057/jors.2010.16
24. Parvin, H, Van Oyen, M., Pandelis, D., Williams, D., Lee, J.: Fixed task zone chaining: worker coordination and zone design for inexpensive cross-training in serial CONWIP lines. IIE Trans. **44**, 1–21 (2012). https://doi.org/10.1080/0740817x.2012.668264
25. Deng, T., Shen, Z.: Process flexibility design in unbalanced networks. Manuf. Serv. Oper. Manage. **15**(1), 24–32 (2013). https://doi.org/10.1287/msom.1120.0390
26. Paul, J., MacDonald, L.: Modeling the benefits of cross-training to address the nursing shortage. Int. J. Prod. Econ. **150**, 83–95 (2014). https://doi.org/10.1016/j.ijpe.2013.11.025

27. Gnanlet, A., Gilland, W.: Impact of productivity on cross-training configurations and optimal staffing decisions in hospitals. Eur. J. Oper. Res. **238**(1), 254–269 (2014). https://doi.org/10.1016/j.ejor.2014.03.033
28. Cuevas, R., Ferrer, J.C., Klapp, M., Muñoz, J.C.: A mixed integer programming approach to multi-skilled workforce scheduling. J. Sched. **19**(1), 91–106 (2016). https://doi.org/10.1007/s10951-015-0450-0
29. Liu, H.: The optimization of worker's quantity based on cross-utilization in two departments. Intell. Decis. Technol. **11**(1), 3–13 (2017). https://doi.org/10.3233/IDT-160273
30. Agrali, S., Caner, Z., Tamer, A.: Employee scheduling in service industries with flexible employee availability and demand. J. Oper. Res. Soc. **66**, 159.169 (2017). https://doi.org/10.1016/j.omega.2016.03.001
31. Taskiran, G., Zhang, X.: Mathematical models and solution approach for cross-training staff scheduling at call centers. Comput. Oper. Res. **87**, 258–269 (2017). https://doi.org/10.1016/j.cor.2016.07.001
32. Mac-Vicar, M., Ferrer, J.C., Muñoz, J.C., Henao, C.A.: Real-time recovering strategies on personnel scheduling in the retail industry. Comput. Ind. Eng. **113**, 589–601 (2017). https://doi.org/10.1016/j.cie.2017.09.045
33. Birge, J., Louveaux, F.: Introduction to Stochastic Programming. Springer, New York (2011)
34. Porto, A.F., Henao, C.A., López-Ospina, H., González, E.R., González, V.I.: Dataset for solving a hybrid flexibility strategy on personnel scheduling problem in the retail industry. Data Brief **32**, 106066 (2020). https://doi.org/10.1016/j.dib.2020.106066

# A Genetic Algorithm for Flexible Job Shop Scheduling Problem with Scarce Cross Trained Setup Operators

Dolapo Obimuyiwa and Fantahun Defersha(✉)

School of Engineering, University of Guelph, 50 Stone Road East, Guelph, Ontario N1G 2W1, Canada
fdefersh@uoguelph.ca
http://www.defersha.ca/

**Abstract.** Many researchers developed algorithms for a dual-resource constrained flexible job shop (DRC-FJSP) where both machines and workers need to be simultaneously scheduled. In those models and algorithms in the literature, the authors assumed that workers are machine operators responsible for performing the production process steps from the beginning to the end of the operation. However, because of increased automation and the adoption of numerically controlled machines, workers become machine tenders and should not be bottleneck and constraining resources. On the other hand, skilled setup operators remain being constraining limited resources in industries. Unlike machine tenders, a setup operator can leave the machine once setup is completed and take on another setup operation on another machine. In this paper, for the first time, we develop a genetic algorithm for a new DRC-FJSP where setup operators and machine tools are constraining resources. Numerical examples of varying problem sizes are presented to show the performance of the algorithm.

**Keywords:** Flexible job-shop scheduling · Dual resource constrained · Scarce setup operators · Detached setup · Genetic algorithm

## 1 Introduction

Flexible job-shop scheduling problem (FJSSP) is an expansion of the job shop scheduling problem (JSSP), which is one of the combinatorial optimization problems considered to be NP-hard [2,3,5]. The consideration of machines and the human resource component in the scheduling environment is known as a Double-Resource, or Dual-Resource constrained (DRC) system. Systems centered around human resources are viewed across literature under DRC, where the degree of versatility and workforce assignment could affect the system's performance [6]. Dual-resource constrained systems are more complex than their counterparts, single resource constrained (SRC), and they pose extra challenges that need to be addressed during resource scheduling [14]. The primary aim of human-resource

© Springer Nature Switzerland AG 2021
D. A. Rossit et al. (Eds.): ICPR-Americas 2020, CCIS 1407, pp. 142–155, 2021.
https://doi.org/10.1007/978-3-030-76307-7_11

scheduling in DRC systems is to have the right amount of eligible people to perform the required task at the right time. Human operators that are not cross-trained cannot address situations where there are unexpected work-orders and unbalanced workloads [8,9]. While the DRC systems' efficiency increases with workers' versatility [1]; however, according to [7], having a workforce system that is fully cross-trained may not be attainable due to the high cost of training. Previous studies have been scheduling workers as operators whose duty is to operate the machines with no experience or training on machine setups. It is assumed that this category of machine operators can be acquired at low cost (i.e., cheap labor that can be hired and fired at will). Also, the machine operators must stay and tend the assigned machines for the entire duration of work (i.e., they are not free to move in between machine operations).

However, in setting up complex and heavy-duty machines, there's a need to be prudent in hiring highly skilled personnel who are cross-trained in machine setups, which comes at a great cost. This category of people is considered costly labor, which might not seem reasonable to hire and fire at will due to their scarcity and cost of hiring. Furthermore, these setup-operators are free to move in between setup operations (i.e., they can move after completing initial setups to subsequent ones) and are not required to stay for the entire duration of the work schedule. Hence, the need for the effective optimization of these limited resources should not be neglected. To this end, in this paper, we propose a new DRC-FJSSP problem that incorporates a limited number of cross-trained skilled setup-operators. For more information on the DRC systems, see other recent publications by [11,13,15], and the recent survey by [4]. The rest of this paper is organized as follows: Sect. 2 presents the problem to be addressed, its assumptions, and the corresponding notations. The proposed genetic algorithm (GA) and its components are presented in Sect. 3. Numerical examples are presented in Sect. 4 to illustrate the problem addressed in this paper and show the computational efficiency of the proposed algorithm. Finally, conclusions are given in Sect. 5.

## 2 Problem Statement, Assumptions and Notations

This section presents the problem to be addressed, its assumptions, and the corresponding notations. Generally, SRC-FJSSP consists of two major subproblems: sequencing of operations and the selection of machines. In contrast, DRC-FJSSP consists of the two major sub-problems of an SRC-FJSSP, with the addition of one more problem referred to as the selection of workers. Therefore, the DRC problem in this paper consists of three (3) subproblems, namely: (1) the selection of a machine for each operation, (2) the sequence of operations on selected machines, and (3) the selection of an eligible setup operator for setting up a machine. For this paper, the DRC-FJSSP in the presence of a scarce number of skilled cross-trained setup operators (SSO) will be hereafter denoted as "DRC-FJSSP-SSO."

### 2.1 Problem Statement

Consider a job-shop consisting of $M$-machines where certain machines $m$ are similar or have some common functionalities. Also, the schedule of the system includes a set of $J$-independent jobs $j$ with a number of operations $O_j$ in a predefined arrangement where each operation $o$ of job $j$ can be processed on a set of alternative and eligible machines. The total time spent by an operation $o$ of job $j$ on machine $m$ is defined as the setup-time denoted by $(S_{o,j,m})$ and the processing time (batch size $B_j$ multiplied by unit processing time $T_{o,j,m}$). The setup of this operation $o$ can be started once the assigned machine completes its previous operation, and an eligible setup operator is available. Furthermore, there are $N$ setup-operators where each setup operator $n$ (where $n = 1, \ldots, N$) is trained to perform setup of operation $o$ on a given set of machines $m$. Hence, as seen in Fig. 1, the problem is the assignment of each operation to one of its alternative and qualified machines, determine the sequence and starting time of the operations on each machine and the eligible setup-operator for each machine. Figure 1 is a case of four jobs, three machines (i.e., partial flexibility), two setup-operators, and the setup-operators are trained to set up some of the machines (i.e., partial flexibility). For example, operation 1 of job 4 can be set up on both machine 2 and machine 3 by either setup-operator 1 or setup-operator 2, and operation 2 of job 4 can be setup on machine 1 by setup-operator 1 only, and on machine 3 by both setup-operator 1 and 2. Note that in a shop system, a machine is said to be partially flexible when all job-operations cannot be carried out on all the machine, and a setup-operator is also said to be partially flexible when he/she cannot carry out setup-operations on all machines (i.e., setup-operator 2).

### 2.2 Assumptions

The following assumptions were made for the model in addressing the problem in this paper:

(i) A machine can process at most a single operation at a given period of time (i.e., machine-capacity constraint)
(ii) Once the processing of an operation begins, it cannot be interrupted until it is done (i.e., Pre-emption not allowed).
(iii) Each operation of a job can be assigned to at most a single machine and a single eligible setup-operator at a given period of time (i.e., resource-constraint)
(iv) Modification is not allowed to the predetermined sequence of the operations for each job (i.e., precedence-constraint)
(v) Each setup-operation can be carried out by only one setup-operator (i.e., setup-operator constraint)
(vi) The nature of all setups are in detached mode (i.e., setup of a particular operation can begin before its arrival on the assigned machine)
(vii) The setup-operator's movement time is negligible.

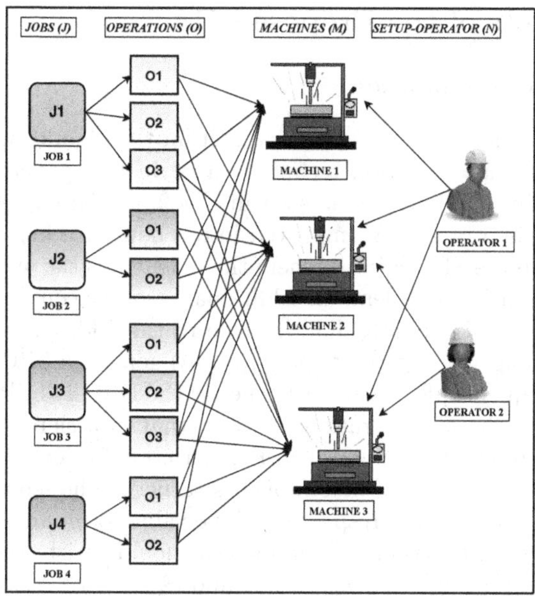


**Fig. 1.** A basic example of the DRC-FJSSP-SSO.

### 2.3 Necessary Notations

The following notations were used in the solution procedure in this paper. Additional notation and MILP formulation can be found in [10].

**Indexes and Parameters:**

| | |
|---|---|
| $m$ | An index for machines, where $m = 1, 2, 3, \ldots, M$, where $M$ is the number of machine; |
| $r$ | An index for machine run, where a run is defined as an processing of an operation by a machine; |
| $n$ | An index of setup operator, where $n = 1, 2, 3, \ldots, N$, where $N$ is the number of setup operators such that $N \ll M$. |
| $k$ | An index of setup operation performed by an operator |
| $P_{o,j,m}$ | A binary variable which takes the value 1 if operation $o$ of job $j$ can be executed on machine $m$, or 0 otherwise; |
| $\Theta_{n,m}$ | A binary variable which takes the value 1 if setup operator $n$ is eligible to perform setup on machine $m$, 0 otherwise |
| $\Omega$ | Large positive number. |

**Decision Variables:**

| | |
|---|---|
| $\widehat{C}_{r,m}$ | Completion time of the $r^{th}$ run of machine $m$; |
| $s_{k,n}$ | Completion time of the $k^{th}$ setup of setup-operator $n$; and |
| $C_{max}$ | Completion time of all jobs ($Makespan$); |

# 3 Components of the Proposed GA

## 3.1 Solution Representation

Solution representation in GA is a critical step in the algorithm's implementation. The solution representation technique used in this paper follows a similar concept used in [5]. However, as seen in Fig. 2(a), the triplet-gene solution encoding scheme used in [5] was improved into a new encoding scheme as seen in Fig. 2(b). Unlike the SRC-FJSSP where the machine is the only resource considered in the scheduling problem, an additional resource constraint is added to the SRC-FJSSP to make it DRC-FJSSP. Due to the addition of an extra constraint (i.e., setup-operators), the triplet($j$, $o$, $m$) gene of a SRC-FJSSP is now augmented into a quadruple($j$, $o$, $m$, $n$) gene for the DRC-FJSSP-SSO as shown in Fig. 2(b), where the $o^{th}$ operation of job $j$ that is assigned to the eligible machine $m$ is setup by setup-operator $n$. Furthermore, for any given $j$, if $o < o'$, a gene with a value ($j$, $o$, $m$, $n$) must appears earlier in the sequence before the gene with value ($j$, $o'$, $m'$, $n$) irrespective of the values of $m$ and $m'$. For a given gene value $(j, o, m, n)$, the value of the indices should satisfy the condition set by eligibility indicator data, i.e., $P_{o,j,m} = 1$ and $\Theta_{n,m} = 1$.

## 3.2 Evaluation

Based on the assignment and sequencing information obtained from a solution string, the resulting starting and completion times of jobs and, hence the makespan of the schedule can be determined. The makespan is used to evaluate how fit the individual is. In calculating the makespan, we take into account the following: (1) the machine setup time by setup-operators, (2) the detached nature of setup, and (3) the cross-training ability of setup-operators. The procedure for this calculation is outlined below:

**Step 1.** Set $l = 1$

**Step 2.** For the particular solution-string under consideration, set the values of indices $j$, $o$, $m$ and $n$ as secured from gene at location $l$.

**Step 3.** Calculate the completion times $s_{k,n}$ and $\widehat{C}_{r,m}$

- If operation $o$ of job $j$ is the first operation assigned to run $r$ of machine $m$ and the first setup-operation $k$ to be performed by setup-operator (n) on the same machine (i.e., $r = 1$, $o = 1$, $k = 1$) then: $s_{k,n} = S_{o,j,m}$ and $\widehat{C}_{r,m} = S_{o,j,m} + B_j \cdot T_{o,j,m}$
- If (a) operation $o > 1$ of job $j$ is the operation assigned to run $r$ of machine $m$ and first setup-operation $k$ to be performed by setup-operator (n) on the same machine (i.e., $r = 1$, $o > 1$, $k = 1$) and (b) operation $o - 1$ was assigned to run $r'$ on machine $m'$, then: $s_{k,n} = S_{o,j,m}$ and $\widehat{C}_{r,m} = \max\{S_{o,j,m};\quad \widehat{C}_{r',m'}\} + B_j \cdot T_{o,j,m}$.

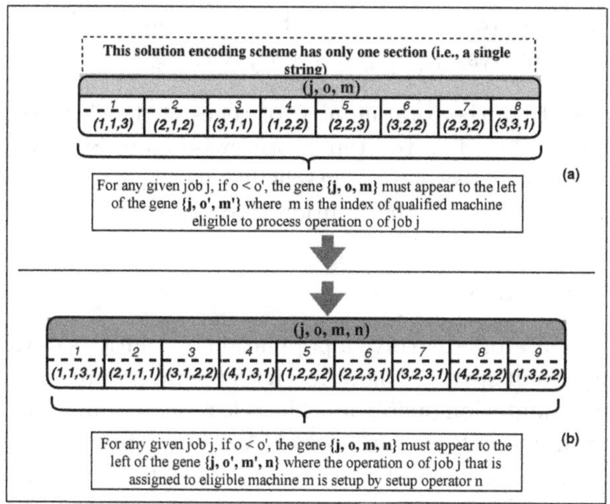


**Fig. 2.** Solution encoding schemes for (a) SRC-FJSSP and (b) DRC-FJSSP-SSO

- If operation $o$ of job $j$ is the operation to be processed on run $r > 1$ machine $m$ and the setup-operation $k$ is the first to be performed by setup-operator (n) machine $m$ (i.e., $r > 1$, $o = 1$, $k = 1$) then: $s_{k,n} = \widehat{C}_{r-1,m} + S_{o,j,m}$ and $\widehat{C}_{r,m} = s_{k,n} + B_j \cdot T_{o,j,m}$
- If (a) operation $o > 1$ of job $j$ is the operation to be processed on run $r > 1$ and the setup-operation $k$ is the first to be performed by setup-operator (n) machine $m$ (i.e., $r > 1$, $o > 1$, $k = 1$) and (b) operation $o - 1$ was assigned to run $r'$ on machine $m'$, then: $s_{k,n} = \widehat{C}_{r-1,m} + S_{o,j,m}$. and $\widehat{C}_{r,m} = \max\{s_{k,n};\ \widehat{C}_{r',m'}\} + B_j \times T_{o,j,m}$.
- If operation $o$ of job $j$ is the first operation assigned to run $r$ on machine $m$ and $k$ is not the first setup to be performed by setup-operator (n) (i.e., $r = 1$, $o = 1$, $k > 1$) then: $s_{k,n} = s_{k-1,n} + S_{o,j,m}$. and $\widehat{C}_{r,m} = s_{k,n} + B_j \cdot T_{o,j,m}$
- If (a) operation $o$ of job $j$ is the first operation to be processed and $k$ is not the first setup to be performed by setup-operator (n) on machine $m$ (i.e., $r = 1$, $o = 1$, $k > 1$) and (b) operation $o - 1$ was assigned to run $r'$ on machine $m'$, then: $s_{k,n} = s_{k-1,n} + S_{o,j,m}$. and $\widehat{C}_{r,m} = \max\{s_{k,n};\ \widehat{C}_{r',m'}\} + B_j \times T_{o,j,m}$.
- If $k$ is not the first setup to be performed by setup-operator (n) on machine $m$ for operation $o = 1$ of job $j$ (i.e., $k > 1$, $r > 1$, $o = 1$) then: $s_{k,n} = \max\{s_{k-1,n};\ \widehat{C}_{r-1,m}\} + S_{o,j,m}$ and $\widehat{C}_{r,m} = s_{k,n} + B_j \cdot T_{o,j,m}$
- If (a) $k$ is not the first setup to be performed by setup-operator (n) on machine $m$ for operation $o$ of job $j$ (i.e., $r = 1$, $o > 1$, $k > 1$) and (b) operation $o - 1$ was assigned to run $r'$ on machine $m'$, then:

$$s_{k,n} = \max\{s_{k-1,n};\quad \widehat{C}_{r-1,m}\} + S_{o,j,m} \text{ and}$$
$$\widehat{C}_{r,m} = \max\{s_{k,n};\quad \widehat{C}_{r',m'}\} + B_j \times T_{o,j,m}.$$

**Step 4.** If $l$ is less than the total number of genes of the chromosome, increase its value by 1 and go to Step 2; otherwise go to Step 5

**Step 5.** Calculate the makespan of the schedule as $c_{max} = \max\ \widehat{C}_{r,m};\ \forall(r,m)\}$ and set the fitness of the solution to $c_{max}$.

### 3.3 Genetic Operators

GAs uses genetic operators to evolve a population of solutions towards a promising region of the search space. Generally, these operators are classified as selection, crossover, and mutation operators. The selection operator chooses the chromosomes for reproduction. In the proposed GA, a $k$-way tournament (KWT) and a roulette wheel selection (RWS) method was used to select individuals. The KWT selection method is used to first select $k$ individuals randomly. Then, the individual with the highest fitness (smallest makespan) is declared the winner. This process of selection is repeated with replacement until the number of the new set of individuals is the same as the population size. In an RWS, the wheel is divided into sections according to the fitness of each solution-string and is rolled $n$-number of times where $n$ is the same as the population size. In each spin, the solution-strings with the most extensive section on the wheel have a higher probability of been selected based on a higher fitness value. Once the individuals have been selected for reproduction, crossover and mutation operators are applied to the offspring. These operators can be classified as assignment and sequencing operators. The assignment operator only alters the assignment property of the parent chromosomes, while sequencing operators only alters the sequence of the operations in the parent chromosomes. The assignment operators used in this paper are (1) machine-operation assignment crossover, (2) machine-operation assignment mutation, (3) single-point crossover, and (4) setup-operator assignment mutation.

Assignment crossover operator is applied with probability $\alpha_1$, and it produces offsprings by exchanging the assignment of arbitrarily chosen operations between two parents, as shown in Fig. 3. The assignment mutation operator is applied with probability $\alpha_2$. It alters the assignment of a few operations of a given chromosome. In a single-point crossover, a point is arbitrarily selected on both parent 1 and parent 2, and labeled as the crossover point. The arbitrarily selected point is chosen with a probability of $\alpha_3$, and all the genetic information beyond the selected point is exchanged between the parent-strings to produce new offsprings. Setup-operator mutation is applied with probability $\alpha_4$ on the randomly selected solution-string. As seen in Fig. 4, it alters the setup-operator assignment of the solution-string by assigning the setup-operator to an alternative machine that he/she is qualified to setup. The sequencing operators used in this paper are (1) job-operation sequence-order crossover and (2) operation sequence-swap mutation. With probability $\alpha_5$, the job-operation sequence-order crossover operator is applied by randomly selecting one operation each from the

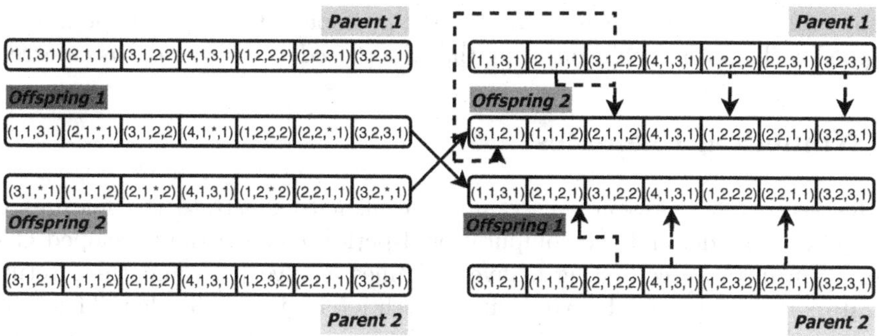


**Fig. 3.** Machine-operation assignment crossover.

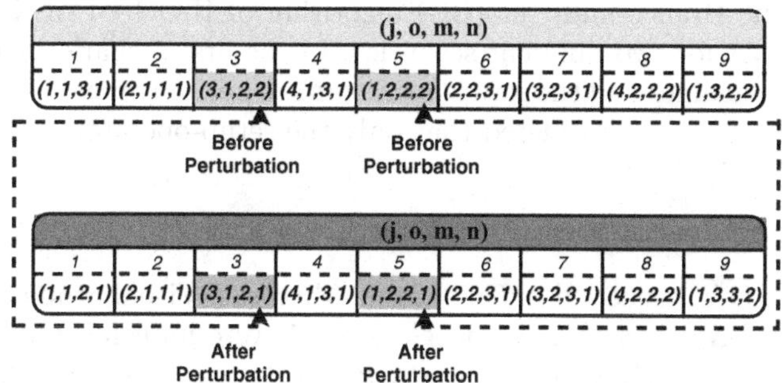


**Fig. 4.** Setup-operator assignment mutation.

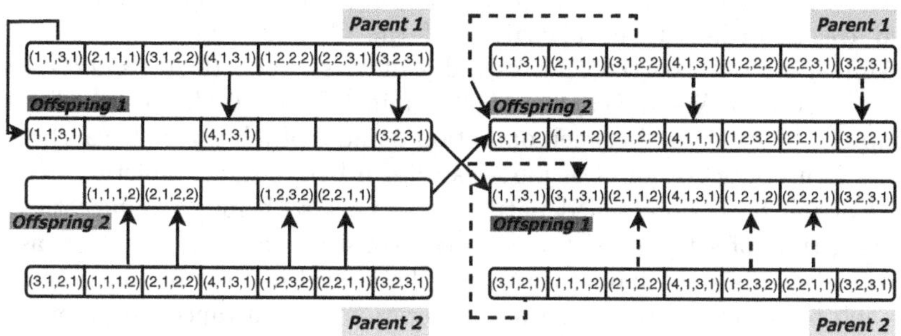


**Fig. 5.** Job-operation sequence-order crossover.

parent chromosomes, copies all the operations of the job to which the selected operation belongs into the corresponding offsprings as illustrated in Fig. 5. It proceeds by completing the new offspring with the remaining operations in the same pattern of the second parent while maintaining the assignment structure of the operations in the first parent. This operator respects the precedence constraints among operations of the same job and ensuring it is not violated. The operation sequence-swap mutation operator is applied with a probability of $\alpha_6$

on the solution-string in such a way that the position of the selected operation is swapped with another while ensuring that the precedence constraint is enforced.

# 4 Numerical Examples

In this section, we present an example problem to illustrate the features of the proposed model and the computational performance of the developed GA. A problem instance consisting of five machines, twelve jobs, and three setup-operators is considered. The batch size of each job is given in Table 1. This table contains, the number of operations in each job, the possible alternative machine routes for each operation, the eligible setup-operators for the machine, and the corresponding unit processing times and setup times. This example problem was solved using the Branch-and-Cut (BC) algorithm of ILOG CPLEX solver and the proposed GA for two different scenarios. In the first scenario, it is assumed that both the machines and the setup-operators are partially flexible. In contrast, in the second case, it is considered that only the setup-operators are flexible.

## 4.1 Solution Illustration

The developed GA was coded in C++ using Microsoft Visual Studio 2017. Both the ILOG CPLEX solver and the developed GA were implemented on a 64-bit operating system, 3.5 GHz Intel (R), 16GB RAM computer. The Gantt charts of the resulting schedules, as solved by the developed GA for the machines, is given in Fig. 6, and the setup-operators is given in Fig. 7. From Fig. 8, it can be concluded that the developed GA ($makespan$ = 1605.0 mins) outperformed the Branch-and-Cut algorithm of the CPLEX solver ($makespan$ = 1927.5 mins) for both partial and total flexibility scenario ($GA - makespan$ = $1605.0 mins, CPLEX - makespan$ = 2065.0 mins). In Fig. 6, the detached nature of setup can be observed between operation 2 and 3 of job $j = 8$. It can be seen that the setup of operation 3 of job 8 can begin before the completion of operation 2 of job 8. The detached nature of setup also helps to reduce the idle or waiting time of setup-operators between consecutive machine setups as the setup-operator does not have to wait for the arrival of the previous operation of the job to begin setup for the next operation. For instance, setup-operator $n = 3$ would be able begin the setup of operation $o = 3$ of job $j = 8$ immediately after his last setup of operation $o = 2$ of job $j = 3$ at 367.5 due to the detached nature of setup, otherwise, the setup-operator would have to wait until operation $o = 2$ of job $j = 8$ is completed at 563.0 which would increase the potential idle-time by 35%. Furthermore, since there was no change in the makespan value for the two flexibility scenarios when solving with GA, it may be reasonable to say that for more work and cost-effectiveness, setup-operators can be trained on some group of machines at a particular workstation than on all the machines in the entire shop system. This would also reduce the movement of setup-operators from one setup operation to another, resulting in a more cost effective and leaner manufacturing environment.

## 4.2 Algorithm Performance

Larger problem instances were developed to evaluate the performance of the genetic algorithm and the behavior of the algorithm to changes in genetic parameters. The case examples used in this study were generated and solved with different genetic problem-parameters for 10 test runs based on the provided procedures in [12] and in other GA published in the literature. Table 2 describes the general features of the case-examples, such as the problem instances, the total number of jobs, operations, machines, setup-operators, and alternative routes. The problem sizes used in this study are much larger than most of the ones used in literature. Figure 9 shows how the algorithm converges during iterations and how it responds to the tuning of genetic parameters for problems 3, 4, 5, and 6 after ten experimental runs. Problems 3 and 4 are considered as medium-sized problems, while problems 5 and 6 are regarded as large-size problems based on the total computational cost required during the iterations by the genetic algorithm. The genetic algorithm can converge during its iteration through a large generation in less computational time for medium-size problems while it takes more computational time to converge and progress from a certain number of generation to another in large-size problems.

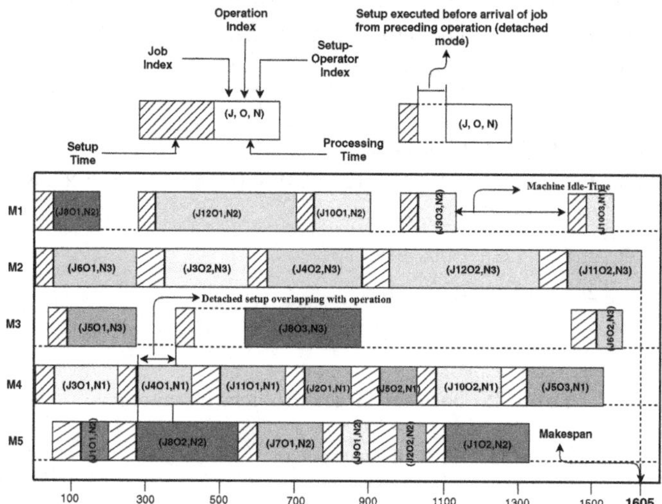


**Fig. 6.** Machines schedule for the example problem.

**Table 1.** Data for the twelve jobs, five machines, and three setup-operators

| | | | Alternative routes, $(m,\, n,\, T_{o,j,m},\, S_{o,j,m})$ | | |
|---|---|---|---|---|---|
| $j$ | $B_j$ | $o$ | 1 | 2 | 3 |
| 1 | 50 | 1 | (2, (2,3), 1.25, 40) | (5, 2, 1.5, 60) | |
| | | 2 | (5, 2, 4.50, 60) | | |
| 2 | 35 | 1 | (3, (2,3), 3.75, 40) | (4, 1, 3.50, 60) | |
| | | 2 | (1, (1,2), 1.75, 40) | (4, 1, 1.75, 60) | (5, 2, 1.50, 80) |
| 3 | 30 | 1 | (2, (2,3), 6.75, 60) | (4, 1, 6.50, 40) | |
| | | 2 | (2, (2,3), 7.00, 80) | | |
| | | 3 | (1, (1,2), 3.50, 40) | (2, (2,3), 3.25, 60) | (5, 2, 4.00, 40) |
| 4 | 40 | 1 | (2, (2,3), 4.00, 60) | (4, 1, 4.00, 40) | |
| | | 2 | (1, (2,3), 2.50, 60) | (2, (2,3), 2.60, 80) | |
| 5 | 30 | 1 | (1, (1,2), 6.25, 40) | (3, (2,3) 6.75, 40) | (5, 2, 7.00, 40) |
| | | 2 | (1, (1,2), 2.50, 60) | (3, (2,3), 2.75, 80) | (5, 2, 2.75, 80) |
| | | 3 | (3, (2,3), 2.75, 80) | (4, 1, 3.00, 80) | (5, 2, 2.75, 40) |
| 6 | 45 | 1 | (2, (2,3), 5.50, 40) | | |
| | | 2 | (3, (2,3), 1.25, 60) | | |
| 7 | 45 | 1 | (5, 2, 4.00, 40) | | |
| 8 | 55 | 1 | (1, (1,2), 2.00, 60) | (4, 1, 1.25, 60) | |
| | | 2 | (1, (1,2), 5.75, 40) | (5, 2, 5.25, 80) | |
| | | 3 | (3, (2,3), 6.00, 60) | (5, 2, 6.00, 80) | |
| 9 | 30 | 1 | (1, (1,2), 3.00, 60) | (3, (2,3), 2.75, 80) | (5, 2, 2.75, 40) |
| 10 | 40 | 1 | (1, (1,2), 4.00, 40) | | |
| | | 2 | (1, (1,2), 6.75, 60) | (4, 1, 7.00, 40) | |
| | | 3 | (1, (1,2), 1.25, 60) | | |
| 11 | 50 | 1 | (1, (1,2), 3.25, 40) | (2, (2,3), 4.00, 80) | (4, 1, 3.25, 80) |
| | | 2 | (2, (2,3), 3.25, 80) | | |
| 12 | 55 | 1 | (1, (1,2), 7.00, 40) | | |
| | | 2 | (2, (2,3), 7.00, 80) | | |

**Table 2.** Data for larger problem instances.

| Problem no. | Machines | Jobs | Total number of operations for the Jobs | Setup-operators | Alternative routes for the operation |
|---|---|---|---|---|---|
| 3 | 9 | 15 | 34 | 4 | 97 |
| 4 | 13 | 20 | 81 | 6 | 190 |
| 5 | 10 | 35 | 154 | 4 | 462 |
| 6 | 15 | 50 | 243 | 7 | 495 |

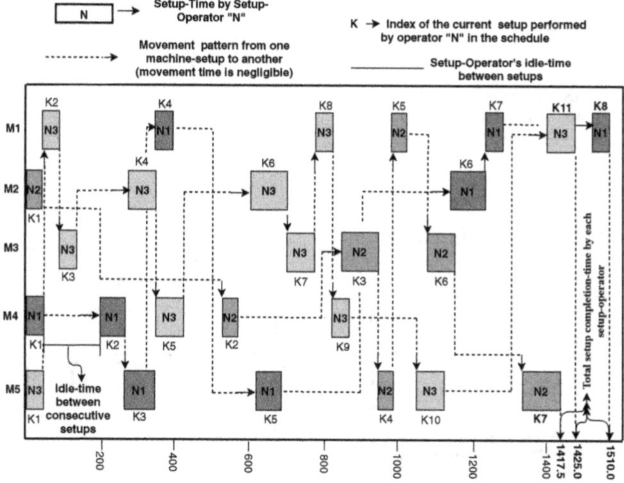


**Fig. 7.** Setup-operators schedule for the example problem.

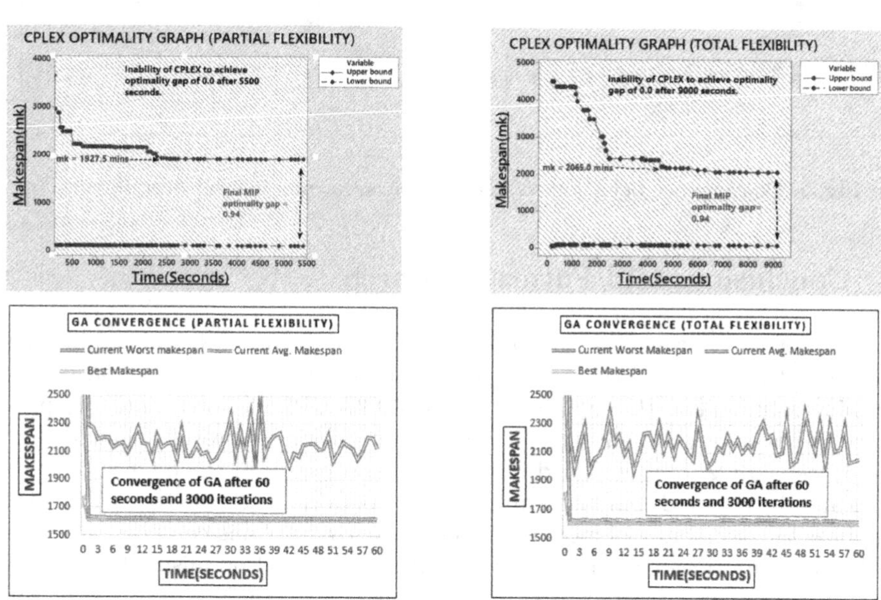


**Fig. 8.** Convergence behaviour of the GA and CPLEX solver.

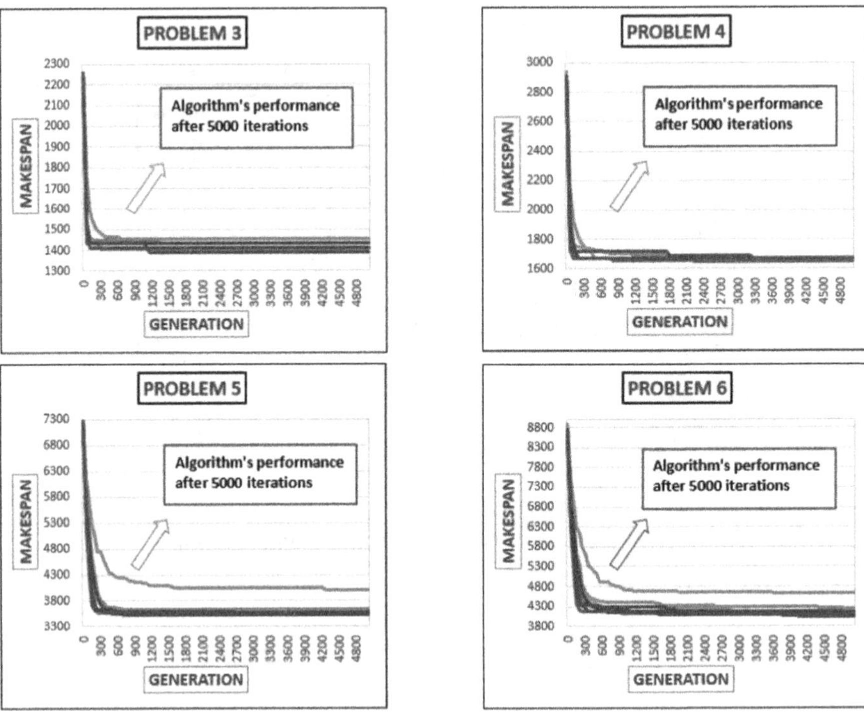


**Fig. 9.** Convergence behaviour of the GA for larger problem set over 10 runs.

## 5 Conclusions and Future Research

In this paper, we considered a DRC system that incorporates an additional resource of scarce cross-trained setup-operators with the detached nature of machine setups. The BC algorithm of the ILOG CPLEX solver and the proposed GA was used in addressing an example problem. The results showed that the developed GA outperformed the BC algorithm both in the quality of solutions and the required computational cost. The performance from the GA further amplifies the dominance of metaheuristics approach to solving complex problems with less computing cost, and it also provides a good and promising foundation for further studies in limited cross-trained human-resource DRC systems. In our future research, we plan to extend the model and the solution procedure to consider multiple objectives such as machine workload-balancing, due dates, and other factors such as machine release dates, sequence-dependency setup times, and transportation time of setup-operators between setups.

**Acknowledgement.** The authors would like to than the Natural Science and Engineering Research Council of Canada (NSERC) for the financial support in conducting this research.

## References


1. Azizi, N., Zolfaghari, S., Liang, M.: Modeling job rotation in manufacturing systems: the study of employee's boredom and skill variations. Int. J. Prod. Econ. **123**(1), 69–85 (2010)
2. Caldeira, R., Gnanavelbabu, A.: Solving the flexible job shop scheduling problem using an improved Jaya algorithm. Comput. Ind. Eng. **137** (2019)
3. Defersha, F.M., Rooyani, D.: An efficient two-stage genetic algorithm for a flexible job-shop scheduling problem with sequence dependent attached/detached setup, machine release date and lag-time. Comput. Ind. Eng. **147** (2020)
4. Dhiflaoui, M., Nouri, H., Driss, O.: Dual-resource constraints in classical and flexible job shop problems: a state-of-the-art review. Procedia Comput. Sci. **126**, 1507–1515 (2018)
5. Fantahun M. Defersha, Chen, M.: A parallel genetic algorithm for a flexible job-shop scheduling problem with sequence dependent setups. Int. J. Adv. Manuf. Technol. **49**(1–4), 263–279 (2010)
6. Fruggiero, F., Riemma, S., Ouazene, Y., Macchiaroli, R., Guglielmi, V.: Incorporating the human factor within manufacturing dynamics. IFAC-PapersOnLine **49**(12), 1691–1696 (2016)
7. Gel, E.S., Hopp, W.J., Van Oyen, M.P.: Hierarchical cross-training in work-in-process-constrained systems. IIE Trans. **39**(2), 125–143 (2007)
8. Givi, Z., Jaber, M., Neumann, W.: Production Planning in DRC systems considering worker performance. Comput. Ind. Eng. **87**, 317–327 (2015)
9. Nembhard, D.A., Nembhard, H.B., Qin, R.: A real options model for workforce cross training. Eng. Econ. **36**(10), 919–940 (2005)
10. Obimuyiwa, D.: Solving Flexible Job Shop Scheduling Problem in the Presence of Limited Number of Skilled Cross-Trained Setup Operators. Masc, University of Guelph (2020). http://www.defersha.ca/thesis/DolapoThesiss.pdf
11. Peng C., Fang Y., Lou P., Yan, J.: Analysis of double-resource flexible job shop scheduling problem based on genetic algorithm. In: Proceedings of the 15th International Conference on Networking, Sensing and Control, ICNSC '18, pp. 1–6 (2018)
12. Pezzella, F., Morganti, G., Ciaschetti, G.: A genetic algorithm for the Flexible Job-Shop Scheduling Problem. J. Comput. Oper. Res. **35**(10), 3202–3212 (2008)
13. Wu, R., Li, S., Guo, S., Xu, W.: Solving the dual-resource constrained flexible job shop scheduling problem with learning effect by a hybrid genetic algorithm. Adv. Mech. Eng. **10**(10), 1–14 (2018)
14. Xu, J., Xu, X., Xie, S.Q.: Recent developments in dual resource constrained (DRC) system research. Eur. J. Oper. Res. **215**(2), 309–318 (2011)
15. Zhang, J., Wang, W., Xu, X.: A hybrid discrete particle swarm optimization for dual-resource constrained job shop scheduling with resource flexibility. J. Intell. Manuf. **28**(8), 1961–1972 (2015). https://doi.org/10.1007/s10845-015-1082-0

# Using the k-Chaining Approach to Solve a Stochastic Days-Off-Scheduling Problem in a Retail Store

María Alejandra Abello, Nicole Marie Ospina, Julia Margarita De la Ossa, César Augusto Henao(✉), and Virginia I. González

Universidad del Norte, Barranquilla, Colombia
{abelloam,nulloque,jdelaossam,cahenao,vvirginia}@uninorte.edu.co

**Abstract.** The present study addresses the personnel assignment and scheduling problems through the multiskilling flexibility strategy with the aim of reducing the negative impacts on a company's cost structure that come as a result of demand uncertainty. The personnel assignment matter is handled using the chaining approach and it is evaluated whether a multiskilled workforce that minimizes the expected costs associated with training and under/overstaffing can be planned while taking into account demand uncertainty and using the k-chaining policy for a days-off scheduling problem. Until now, literature's main focus has been 2-chaining policies, where every multiskilled employee is at most trained in two departments, i.e., $k = 2$. Hence, this research intends to determine scenarios where a k-chaining policy with $k \geq 2$ offers more cost-effective multiskilling structures than a 2-chaining one. Initially, a deterministic mixed-integer linear optimization model is developed and then modified into a two-stage stochastic optimization model in order to add demand uncertainty. Such methodology is subsequently applied to a Chilean retail store where a Monte Carlo simulation is used to generate different daily demand scenarios. The results obtained suggest that with high levels of demand variability k-chaining policies seem favorable. Nonetheless, none of the employees evaluated were able to work in more than three departments, which leads to the conclusion that it does not appear to be advantageous to implement excessive flexibility levels.

**Keywords:** Personnel scheduling · Multiskilling · Chaining · Labor flexibility · Retail

## 1 Introduction

An adequate personnel scheduling strategy is not only essential to a company in terms of securing a competitive advantage and thus acquiring means to outperform its competitors within a continuously growing market, but also in regard to the decrease of workforce related costs that both manufacturing and service companies must discharge.

Demand seasonality is a predictable phenomenon typically faced by the service sector that affects a company's usual personnel scheduling system since special needs

© Springer Nature Switzerland AG 2021
D. A. Rossit et al. (Eds.): ICPR-Americas 2020, CCIS 1407, pp. 156–170, 2021.
https://doi.org/10.1007/978-3-030-76307-7_12

must be met. Formulating such system becomes a challenging activity for the retail trade especially when unpredictable phenomena such as demand uncertainty and staff absenteeism are taken into account, for the mismatch between demand and supply ought to be reduced as much as possible [1–3]. As a result of the service industries' inherent variability, firms eventually face situations where the number of employees is below staff requirements (understaffing) or where there are more employees than the actual amount needed (overstaffing).

Therefore, in order to protect themselves from uncertainty and hence reduce the mismatch between employee supply and demand, enterprises apply several flexibility strategies such as a multiskilled staff [4–7]. By implementing this strategy in the retail industry, the operational manager of a retail store is able to transfer multiskilled employees from overstaffed departments to those understaffed. However, since training employees is costly, it is necessary to identify efficient multiskilling designs in order to obtain a maximum of benefits from this approach at a reasonable cost.

Assigning multiskilled staff to different activities can be modeled as an assignment problem in a bipartite graph where the supply nodes are employees, and the demand nodes are the departments of a store. These types of designs are known as chained multiskilling designs where different multiskilling structures can be arranged. Amongst the most evaluated structures by previous works in the literature we can find closed long chains (CLC) and closed short chains (CSC). A CLC is a bipartite graph where there is a multiskilled employee in each department so that its arcs form exactly one undirected cycle, while a CSC is an induced bipartite subgraph that forms an undirected cycle since there is a subset of multiskilled employees that only connects a subset of departments.

Particularly, the 2-chaining policy, where certain employees are trained in two departments, i.e., k = 2, has been acknowledged by many studies as the most popular and endorsed chaining policy [4, 6, 8–11]. However, it would be interesting to analyze scenarios where a certain number of employees are able to work in k departments with k ≥ 2. This type of policy is known as k-chaining and only a few studies have evaluated its benefits before. Figure 1 shows a comparison between a 2-chaining structure with one CLC and one CSC, and a k-chaining structure with k = 3.

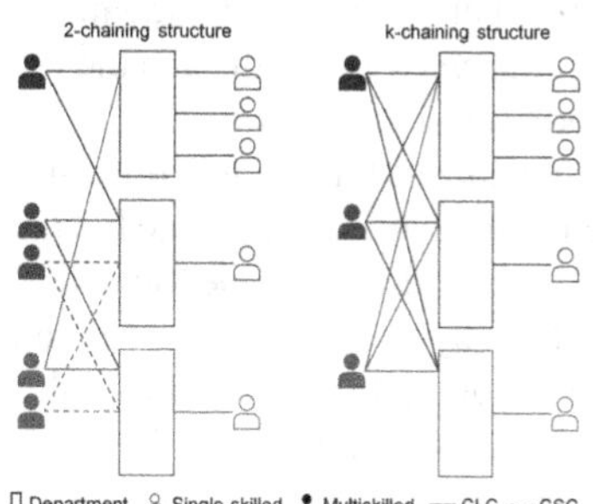


**Fig. 1.** Comparison between a 2-chaining structure with one CLC and one CSC, and a k-chaining structure with k = 3.

Therefore, the present study evaluates if it is possible to plan a multiskilled staff, using the k-chaining approach and considering demand uncertainty, that minimizes the expected costs associated with training and under/overstaffing for a days-off scheduling problem. A deterministic mixed-integer Linear Programming (MILP) model is initially proposed as a foundation for the stochastic analysis. Subsequently, a Two Stage Stochastic Optimization (TSSO) MILP model is developed and thus applied to a case study of a Chilean retail store.

## 2 Literature Review

Previous studies have analyzed the multiskilling as a useful strategy to address personnel scheduling problems with the aim of minimizing the gap between demand and supply. A series of these articles is presented in Table 1. The following symbology is introduced to contextualize the main aspects in consideration:

1. *Decision level of Human Resources studied (ND-RH):* The ND-RH column indicates which decision problems were covered in each study. At the planning level, the main decision problem would be to determine the staffing by type of task and contract (S). At the scheduling level, the decision problems are: (i) shift scheduling (SS), that is, assigning employees to daily shifts; (ii) days-off scheduling (DOS), that is, assigning employees to days off between their working days over a given planning horizon; and (iii) Tour scheduling (TS), that is, the simultaneous assignment of days off and shifts. Finally, the assignment level (A) simply consists of assigning employees to particular types of tasks regardless of their corresponding shifts.
2. *Chaining (CH):* Indicates if any sort of chaining application is present in the model, if there is not (–), if there is chaining in maximum two types of tasks (2-chaining), if there is chaining in three or more task types (k-chaining with k > 2).
3. *Multiskilling (MS):* Indicates whether the study uses a parameter (Par) representing an employee's fixed skills or a variable (Var) representing the decision as to whether or not an employee is trained in a given additional task type.
4. *Constraint of maximum number of skills in the model (MSP):* Indicates the maximum number of skills per employee that the model allows for the solution.
5. *Number of skills in the solution (NSS):* Indicates the resultant value of the number of skills per employee present in the solution.
6. *System type (ST):* (i) homogeneous (Hom), meaning that all task types, departments or resources in the system are identical, each one having the same supply level and installed capacity as well as possessing an identical demand and level of productivity, multiskilling and probability of absenteeism; (ii) heterogeneous (Het), meaning that not all the task types or departments are identical.
7. *Demand uncertainty (DU):* Indicates whether or not the problem considers variability in the staff demand.
8. *Supply uncertainty (SU):* Indicates whether or not the problem considers variability of supply or staff absenteeism.
9. *Method:* Indicates the solution method used, may be: (i) heuristics (H); (ii) simulation (S); (iii) optimization (OPT); (iv) Markov processes (MP); (v) analytic (AN); (vi) queuing theory (QT); (vii) algorithm (AL); (viii) genetic algorithm (GA); (ix)

constraint-based selection procedure (CODEMI); (x) constraint scheduling (CS) and (xi) statistical analysis (SA).

10. *Application (AP):* The industry or sector where the proposed model was applied in the study, which may be (i) manufacturing (M); (ii) services (S); (iii) call centers (CC); (iv) health (H); (v) retail (R); (vi) Power plant (CE); (vii) Cyber Security Operations Center (CS); (viii) Construction (C).

**Table 1.** Characteristics of published studies related to chaining and multiskilling.

| Reference | ND-RH | CH | MS | MSP | NSS | ST | DU | SU | Method | AP |
|---|---|---|---|---|---|---|---|---|---|---|
| Brusco and Johns [12] | S+SS | k-cha | Par | ≤3 | ≤3 | Het | No | No | OPT | S |
| Cai and Li [13] | SS+DOS+A | – | Par | ≤2 | ≤2 | Het | No | No | GA, H | S |
| Gomar et al. [14] | A | – | Par | – | ≤3 | Het | No | No | S | C |
| Eitzen et al. [15] | DOS+SS | – | Par | ≤9 | ≤3 | Het | No | No | AN | CE |
| Inman et al. [16] | S+A | 2-cha | Par | ≤2 | ≤2 | Hom | No | Yes | S | M |
| Iravani et al. [17] | A | k-cha | Par | ≤6 | ≤2 | Het | Yes | Yes | S | CC |
| Yang [18] | A | k-cha | Par | – | ≤2 | Het | No | Yes | S | M |
| Simchi-Levi and Wei [8] | A | 2-cha | Par | ≤2 | ≤2 | Het | Yes | No | AN | M |
| Deng and Shen [19] | A | 2-cha | Par | ≤2 | ≤2 | Het | Yes | No | OPT, S, AN | M |
| Paul and MacDonald [20] | S+A | 2-cha | Var | ≤2 | ≤2 | Het | Yes | No | H, OPT, AN | H |
| Olivella and Nembhard [21] | A | – | Var | – | ≤3 | Hom | Yes | Yes | CODEMI | M |
| Henao et al. [9] | A | 2-cha | Var | ≤2 | ≤2 | Hom | Yes | No | OPT, H, S | R |
| Ahmadian Fard Fini et al. [22] | S+A | – | Var | ≤3 | ≤3 | Het | No | No | CS, GA | C |
| Ağralı et al. [23] | SS | – | Par | ≤10 | ≤10 | Het | No | No | OPT | S |
| Liu [24] | D+A | 2-cha | Par | ≤2 | ≤2 | Het | Yes | No | OPT, S | M |
| Taskiran and Zhang [25] | S+TS | – | Var | ≤4 | ≤4 | Het | No | No | OPT | CC |
| Henao et al. [10] | A | 2-cha | Var | ≤2 | ≤2 | Hom | Yes | No | AN+S | R |
| Porto et al. [4] | TS+S | 2-cha | Var | ≤2 | ≤2 | Hom | No | No | OPT | R |
| Altner et al. [26] | DOS | – | Var | – | ≤6 | Hom | Yes | No | OPT, H | CS |

According to Table 1, in the vast majority of the studies the multiskilling effect it is entered as a fixed data input, which restricts the possibilities of obtaining better solutions. This issue could be solved by using a decision variable instead.

Subsequently, it can be noticed that it is frequently concluded that by training the personnel in a maximum of one additional skill, optimal results are obtained in terms of budget and level of service. However, it is not common to execute chaining methods in the solution procedures, which are considered of utmost importance to evaluate the impact that its application may possibly cause in this type of systems in relation to a more accurate approach to the optimal number of additional skills required per employee.

In addition, we identify that in the scarcely articles where chaining is contemplated, the 2-chaining method is usually applied (k = 2), only few studies implement multiskilling as an unrestricted decision variable for the purpose of developing k-chaining setups. It is also evident that a large section of articles does not evaluate demand uncertainty.

In summary, this study aims to fill the gaps identified from the literature review by solving a days-off-scheduling problem that minimizes the under/overstaffing costs, using a multiskilled staff with a k-chaining approach, and taking into account the demand uncertainty.

## 3 Problem Description

The problem under study consists in planning a multiskilled workforce, over a two-week horizon, that helps mitigate the negative impacts of the inherent uncertainties involved in the personnel scheduling process. Specifically, the problem we propose addresses the retail industry and seeks to minimize the costs associated with training and personnel shortage/surplus. In addition, this research aims to define robust multiskilling structures through k-chaining policies with $k \geq 2$, involving CLCs and/or CSCs, for a days-off scheduling human resources decision level. Since training employees is costly, the solution of this problem will define the most cost-effective multiskilling structure that not only minimizes the training cost but also guarantees robusticity against different levels of demand variability.

Briefly, the solution of this problem will determine: (i) how many employees should be single-skilled and how many should be multiskilled; (ii) the amount of additional skills each multiskilled employee should have; and (iii) under what levels of demand uncertainty a 2-chaining policy or a k-chaining policy with $k \geq 2$ is required. The solution will also assign rest days between working days for each employee over the two-week planning horizon.

In order to obtain those results, we formulate deterministic (see Sect. 4.1) and stochastic (see Sect. 4.2) models that consider the following assumptions: (i) Training costs are included in the total cost function and assumed to be the same for each department. (ii) The staff shortage/surplus cost per person-hour is assumed to be the same for each department and independent of the size of the staff shortage/surplus. (iii) The hours available for assignment in any department at the start of the planning horizon are exactly equal to the value in hours of that department's mean demand. (iv) There is no uncertainty of supply, that is, no staff absenteeism. (v) All employees have the same contract and therefore work the same number of weekly hours. (vi) Each employee should get at least one weekend off during the two-week planning period. (vii) At the start of the planning horizon, all employees are single-skilled and, thus, they can work only in a single department. (viii) Employees are homogeneous, i.e. all equally productive.

## 4 Methodology

The objective of the present section is to structure a workforce using the k-chaining approach for a days-off scheduling problem while considering uncertain demand. In

Subsect. 4.1, a deterministic optimization model is proposed for solving the addressed problem. Then, in Subsect. 4.2, we present a two-stage stochastic optimization model for determining the robust multiskilling levels that minimize the expected under/overstaffing and training costs.

## 4.1 Deterministic Optimization Model

In this subsection we present a deterministic version of the mixed-integer linear programming model useful to solve the personnel days-off scheduling problem with k-chaining. Later, in Subsect. 4.2, we will introduce the necessary modifications to incorporate demand uncertainty. First, let us define the model's sets, parameters and variables and their respective notation.
Model sets:

$I$: Store employees, indexed by $i$.
$L$: Store departments, indexed by $l$.
$I_l$: Store employees under contract in department $l$, $I_l \subseteq I, \forall l \in L$
$S$: Weeks, indexed by $s$. $\{1, 2\}$.
$D$: Days of the two-week period, indexed by $d$ $\{1, 2, \ldots, 13, 14\}$.
$D_S$: The set of days during week $s$ of the two-week period.
$D5_d$: Next five consecutive days in the two-week period after day $d$. For example, $D5_2 = \{3, 4, 5, 6, 7\}$, where day 2 corresponds to the first week's Tuesday and day 7 corresponds to the first week's Sunday.
$D^+$: Maximum consecutive days, indexed by $d$ $\{1, 2, \ldots, 8, 9\}$.

Model parameters:

$c$: Cost of training an employee to work in any department; [$-biweek/employee].
$u$: Staff shortage cost per hour (equivalent to the expected cost of lost sales); [$/Hour].
$o$: Staff surplus cost per hour (equivalent to the opportunity cost of idle employee-hours); [$/Hour].
$r_{ld}$: Number of hours required in department $l$ on day $d$, $\forall l \in L, d \in D$.
$t$: Number of daily hours employees must work under store employment contract.
$m_i$: Department for which employee $i$ was initially trained, $\forall\, i \in I$.

Model decision variables:

$x_{il}$: Equal to 1 if employee $i$ is trained for department $l$, otherwise 0, $\forall i \in I, l \in L$.
$y_{id}$ Equal to 1 if employee $i$ works on day $d$, otherwise 0, $\forall i \in I, d \in D$.
$\omega_{ild}$: Number of working hours assigned to employee $i$ on day $d$ in department $l$, $\forall\, i \in I, l \in L, d \in D$.
$z_i$: Equal to 1 if employee $i$ does not work on the first weekend (day 6, day 7) and 0 if employee $i$ does not work on the second weekend (day 13, day 14) $\forall i \in I$.
$v_i$ Number of additional skills for employee $i$, $\forall i \in I$.
$k_{ld}$: Staff shortage in daily hours in department $l$ on day $d$, $\forall l \in L, d \in D$.
$\delta_{ld}$: Staff surplus in daily hours in department $l$ on day $d$, $\forall l \in L, d \in D$.

The deterministic mixed-integer linear programming (MILP) model can now be formulated as follows:

$$Min \underbrace{\sum_{i\in I} \sum_{l\in L:l\neq m_i} cx_{il}}_{(a)} + \underbrace{\sum_{d\in D} \sum_{l\in L} (uk_{ld} + o\delta_{ld})}_{(b)} \tag{1}$$

*s.t.*

$$\sum_{i\in I} \omega_{ild} + \kappa_{ld} - \delta_{ld} = r_{ld} \qquad \forall l \in L, d \in D \tag{2}$$

$$\omega_{ild} \leq tx_{il} \qquad \forall i \in I, l \in L, d \in D \tag{3}$$

$$\sum_{l\in L} \omega_{ild} = ty_{id} \qquad \forall i \in I, d \in D \tag{4}$$

$$\sum_{d\in D_S} y_{id} = 5 \qquad \forall i \in I, s \in S \tag{5}$$

$$y_{id} + \sum_{\overline{d}\in D5_d} y_{i\overline{d}} \leq 5 \qquad \forall i \in I, d \in D^+ \tag{6}$$

$$y_{id} \leq 1 - z_i \qquad \forall i \in I, d \in \{6, 7\} \tag{7}$$

$$y_{id} \leq z_i \qquad \forall i \in I, d \in \{13, 14\} \tag{8}$$

$$x_{il} = 1 \qquad \forall i \in I, l \in L : l = m_i \tag{9}$$

$$v_i = \sum_{l\in L:l\neq m_i} x_{il} \qquad \forall i \in I \tag{10}$$

$$\sum_{i\in I_l} v_i = \sum_{i\in[I-I_l]} x_{il} \qquad I_l \subseteq I, \forall l \in L \tag{11}$$

$$z_i \in \{0, 1\} \qquad \forall i \in I \tag{12}$$

$$x_{il} \in \{0, 1\} \qquad \forall i \in I, l \in L \tag{13}$$

$$y_{id} \in \{0, 1\} \qquad \forall i \in I, d \in D \tag{14}$$

$$v_i \geq 0 \qquad \forall i \in I \tag{15}$$

$$k_{ld} \geq 0 \qquad \forall l \in L, d \in D \tag{16}$$

$$\delta_{ld} \geq 0 \qquad \forall l \in L, d \in D \tag{17}$$

$$\omega_{ild} \geq 0 \qquad \forall i \in I, l \in L, d \in D \tag{18}$$

In this formulation the objective function (1) minimizes the following costs: (a) training of employees in additional skills; and (b) staff shortage/surplus. Constraints (2), (16) and (17) yield the level of (non-negative) staff shortage/surplus associated with each department. Constraint (3) ensures the employees work the number of hours required by the employment contract only in departments they have been trained for. Constraint (4) guarantees the employees work the number of daily hours required by the employment contract only on the days they have been scheduled to work. Constraints from (5) to (8) implement days off scheduling constraints developed by Altner et al. [26]. Constraint (5) ensures each employee works exactly five days per week. Constraint (6) guarantees each employee works at most five consecutive days. Constraints (7) and (8) ensure each employee gets at least one weekend off. Constraints from (9) to (11) implement k-chaining constraints. Constraint (9) determines the department each employee was originally trained for. Constraint (10) indicates the number of additional departments an employee is trained for. Constraint (11) guarantees that for each department $l$, the number of employees belonging to it that have also been trained to work in other departments, is equal to the number of employees belonging to other departments that have been reciprocally trained to work in department $l$. Constraint (11) also ensures the formation of long or short closed chains. Either case requires that the indegree of each supply node belonging to the chain is equal to the outdegree of its corresponding demand node. Finally, constraints (12)–(18) define the domain of each problem variable.

### 4.2 Two Stage Stochastic Optimization Model

As mentioned before, in this subsection it is intended to describe the modifications applied to the original deterministic model in order to incorporate demand uncertainty in the stochastic model.

The approach that will be used to incorporate the uncertainty assumes that said uncertainty is present in certain parameters of the problem, being denoted by $\varepsilon$. Along with this fact, it also assumes that the probability distribution of these uncertain parameters is perfectly known and deducted from historical data. The two-stage stochastic optimization (TSSO) approach begins by dividing into two sets the decision variables: first-stage and second-stage. The first-stage variables are represented by the vector $x$, and are decisions known as "here-and-now" because they are made before the actual perception of the uncertain parameters becomes available. The second-stage variables, $y(x)$, known as "wait and see" are taken after the realization of uncertainty and in consequence they can be denoted as $y(x, \varepsilon)$.

In addition, the TSSO problems are commonly solved by the sample average approximation (SAA) approach that creates one set of second-stage variables for every possible scenario. The quality of the SAA approach is high when it is generated a large enough, although finite, set of samples from the continuous distribution of the random parameters

$\varepsilon$, represented by $\varepsilon \in \otimes$, where $\otimes$ is the set of scenarios generated. The first-stage variables $x_{il}$, $v_i$ are related to decisions about training employees in a different department. These tactical decisions are made before the random realizations of the weekly demand are known. The second-stage variables, $\omega_{ild}(\varepsilon)$, $k_{ld}(\varepsilon)$, $\delta_{ld}(\varepsilon)$ are the operational adjustments once the uncertain demand in each department, $r_{ld}(\varepsilon)$ is observed. That is, after demands become known, staff shortages or surpluses may occur.

Taking this under appreciation, the two-stage stochastic MILP model can now be formulated as follows:

$$Min \sum_{i \in I} \sum_{l \in L: l \neq m_i} cx_{il} + \frac{1}{|\otimes|} \left[ \sum_{d \in D} \sum_{l \in L} \sum_{\varepsilon \in \otimes} (uk_{ld}(\varepsilon) + o\delta_{ld}(\varepsilon)) \right] \tag{19}$$

*(first-stage)* *second-stage*

*s.t.*

$$\sum_{i \in I} \omega_{ild}(\varepsilon) + \kappa_{ld}(\varepsilon) - \delta_{ld}(\varepsilon) = r_{ld}(\varepsilon) \qquad \forall l \in L, d \in D, \varepsilon \in \otimes \tag{20}$$

$$\omega_{ild}(\varepsilon) \leq tx_{il} \qquad \forall i \in I, l \in L, \varepsilon \in \otimes \tag{21}$$

$$\sum_{l \in L} \omega_{ild}(\varepsilon) = ty_{id} \qquad \forall i \in I, d \in D, \varepsilon \in \otimes \tag{22}$$

$$\sum_{d \in D_S} y_{id} = 5 \qquad \forall i \in I, s \in S \tag{23}$$

$$y_{id} + \sum_{\overline{d} \in D5_d} y_{i\overline{d}} \leq 5 \qquad \forall i \in I, d \in D^+ \tag{24}$$

$$y_{id} \leq 1 - z_i \qquad \forall i \in I, d \in \{6, 7\} \tag{25}$$

$$y_{id} \leq z_i \qquad \forall i \in I, d \in \{13, 14\} \tag{26}$$

$$x_{il} = 1 \qquad \forall i \in I, l \in L : l = m_i \tag{27}$$

$$v_i = \sum_{l \in L: l \neq m_i} x_{il} \qquad \forall i \in I \tag{28}$$

$$\sum_{i \in I_l} v_i = \sum_{i \in [I - I_l]} x_{il} \qquad \forall l \in L \tag{29}$$

$$z_i \in \{0, 1\} \qquad \forall i \in I \tag{30}$$

$$x_{il} \in \{0, 1\} \qquad \forall i \in I, l \in L \tag{31}$$

$$y_{id} \in \{0, 1\} \qquad \forall i \in I, d \in D \tag{32}$$

$$v_i \geq 0 \qquad \forall i \in I \tag{33}$$

$$k_{ld}(\varepsilon) \geq 0 \qquad \forall l \in L, d \in D, \varepsilon \in \otimes \tag{34}$$

$$\delta_{ld}(\varepsilon) \geq 0 \qquad \forall l \in L, d \in D, \varepsilon \in \otimes \tag{35}$$

$$\omega_{ild}(\varepsilon) \geq 0 \qquad \forall i \in I, l \in L, d \in D, \varepsilon \in \otimes \tag{36}$$

In this formulation the objective function (19) minimizes the following costs: First-stage, training of employees in additional skills; and Second-stage, expected staff shortage/surplus. In summary, the constraints (2)–(4) and (16)–(18) were modified to explicitly incorporate uncertainty. Note that all these constraints involve only second-stage variables. Thus, the constraints (5)–(15) did not receive any modification because they are associated with first-stage variables. In addition, note that the interpretation of the constraints (20)–(36) and the constraints (2)–(18) is the same.

## 5 Computational Experiments

The developed methodology was applied to a case study of a Chilean retail store. Regarding the deterministic model (1)–(18), it was written in AMPL and solved using the software CPLEX running through a computer with a random-access memory of 4 gigabytes. Regarding the TSSO model (19)–(36), it was also written in AMPL and solved using the software CPLEX running through NEOS-server, a free internet-based service for solving numerical optimization problems, subjected to a computational restricted time limit of 8 h. The following subsections will elaborately describe the case study considered and the results obtained from the approach used.

### 5.1 Case Study

We use real and simulated datasets to solve the proposed methodology. As to the real dataset, information concerning the size of the departments, initial skills of each employee and type of contract assigned to each employee, was retrieved from the Chilean retail store considered in this study. A total of 30 employees were analyzed within a store with 6 departments, that is, $|I|= 30$ and $|L|= 6$. A 9-h full-time contract was assumed for the whole workforce; hence all employees should work exactly 5 days per week, resulting in 45 h per week.

As for the costs parameters, the cost related to staff shortages is equal to $u = US\$$ $60/\ h$ while staff surpluses represent a cost of $o = US\$\ 15/h$ [27]. The training cost per employee is assumed minimal, that is, $c = US\$$ 1-*biweek/employee*. Since training employees is almost free of charge, the results obtained from this research could be evaluated as an upper bound on the potential beneficence of multiskilling [1, 9, 10].

In this study we considered the correspondent in-sample data, which alludes to the initial information used to attain the Multiskilled Personnel Days-off Scheduling Problem solution with uncertain demand. We used the TSSO approach, which is solved using a SAA method. To obtain this in sample-data we appeal to a classic Monte Carlo simulation to generate 60 scenarios at random which in the same way represent the possible future parameters of the daily demand $r_{ld}(\varepsilon)$ $\forall l \in L, d \in D, \varepsilon \in \otimes$, such that $|\otimes| = 60$.

This exact procedure is repeated as many times as possible to produce proper random demand values for each department under three coefficients of variation (CV) of the demand, which were 10%, 30% and 50%. Followed by these values, we assign a zero-truncated normal distribution to the stochastic demand in each department to avoid negative demand values which are not possible events.

With the purpose of studying the multiskilling levels defined using the k-chaining approach, the following metrics were evaluated.

First, the percentage of multiskilled employees: number of multiskilled employees required in relation to the total number of hired employees.

$$\%ME = \frac{\sum\limits_{l \in L} \lambda_l}{|I|}.100 \tag{37}$$

Second, the percentage of multiskilled employees with $h$ additional skills: number of multiskilled employees with $h$ additional skills in relation to the total number of hired employees. Being $m_{lh}$ the number of multiskilled employees belonging to department $l$ with $h$ additional skills,

$$\%ME_h = \frac{\sum\limits_{l \in L} m_{lh}}{|I|}.100 \qquad \forall h \in \{1, 2, 3, 4, 5\} \tag{38}$$

Third, the percentage of total multiskilling: Number of additional skills that have been trained, relative to the theoretical maximum, in this case 150.

$$\%TM = \frac{\sum\limits_{i \in I} \sum\limits_{l \in L: l \neq m_i} x_{il}}{|I|(|L| - 1)}.100 \tag{39}$$

## 5.2 Results and Discussion

In order to analyze the results obtained from applying the developed methodology to a Chilean retail store, various aspects regarding both the deterministic and stochastic models will be presented. As expected, the computational time was way higher for the TSSO zero-truncated (TSSO-ZT) problem than for the deterministic one (DT). Apropos of the size of each model, the following Table 2 exhibits some important measures.

**Multiskilling Decisions.** Table 3 presents the results obtained once each of the metrics presented in the case study section was evaluated. Figure 2 shows the number of employees trained in one additional skill (k = 2) and two additional skills (k = 3) for each CV.

**Table 2.** Size of the deterministic and stochastic models

| Model | Constraints | Binary variables | Continuous variables |
|---|---|---|---|
| TSSO-ZT | 156 738 | 600 | 161 322 |
| DT | 3 102 | 600 | 2 730 |

**Table 3.** Metrics used to measure the multiskilling requirements.

| %CV | %ME | $\%ME_1$ | $\%ME_2$ | %TM |
|---|---|---|---|---|
| 10 | 47 | 40 | 7 | 11 |
| 30 | 83 | 80 | 3 | 17 |
| 50 | 100 | 77 | 23 | 25 |

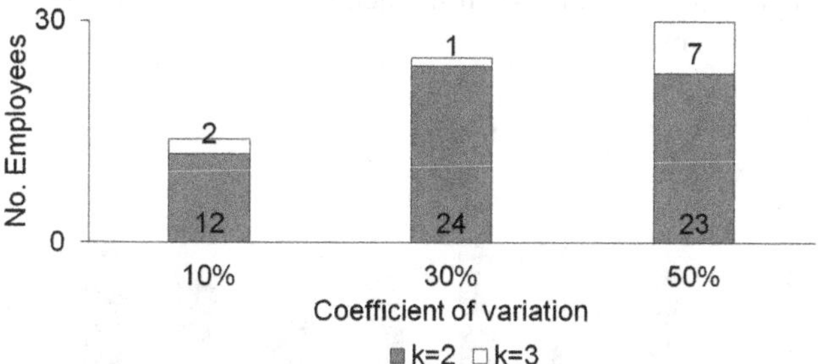


**Fig. 2.** Number of multiskilled employees trained in one additional skill (k = 2) and two additional skills (k = 3).

For a stochastic problem where 60 demand scenarios were generated, the ME, $ME_1$ and TM increase as the coefficient of variation rises. As to the $ME_2$, its greatest and lowest percentage are achieved at the CV = 50% and 30% respectively, behavior that could be explained by the quantity of demand scenarios analyzed.

It is also observed that when the demand variability is the highest (i.e., CV = 50%) the ME reaches its theoretical maximum. Since the problem does not constraint the amount of additional skills an employee can acquire, it is expected that the resulting levels of multiskilling for the CV = 50% are those that not only most likely minimize the under/overstaffing cost, but also work as an upper bound on the potential benefits of multiskilling considering the low training cost considered. The maximum amount of additional skills obtained was two, that is, there are some employees who are able to work in three different departments.

As to the highest level of demand variability, every single employee is multiskilled, just as assured before with the ME metric. In this scenario, 7 out of 30 employees were trained in two additional skills and the remaining 23 were trained in one additional skill.

Although the maximum number of employees trained in more than one additional skill is obtained at the CV = 50%, at both CV = 10% and 30% some employees are able to work in more than 2 departments. These results demonstrate that k-chaining policies with $k \geq 2$ are convenient, especially when the demand variability levels are high. It could also be concluded that flexibility structures with $k > 3$ are not quite attractive since such amount of multiskilling could be impractical. However, a larger amount of demand scenarios should be evaluated in order to make a more objective statement.

**Days-Off-Scheduling Decisions.** Figure 3 shows the number of employees assigned in each day of the second week of the planning horizon, indicating if the employee is single-skilled, trained in one additional skill (k = 2) or trained in two additional skills (k = 3), considering the demand variability with CV = 10%, 30% and 50%. It is observed that when the demand variability is low (CV = 10% and 30%) there are more employees assigned to work on the weekends (Saturday and Sunday), but as the demand variability increases (CV = 50%) the number of employees assigned to work on the weekends is less or equal to any other day of the week. Also, when the demand variability is the lowest (CV = 10%) there are more single-skilled employees, but these decrease in inverse proportion with the CV, and when the demand variability is the highest (CV = 50%) all employees are multiskilled.

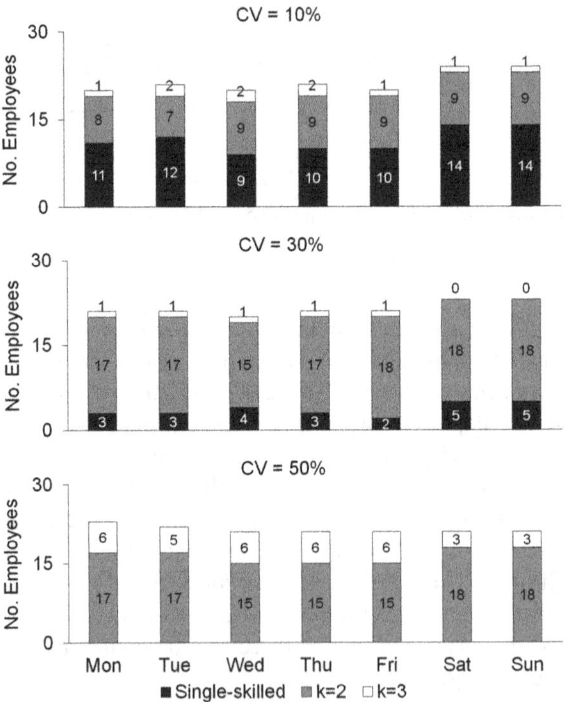


**Fig. 3.** Number of employees assigned in each day of the second week of the planning horizon.

## 6 Conclusions

Through the development of this study we were able to demonstrate how, under a series of realistic conditions of demand uncertainty, it is profitable and advisable to apply multiskilling setups using k-chaining with $k \geq 2$. It was observed that the number of multi-skilled employees needed grows as the demand variability increases. This is explained because it is crucial to count on a strong organization and programing of multiskilled employees that will keep us prepared to cover the majority of attainable scenarios in the practice, this is why the resultant percentage of multiskilled employees with a 50% coefficient of variation in the demand is consequently of the 100%. Among some of the assumptions of the model, we suppose that the training cost is minimal and the productivity of an employee will be identical from one another and independent from the quantity of additional skills that he/she has been trained in. Nevertheless, the study shows that total multiskilling is not required to obtain maximum profits. Finally, the study also proves that k-chaining policies with $k \geq 2$ are convenient, especially when the demand variability levels are high.

## References


1. Henao, C.A., Muñoz, J.C., Ferrer, J.C.: The impact of multi-skilling on personnel scheduling in the service sector: a retail industry case. J. Oper. Res. Soc. **66**(12), 1949–1959 (2015)
2. Álvarez, E., Ferrer, J.C., Muñoz, J.C., Henao, C.A.: Efficient shift scheduling with multiple breaks for full-time employees: a retail industry case. Comput. Ind. Eng. **150**, 106884 (2020)
3. Mac-Vicar, M., Ferrer, J.C., Muñoz, J.C., Henao, C.A.: Real-time recovering strategies on personnel scheduling in the retail industry. Comput. Ind. Eng. **113**, 589–601 (2017)
4. Porto, A.F., Henao, C.A., López-Ospina, H., González, E.R.: Hybrid flexibility strategic on personnel scheduling: retail case study. Comput. Ind. Eng. **133**, 220–230 (2019)
5. Felan, J.T., Fry, T.D., Philipoom, P.R.: Labor flexibility and staffing levels in a dual-resources constrained job shop. Int. J. Prod. Res. **31**(10), 2487–2506 (1993)
6. Hopp, W.J., Tekin, E., Van Oyen, M.P.: Benefits of skill chaining in serial production lines with cross-trained workers. Manage. Sci. **50**(1), 83–98 (2004)
7. Molleman, E., Slomp, J.: Functional flexibility and team performance. Int. J. Prod. Res. **37**(8), 1837–1858 (1999)
8. Simchi-Levi, D., Wei, Y.: Understanding the performance of the long chain and sparse designs in process flexibility. Oper. Res. **60**(5), 1125–1141 (2012)
9. Henao, C.A., Ferrer, J.C., Muñoz, J.C., Vera, J.: Multiskilling with closed chains in a service industry: a robust optimization approach. Int. J. Prod. Econ. **179**, 166–178 (2016)
10. Henao, C.A., Muñoz, J.C., Ferrer, J.C.: Multiskilled workforce management by utilizing closed chains under uncertain demand: a retail industry case. Comput. Ind. Eng. **127**, 74–88 (2019)
11. Fontalvo Echavez, O., Fuentes Quintero, L., Henao, C.A., González, V.I.: Two-stage stochastic optimization model for personnel days-off scheduling using closed-chained multiskilling structures. In: Rossit, D.A., Tohmé, F., Mejía, G. (eds.) Production Research. ICPR-Americas 2020. Communications in Computer and Information Science, vol. 1407, Springer Nature Switzerland AG (2021)
12. Brusco, M., Johns, T.: Staffing a multiskilled workforce with varying levels of productivity: an analysis of cross-training policies. Decis. Sci. **29**(2), 499–515 (1998)

13. Cai, X., Li, K.: A genetic algorithm for scheduling staff of mixed skills under multi-criteria. Eur. J. Oper. Res. **125**(2), 359–369 (2000)
14. Gomar, J., Haas, C., Morton, D.: Assignment and allocation optimization of partially multiskilled workforce. J. Constr. Eng. Manage. **128**(2), 103–109 (2002)
15. Eitzen, G., Panton, D., Mills, G.: Multi-skilled workforce optimisation. Ann. Oper. Res. **127**(1–4), 359–372 (2004)
16. Inman, R.R., Jordan, W.C., Blumenfeld, D.E.: Chained cross-training of assembly line workers. Int. J. Prod. Res. **42**(10), 1899–1910 (2004)
17. Iravani, S., Van Oyen, M., Sims, K.: Structural flexibility: a new perspective on the design of manufacturing and service operations. Manage. Sci. **51**(2), 151–166 (2005)
18. Yang, K.K.: A comparison of cross-training policies in different job shops. Int. J. Prod. Res. **45**(6), 1279–1295 (2007)
19. Deng, T., Shen, Z.J.M.: Process flexibility design in unbalanced networks. Manuf. Serv. Oper. Manage. **15**(1), 24–32 (2013)
20. Paul, J., MacDonald, L.: Modeling the benefits of cross-training to address the nursing shortage. Int. J. Prod. Econ. **150**, 83–95 (2014)
21. Olivella, J., Nembhard, D.: Calibrating cross-training to meet demand mix variation and employee absence. Eur. J. Oper. Res. **248**(2), 462–472 (2016)
22. Ahmadian Fard Fini, A., Rashidi, T., Akbarnezhad, A., Travis Waller, S.: Incorporating multiskilling and learning in the optimization of crew composition. J. Constr. Eng. Manage., **142**(5), 04015106 (2016)
23. Ağralı, S., Taşkın, Z.C., Ünal, A.T.: Employee scheduling in service industries with flexible employee availability and demand. Omega **66**, 159–169 (2017)
24. Liu, H.: The optimization of worker's quantity based on cross-utilization in two departments. Intell. Decis. Technol. **11**(1), 3–13 (2017)
25. Taskiran, G.K., Zhang, X.: Mathematical models and solution approach for cross-training staff scheduling at call centers. Comput. Oper. Res. **87**, 258–269 (2017)
26. Altner, D.S., Mason, E.K., Servi, L.D.: Two-stage stochastic days-off scheduling of multi-skilled analysts with training options. J. Combin. Optim. **38**(1), 111–129 (2018). https://doi.org/10.1007/s10878-018-0368-5
27. Porto, A.F., Henao, C.A., López-Ospina, H., González, E.R., González, V.I.: Dataset for solving a hybrid flexibility strategy on personnel scheduling problem in the retail industry. Data Brief **32**, 106066 (2020)

# Efficient Scheduling of a Real Case Study of the Pharmaceutical Industry Using a Mathematical-Algorithmic Decomposition Methodology

Josías A. Stürtz[1] and Pablo A. Marchetti[1,2(✉)]

[1] Universidad Tecnológica Nacional, Facultad Regional Santa Fe, Lavaisse 610, 3000 Santa Fe, Argentina

[2] Instituto de Desarrollo Tecnológico para la Industria Química (INTEC), UNL-CONICET, Güemes 3450, 3000 Santa Fe, Argentina

pmarchet@intec.unl.edu.ar

**Abstract.** This work presents a mathematical-algorithmic methodology to handle scheduling decisions on multiproduct multistage batch facilities, developed with the aim to solve problems of industrial size. The proposal is based on the iterative construction of a solution by solving a sequence of subproblems associated to each stage, identifying and fixing the critical decisions at each step, and keeping rigorous information on the bounds. The methodology is applied to a real case study obtaining good quality solutions in competitive computing times.

**Keywords:** Multiproduct batch plants · Scheduling · Optimization

## 1 Introduction

In the last 30–40 years, scheduling decision tools have become a critical asset for the success and competitiveness of manufacturing organizations. Scheduling problems have been widely studied, and different mathematical formulations have been proposed that can be broadly classified into two groups: discrete and continuous time representations [1]. When comparing different models the quality of the solution and time required for the computation are critical. In general, a tighter formulation obtains better solutions and more competitive performances by reducing the number of alternatives to be explored by the solver, usually a branch-and-cut algorithm.

Despite countless efforts in formulating and solving scheduling problems, from both researchers in academia and engineers in industry, there is still a significant gap between theory and practice. One of the main causes of this gap is the intrinsic difficulty of solving industrial-scale problems, due to the underlying combinatorial explosion of alternatives. Harjunkoski et al. [2] present an extensive review of the strengths and weaknesses of different scheduling approaches, focusing on the development and deployment of industrial applications. They

© Springer Nature Switzerland AG 2021
D. A. Rossit et al. (Eds.): ICPR-Americas 2020, CCIS 1407, pp. 171–176, 2021.
https://doi.org/10.1007/978-3-030-76307-7_13

argue that current optimization tools can be successfully applied to increase plant floor production efficiency, although they note there is still a large potential for improvement.

This work presents a scheduling methodology for multiproduct multistage batch plants considering parallel equipment, unlimited storage, and sequence-dependent transition times, which is applied to an industrial-scale case study from the pharmaceutical industry [3,4]. The approach is based on the idea of prioritizing the tasks that require the critical resources of the installation, and subsequently accommodating the tasks that have more slack time available. In particular, the existence of a critical stage or "bottleneck" is assumed.

## 2 Mathematical-Algorithmic Proposed Methodology

The proposed method consists of an iterative algorithm that solves a sequence of mixed-integer linear programming (MILP) mathematical models. Each of these models corresponds to a subproblem of the complete problem being tackled, and considers one of the stages together with a subset of the preceding and succeeding tasks required at the associated upstream and downstream stages, respectively, of the process. To model the subproblems a new version of the continuous time representation based on time-slots by Pinto and Grossmann [5] has been developed, in which it is possible to add a "wildcard" slot to a given unit. That is, a multi-valued slot to which an arbitrary number of batches can be assigned.

The algorithm is structured with a main iteration and an inner iteration. The main iteration, which considers all the stages of the process, gradually builds a solution of the complete problem by fixing one by one the solutions of each stage. On each main iteration, the set of pending (or not fixed) stages is analyzed, solving a sequence of subproblems (inner iteration) for each stage. The aim is to identify the critical stage, that is, the one whose solution produces the greatest deterioration to the value of the objective function (makespan). Once this stage is identified, the allocation and sequencing decisions obtained in the corresponding subproblem are fixed, and the set of pending stages is reduced. The main iteration ends when all the stages have been fixed and a complete solution of the problem (schedule) is obtained.

On the other hand, the inner iteration is focused on solving the subproblems corresponding to each stage. In this iteration, an approximate model and an exact model are alternately solved. The approximate subproblem, which includes multi-valued slots in all units, is used to obtain a candidate slot configuration for the stage. Subsequently, the exact subproblem fixes the previously obtained slot configuration to find a feasible solution for the stage. Based on the quality of this solution and the parameters of the algorithm it is decided to end the inner iteration or repeat the process. If it continues, the recently analyzed slot configuration is removed from the following approximate subproblems by adding appropriate integer cuts. The inner iteration returns the best feasible solution found and accurately reports the bounds (MIP solution and best possible solution) to the main iteration.

The parameters taken into account for the inner iteration on a given stage are the following: (a) the number of slots $R$ to be included on the equipment units of the preceding (upstream) and succeeding (downstream) stages of the process, (b) a time limit $T$ to solve each subproblem, (c) a time limit for the inner iteration process, and (d) the maximum number of iterations allowed. The inner iteration concludes after a solution is obtained for an exact subproblem such that its best possible solution cannot be improved, or when the limits set in (c) or (d) are exceeded.

## 3 Results and Discussion

The proposed method has been applied to a real case study of the pharmaceutical industry (see [4]) comprising a multiproduct batch installation with 17 processing units running in parallel and distributed in 6 stages. One of the main complexities of the problem is the presence of sequence-dependent transition times, which are larger than the processing times in some stages. Based on the 30 batches problem, instances including the first 10, 15, 20, 25 and 30 batches have been evaluated. Different alternatives were considered for the parameters (a) with values $R = 2$ and $R = 3$, and (b) with values $T = 300$, 600, 1200 s. Besides, a maximum of 1 h and a limit of 10 cycles were set for the parameters (c) and (d) of the inner iteration.

The proposed methodology has been implemented in GAMS, using the MILP solver GUROBI 7.5 for the inner iteration subproblems. In order to compare the computational performance and the quality of the solutions obtained (objective function value and integrality gap), two formulations of the complete problem (i.e., considering all stages) were also solved, setting in this case a time limit of 5 h. On the one hand, a model with a fixed number of slots on each unit, which is called FULL FIXED, has been evaluated. Here, the number of slots is equivalent to the slot configuration of the best solution obtained with the proposed method. On the other hand, a model that sets the number of slots based on the maximum number of batches that can be processed on each unit, which is denoted FULL, has been solved. All computations were performed on a generic PC with an Intel Core i7 3.2 GHz processor and 16 GB of RAM.

A comparison of the solutions obtained is presented in Fig. 1 where, for each instance considering a given number of batches, the first column shows the best solution obtained with the proposed methodology, and the remaining columns the solutions of the complete formulations. The total height of each column corresponds to the optimal MIP value found, and is composed by the best possible solution in blue and the remaining gap to prove optimality in red. Figure 1 shows how, when the number of batches increases, the proposed methodology allowed to obtain better quality solutions with lower bounds similar to the complete problems (FULL and FULL FIXED), even though it only solves subproblems.

The model sizes and CPU times for the 30 batch problem are compared in Table 1. Regarding the sizes of the models, it is observed that the FULL FIXED alternative reduces more than by half the number of binary variables of the FULL model and, however, the improvement in the solution is not substantial. In contrast, given that partial subproblems are solved, for the proposed methodology the minimum and maximum values of these statistics are shown. The largest evaluated subproblem includes 1556 binary variables compared to 4090 for the FULL FIXED formulation. Besides, in this case the CPU time required to complete the algorithm is 1 h 27 min, compared to 5 h for the complete models. Finally, the Gantt chart of the best solution found is presented in Fig. 2.

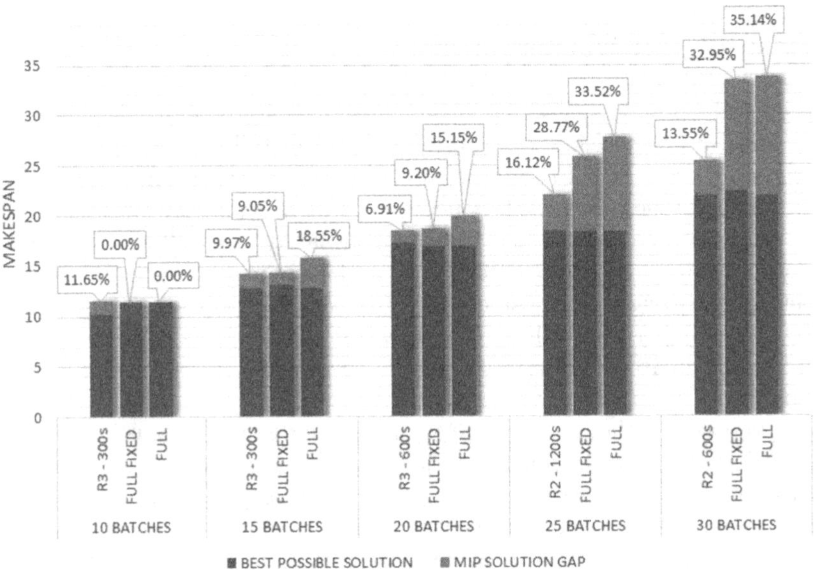


**Fig. 1.** Comparison of the best solutions obtained with the proposed methodology and the solutions of complete formulations with fixed (FULL FIXED) or maximum possible (FULL) number of slots.

**Table 1.** Comparison of model sizes and CPU times for the 30 batch problem.

| | | Binary vars | Cont. Vars | Equations | Time (s) |
|---|---|---|---|---|---|
| $R = 2$, $T = 600$ s | min | 120 | 395 | 2935 | 5221 |
| | max | 1556 | 775 | 19583 | |
| FULL FIXED | | 4090 | 775 | 19847 | 18000 |
| FULL | | 10264 | 1384 | 50190 | 18000 |

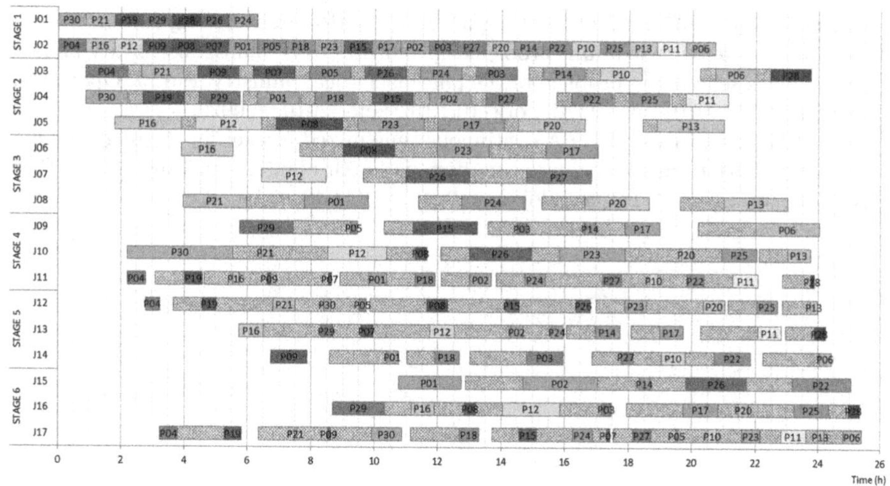


**Fig. 2.** Gantt chart of the best solution obtained (makespan: 25.38 h) for the 30 batch problem, with parameters $R = 2$ and $T = 600\,\text{s}$.

# 4 Conclusions

A mathematical-algorithmic methodology for multiproduct multistage batch plant scheduling that relies on solving a sequence of subproblems has been presented. The proposed method has been applied to an industrial size case study from the pharmaceutical industry, allowing not only to obtain good quality solutions in competitive times but also to provide rigorous information on the associated integrality bounds.

**Acknowledgments.** The authors gratefully acknowledge financial support from Universidad Tecnológica Nacional through grants PID 4932 and PID 7768.

# References

1. Méndez, C.A., Cerdá, J., Grossmann, I.E., Harjunkoski, I., Fahl, M.: State-of-the-art review of optimization methods for short-term scheduling of batch processes. Comput. Chem. Eng. **30**, 913–946 (2006). https://doi.org/10.1016/j.compchemeng.2006.02.008
2. Harjunkoski, I., et al.: Scope for industrial applications of production scheduling models and solution methods. Comput. Chem. Eng. **62**, 161–193 (2014). https://doi.org/10.1016/j.compchemeng.2013.12.001
3. Castro, P.M., Harjunkoski, I., Grossmann, I.E.: Optimal short-term scheduling of large-scale multistage batch plants. Ind. Eng. Chem. Res. **48**, 11002–11016 (2009). https://doi.org/10.1021/ie900734x

4. Kopanos, G.M., Méndez, C.A., Puigjaner, L.: MIP-based decomposition strategies for large-scale scheduling problems in multiproduct multistage batch plants: a benchmark scheduling problem of the pharmaceutical industry. Eur. J. Oper. Res. **207**, 644–655 (2010). https://doi.org/10.1016/j.ejor.2010.06.002
5. Pinto, J.M., Grossmann, I.E.: A continuous time mixed integer linear programming model for short term scheduling of multistage batch plants. Ind. Eng. Chem. Res. **34**, 3037–3051 (1995). https://doi.org/10.1021/ie00048a015

# Solving Order Batching/Picking Problems with an Evolutionary Algorithm

Fabio M. Miguel[1], Mariano Frutos[2](✉), Máximo Méndez[3], and Fernando Tohmé[4]

[1] Universidad Nacional de Río Negro, Sede Alto Valle y Valle Medio, Villa Regina, Argentina
fmiguel@unrn.edu.ar

[2] Departamento de Ingeniería, Universidad Nacional del Sur e IIESS UNS-CONICET, Bahía Blanca, Argentina
mfrutos@uns.edu.ar

[3] Instituto Universitario de Sistemas Inteligentes SIANI, Universidad de Las Palmas de Gran Canaria ULPGC, Las Palmas, Spain
maximo.mendez@ulpgc.es

[4] Departamento de Economía, Universidad Nacional del Sur e INMABB UNS-CONICET, Bahía Blanca, Argentina
ftohme@criba.edu.ar

**Abstract.** We present an evolutionary algorithm to solve a combination of the Order Batching and Order Picking problems. This integrated problem consists of selecting and picking up batches of various items requested by customers from a storage area, given a deadline for finishing each order according to a delivery plan. We seek to find the plan that minimizes the total cost of picking the goods, proportional to the time devoted to traverse the storage facility, grabbing the good and leaving it at the dispatch area. Earliness and tardiness induce inefficiency costs due to the excess use of space or breaching the delivery contracts. The results of running the algorithm compare favorably to those reported in the literature.

**Keywords:** Order Batching · Order Picking · Evolutionary algorithm

## 1 Introduction

A critical factor in the operational performance of the internal and external logistics of a firm involves the optimization of processes in distribution centers. Moving the items inside those centers, receiving, locating, selecting and collecting them are some of those processes [1, 2]. The selection and collection of items are the main activities carried out in storage facilities. They amount to picking up the right number of items requested by the customers and then taking them to the area in which orders are prepared [3]. The goods are classified, regrouping the units in each batch to prepare the individual orders [4, 5]. Most cases involve also marking, labelling and boxing up the goods into indivisible parcels. The process finishes once each of those parcels are checked out, loaded on the delivery trucks and finished the necessary documents. In this paper we focus on the optimization of the selection and collection of requested items.

© Springer Nature Switzerland AG 2021
D. A. Rossit et al. (Eds.): ICPR-Americas 2020, CCIS 1407, pp. 177–186, 2021.
https://doi.org/10.1007/978-3-030-76307-7_14

## 2 Problem Description and Literature Review

We can distinguish three problems concerning the operations in storage facilities. The first one is the allocation of goods to different storage positions. The second involves the grouping of items in batches for their collection. The third problem is that of scheduling the sequence of pick-ups of goods and taking them to the dispatch area [6–8]. In this paper we focus on the integrated treatment of the second and third problems, critical for the efficiency of operations since they generate most of the costs of the operations on the floor of the storage facility, being intensive in the use of manpower [9, 10]. This combined problem starts with the arrival of the orders from different customers, detailing the amounts, specifications and availability dates at the merchandise dispatch area. The whole procedure is optimized by choosing the plan that minimizes the operational costs of the operations leading to satisfy the requests in due time of the batches, consisting of different goods for disparate customers [11–13]. Delays in complying with the plan create costs of breaching the contracts with customers. On the other hand, an early finishing of the plan creates costs of crowding the dispatch area, blocking flows of activity and increasing processing times for new orders [11]. Formally, the problem integrates the Order Batching Problem (OBP) and the Order Picking Problem (OPP) [14–16]. OBP consists in finding the amount and size of the batches of items, readying them for the pick-up team [13]. This requires to take into account the capacity of the team and the time at which each item must be available for finishing at the delivery area [17–19]. OPP consists in identifying optimal navigation plans around the sites where the items are stored [20–22]. From now on we denote this integrated problem as OBP+OPP.

A thorough review of the literature on OBP and OPP can be found in [2], while [1, 21, 23] review the heuristics for solving OPP. [8] presents two ways of solving OPP, using Ant Colony Optimization and Iterated Local Search. Other meta-heuristics applied to related routing problems can be found in [15]. [4] uses a clustering approach to solve OBP taking into account demand patterns instead of the distances covered by each sequence of visits. Other heuristics for OBP can be found in [7]. An integer programming approach to OBP is presented in [13] were the visit sequences are estimated and the problem is solved with a heuristic based on fuzzy logic. In turn [11] uses a multiple genetic algorithm for OBP+OPP. This latter contribution uses flexible time windows for the delivery time of each order. We adopt this methodology, but using an evolutionary algorithm with a specific chromosome covering different batches. This algorithm yields results that improve over those obtained at the instances and lay-out of [11].

## 3 Formulation of the Problem

Let $\mathcal{P} = \{1, \ldots, nIt\}$ be the set of items, where $nIt$ is the amount of different goods. Each item has a unit weight, defining a corresponding class $\mathcal{W} = \left\{w_1, \ldots, w_p, \ldots, w_{nIt}\right\}$. Each customer $i$ makes a single request, involving a list of items $\mathcal{P}_i$. Then, the number of customers is the same as the number of requests, $nReq$, being the set of customers $\mathcal{I} = \{1, \ldots, i, \ldots, nC\}$. Each request has a finishing time, defining a class $\mathcal{T} = \left\{t_1, \ldots, t_i, \ldots, t_{nReq}\right\}$. Let $\mathcal{L} = \left\{\ell_0, \ell_1, \ldots, \ell_p, \ldots, \ell_{nIt}\right\}$ be the positions on the floor of the store, where $\ell_0$ is the dispatch area while the others correspond to the placements of the different items. So, for an item $p \in \mathcal{P}$, its position $\ell_{\mathrm{p}} = (x_p, y_p)$ corresponds to its coordinates in the floor. Given a pair of positions $\ell_{\mathrm{p}}$ and $\ell_{\mathrm{p'}}$, if there is an open direct path between them, we define a distance $D_{lp,lp'}$. We denote with $\mathcal{P}_{\mathrm{r}}$ the class of items in a batch $r$ (the goods in $\mathcal{P}_{\mathrm{r}}$ may or not correspond to the requests of a single customer). $\mathcal{R} = \{1, .., \mathrm{r}, \ldots, |\mathcal{R}|\}$ is the class of total batches to be picked-up. $\mathcal{S}_r = \mathrm{s}_1, \ldots, \mathrm{s}_{\mathrm{u}}, \ldots \mathrm{s}_{|\mathcal{S}_r|}$ is the sequence of positions to be visited to complete batch $r$, where $s_u$ is the $u$-th position to be visited and $|\mathcal{S}_{\mathrm{r}}|$ is the amount of different items in batch $r$. $q_{i,p} \in \mathcal{Q}$ denotes the total number of units of item $p$ requested by customer $i$. Then $\mathcal{Q}_i = \sum_{p \in \mathcal{P}_i} q_{i,p}$ is the total number of units in the request of customer $i$ and $\mathcal{Q}_p = \sum_{i \in \mathcal{I}} q_{i,p}$ the total number of units requested of item $p$. Analogously, we define $\mathcal{Q}_{\mathrm{r}}$ to be the total number of units in batch $r$. $\mathcal{K} = \{1, \ldots, |\mathcal{K}|\}$ denotes the class of pick-up teams. Each team as a maximum carrying capacity $Cap$. Then, the integrated problem OBP+OPP is defined on an undirected graph $\mathcal{G} = (\mathcal{V}, \mathcal{A})$, where $\mathcal{V}$ are nodes denoting the storage positions of the items in $\mathcal{P}$, plus two copies of the initial node (the position of the dispatch area). In turn, $\mathcal{A}$ represents all the feasible direct paths between nodes in $\mathcal{V}$. Each such edge $(h, l) \in \mathcal{A}$ has an associated time $t_{hl}$ defined as the length of the distance between positions $h$ and $l$ divided by the speed of the pick-up team $v$ (i.e. $t_{hl} = D_{h,l}/v$). Each edge has also an associated monetary cost per time unit $\varsigma$. $t_{pick}$ is the mean time to pick any item once reached its corresponding position. A Boolean variable $x_{hlkr}$ is 1 if and only if item $h$ is picked up right before item $l$ by the picking team $k$ in the sequence for batch $r$, where $h, l \in \mathcal{V}$, $k \in \mathcal{K}$ and $r \in \mathcal{R}$. Another Boolean variable is $y_{hkr} = 1$ if and only if the pick-up team $k$ grabs item $h$ for batch $r$, where $h \in \mathcal{V}$, $k \in \mathcal{K}$ and $r \in \mathcal{R}$. Then the formal presentation of OBP+OPP is as follows [3, 11]:[1]

$$\min C_{Total}: \left[ \frac{\sum_{h \in \mathcal{V}} \sum_{l \in \mathcal{V}} D_{h,l} \cdot \sum_{k \in \mathcal{K}} \sum_{r \in \mathcal{R}} x_{hlkr}}{v} + \sum_{\substack{p \in P \\ q \in Q}} q_p \cdot t_{pick} \right] \cdot \varsigma + \sum_{i \in \mathcal{I}} (\alpha \cdot E_i + \beta \cdot T_i) \tag{1}$$

s.t.:

$$\sum\nolimits_{p \in \mathcal{P}_r} (q_p \cdot w_p) \cdot y_{pkr} \leq Cap, \ \forall k \in \mathcal{K}, \ r \in \mathcal{R} \tag{2}$$

$$\sum\nolimits_{r \in \mathcal{R}} y_{hkr} = 1, \forall h \in \mathcal{P}, \forall k \in \mathcal{K} \tag{3}$$

$$\sum\nolimits_{k \in \mathcal{K}} y_{hkr} = |\mathcal{K}|, \forall h \in \{0, n+1\}, r \in \mathcal{R} \tag{4}$$

$$\sum\nolimits_{h \in \mathcal{V}} x_{hlkr} = y_{lkr}, \forall l \in \mathcal{V} \backslash \{0\}, k \in \mathcal{K}, \ r \in \mathcal{R} \tag{5}$$

$$\sum\nolimits_{l \in \mathcal{V}} x_{hlkr} = y_{hkr}, \forall h \in \mathcal{V} \backslash \{n+1\}, k \in \mathcal{K}, \ r \in \mathcal{R} \tag{6}$$

[1] We denote with $\{0, n+1\}$ the start and end at the dispatch area.

$$\sum_{i\in\mathcal{I}}\sum_{k\in\mathcal{K}}\sum_{r\in\mathcal{R}} q_{i,p}\cdot y_{pkr} = \mathcal{Q}_p, \forall p \in \mathcal{P} \tag{7}$$

$$\sum_{p\in\mathcal{P}}\sum_{k\in\mathcal{K}}\sum_{r\in\mathcal{R}} q_{i,p}\cdot y_{pkr} = \mathcal{Q}_i, \forall i \in \mathcal{I} \tag{8}$$

$$x_{hlkr} \in \{0, 1\}, \forall h, l \in \mathcal{V},\ k \in \mathcal{K},\ r \in \mathcal{R} \tag{9}$$

$$y_{hkr} \in \{0, 1\}, \forall h \in \mathcal{V},\ k \in \mathcal{K},\ r \in \mathcal{R} \tag{10}$$

The goal (1) is the minimization of the total money cost of the time spent in collecting the batches plus a penalty for failing to meet the agreed-on deadlines for finishing the requests. The first term represents the cost of the time devoted to picking up the goods and taking them to the dispatch area. A penalization to earliness or tardiness in getting the items in time to the dispatch area is defined as follows. $\alpha$ is the earliness penalty per unit time while $\beta$ is the corresponding unit time fine for tardiness. $E_i$ is the time length of earliness in the fulfillment of the request of customer $i$ and $T_i$ corresponds to tardiness: $E_i = max\{0, t_i - c_i\}$ and $T_i = max\{0, c_i - t_i\}$, where $c_i$ is the actual finishing time of the request of $i$ (i.e. the time at which all the items in the request are finally prepared in the dispatch area). The list of constraints (2) indicates that the total weight carried by a team cannot exceed its capacity. The restrictions in (3) mean that no storage position should be visited more than once for the preparation of a given batch. (4) ensures that all pick-up teams start and end at the dispatch area. Restrictions (5) and (6) preserve the orderly sequence of pick-ups. (7) means that the requested amounts of each item $p$ are picked-up. Analogously, (8) indicates that the amounts requested by customer $i$ are picked-up. Finally, (9) and (10) indicate that variables $x_{hlkr}$ and $y_{hkr}$ are Boolean.

## 4 Solution Method

The problem presented in the previous section belongs to the NP-Hard complexity class. This means that, in practice, it can be solved analytically only in very small instances. Therefore, to find solutions in polynomial time we need to use heuristic methods. We propose here an evolutionary algorithm based on the usual integer representation used to solve combinatorial problems. In this representation we consider a chromosome consisting of two genomes. We can see how this works in a very simple instance, in which we codify the solution for three requests of four items. Table 1 shows that request 1 involves 3 units (one unit of item A and two units of C); request 2 asks for one unit of A, one of B, two of C and one of D while request 3 demands two units of B and two of D.

To codify this we consider two rows, one for items (identifying to which requests they belong) and another indicating cumulative amounts (Table 2).

A chromosome with two genomes captures the codification in Table 2 (see Table 3).

Chromosome: {[G1] [G2]}

Genomes:

[G1]: Sequence in which items will be picked up.

[G2]: Amount of items picked up in [G1] for each batch.

Number of entries:

**Table 1.** Example for OBP + OPP.

| Request | Items | | | | Total |
|---|---|---|---|---|---|
| | A | B | C | D | |
| 1 | 1 | 0 | 2 | 0 | 3 |
| 2 | 1 | 1 | 2 | 1 | 5 |
| 3 | 0 | 2 | 0 | 2 | 4 |
| Total amounts | 2 | 3 | 4 | 3 | 12 |

**Table 2.** Codification.

| Items (corresponding request) | | | | | | | | | | | |
|---|---|---|---|---|---|---|---|---|---|---|---|
| A(1) | A(2) | B(2) | B(3) | B(3) | C(1) | C(1) | C(2) | C(2) | D(2) | D(3) | D(3) |
| ↑ | ↑ | ↑ | ↑ | ↑ | ↑ | ↑ | ↑ | ↑ | ↑ | ↑ | ↑ |
| 1 | 2 | 3 | 4 | 5 | 6 | 7 | 8 | 9 | 10 | 11 | 12 |

**Table 3.** Chromosome = Genome 1 + Genome 2

| Cumulative pick ups → | 1 | 2 | 3 | 4 | 5 | 6 | 7 | 8 | 9 | 10 | 11 | 12 |
|---|---|---|---|---|---|---|---|---|---|---|---|---|
| Genome 1 | | | | | | | | | | | | |
| [G1] → | 6 | 7 | 9 | 10 | 11 | 1 | 3 | 2 | 12 | 4 | 8 | 5 |
| | batch 1 | | | | | batch 2 | | batch 3 | | | | |
| Genome 2 | | | | | | | | | | | | |
| [G2] → | 5 | 7 | 12 | – | – | – | – | – | – | – | – | – |

[G1]: Total number of requested items $\sum_{i \in \mathcal{I}} \mathcal{Q}_i$.

[G2]: Total number of requested items $\sum_{i \in \mathcal{I}} \mathcal{Q}_i$ (non-null entries correspond to batches).

The sequence in which items of each batch are picked up is the following. Batch 1: $\{6 \rightarrow 7 \rightarrow 9 \rightarrow 10 \rightarrow 11\}$, Batch 2: $\{1 \rightarrow 3\}$ and Batch 3: $\{2 \rightarrow 12 \rightarrow 4 \rightarrow 8 \rightarrow 5\}$. This means that, for instance for Batch 2, that the pick-up team has to go to the first position on Table 2 (A(1)) and then to third position (B(2)) to finally take to the dispatch area the two picked up units of items A and B. Figure 1 shows the navigation path of the three pick-up teams.

This visual representation ensures to warrant the satisfaction of constraints (2)–(8). In the initial stage of the algorithm a population is generated at random. The procedure iterates, by selecting the fittest individuals, the winners in repeated tournaments among $k$

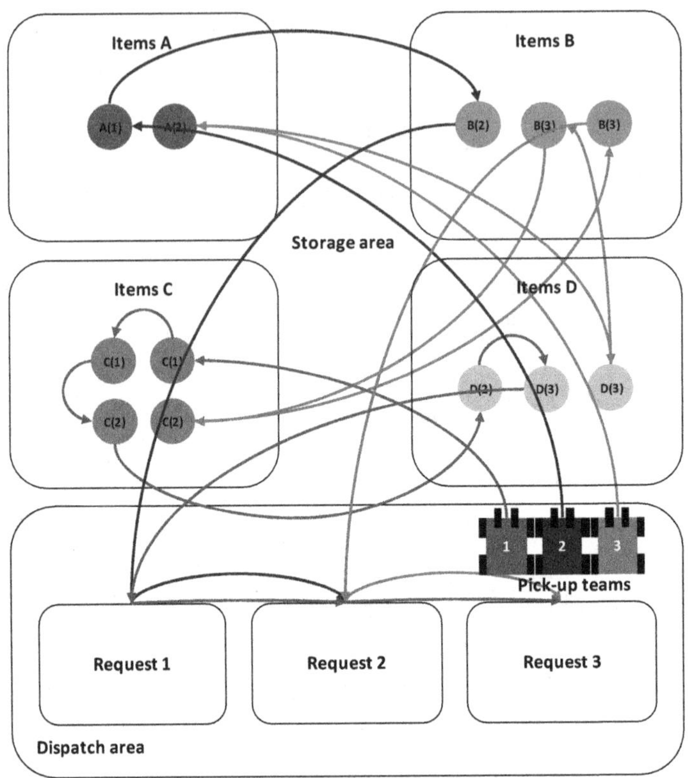


**Fig. 1.** Navigation paths of the pick-up teams

individuals chosen at random from the current population [24]. Then, the *edge recombination* Operator [25] and the *mutation of insertion* operator [26] are applied on Genome 1. Genome 2 is completed according to the capacity of pick-up teams. A cost criterion caps the number of iterations. Figure 2 presents the pseudo-code of the algorithm.

## 5 Computational Experiment

In order to evaluate the quality of the solutions and the performance of the algorithm we run it on the instances presented in [11]: a medium size instance (DS1/M1), a large one (DS2/M1) and a very large instance (DS3/M1) (Table 4).

We assume that the number of units of an item $p$ requested by a customer $i$ follows a uniform distribution between 1 and 10, i.e. $q_{i,p} \sim U(1, \ldots, 10)$.The number of different items requested by a customer $i$ is assumed to obey to a normal distribution with mean 10 and standard deviation 5, that is $|\mathcal{P}_i| \sim N(10, 5)$. The finishing time of a request of customer $i$, follows a uniform distribution, on discrete periods of time measured in seconds between 10:00 am and 06:00 pm, that is, $t_i \sim U(36000, \ldots, 64800)$. The unit weight of each item $p$, obeys also to a uniform distribution ranging between 8 and 24 kg., i.e. $w_p \sim U(8, \ldots, 24)$. With respect to the pick-up teams we assume that their average

```
Load Input % information of requests, lay-out and parameters of the algorithm.
nLot ← nLotMin
while nLot < nLotMax
 t ← 0;
 P(t) ← InitPop(Entrada);
 FitP(t) ← EvalPop(P(t));
 For t ← 1 a MaxNumGen
     Q(t) ← SelecBreeders (P(t), FitP(t));
     Q(t) ← Crossover(Q(t));
         Q(t) ← Mutation(Q(t));
         FitQ(t) ← EvalPop(Q(t));
         P(t) ← SelSurviv(P(t), Q(t), FitP(t), FitQ(t));
         FitP(t) ← EvalPop(P(t));
         if TermCond(P(t), FitP(t))
             break;
     end
     end
nLot ← nLot +1;
end
```

**Fig. 2.** Pseudo-code of the algorithm

**Table 4.** Features of the instances [11]

| | DS1/M1 | DS2/M1 | DS3/M1 |
|---|---|---|---|
| Number of requests | 40 | 80 | 200 |
| Number of different items | 80 | 160 | 300 |
| Total average weight (kg.) | 13.704 | 37.152 | 158.784 |
| Capacity of pick-up teams (kg.) | 10.000 | 10.000 | 20.000 |

speed is $v = 2m/s$, will the mean grabbing time of items is $t_{pick} = 15$ s while the cost of displacement per unit of time is $\varsigma = \$0,05$ with $Cap$ depending on the instance. For the penalty fees we consider $\alpha = \$0.5$ and $\beta = \$1$. We adopt also the strategy of bounding the search space presented in [11] as to compare our results with those reported in that article. This amounts to define lower and upper bounds on the number of batches: $|\mathcal{R}|_{min} \leq |\mathcal{R}| \leq |\mathcal{R}|_{max}$. These bounds are $|\mathcal{R}|_{min} = \left(\varphi_1 \cdot \sum_{p \in \mathcal{P}} w_p\right)/Cap$ y $|\mathcal{R}|_{max} = \left(\varphi_2 \cdot \sum_{p \in \mathcal{P}} w_p\right)/Cap$ where $\varphi_1$ and $\varphi_2$ are such that $\varphi_2 \geq \varphi_1$. The number of iterations is capped at 500, the size of the population is 150, the number of participants in the tournament is 2, while $\varphi_1 = 2$ and $\varphi_2 = 4$, with a crossover probability of 0.9, a mutation probability of 0.15, and a 5% of the population in the elite. We ran the experiment on a PC with an Intel Core i7 3.00 GHz processor and a RAM of 8 GB.

## 6 Results

We compare the results of running the algorithm on the three instances DS1/M1, DS2/M1 and DS3/M1 to those obtained in [11] (Table 5, $D_{Total}$: total distance in meters; $n_{Batch}$: optimal number of batches; $D_{Average}$: average distance in meters; $D_{\sigma}$: standard deviation in meters; $T_{fail}$: tardiness and earliness in seconds; $C_{Total}$: total cost in $; $T_{CPU\ average}$: running time in seconds). We ran each instance 30 times.

**Table 5.** Results

| | Results from [11] | | | Evolutionary algorithm | | |
|---|---|---|---|---|---|---|
| | DS1/M1 | DS2/M1 | DS3/M1 | DS1/M1 | DS2/M1 | DS3/M1 |
| $D_{Total}$ | 1304.00 | 3569.00 | 16945.00 | 1266.00 | 3215.00 | 14548.00 |
| $n_{Batch}$ | 8.00 | 11.00 | 27.00 | 8.00 | 12.00 | 27.00 |
| $D_{Average}$ | 163.00 | 324.46 | 627.59 | 158.25 | 267.91 | 538.81 |
| $D_{\sigma}$ | 3.70 | 25.17 | 18.69 | 2.90 | 23.41 | 21.75 |
| $T_{fail}$ | 1181.00 | 4704.00 | 15481.00 | 1158.70 | 4801.91 | 15621.31 |
| $C_{Total}$ | 1092.60 | 3104.83 | 9207.13 | 1090.78 | 2961.89 | 8898.01 |
| $T_{CPUaverage}$ | 753.60 | 2629.90 | 5785.50 | 605.70 | 2121.90 | 4474.20 |

We can see that $D_{Total}$ improves with our algorithm in all three instances (DS1/M1, DS2/M1 and DS3/M1, 2.91%, 9.92% and 14.15%, respectively). On instances DS1/M1 and DS3/M1, $n_{Batch}$ was the same as in [11], while for DS2/M1 it was larger. For instance DS1/M1, $T_{fail}$ improves 1.89%, while on DS2/M1 and DS3/M1 it worsens 2.08% and 0.91% respectively. Finally, $T_{CPU\ average}$ was lower in all instances. Figure 3 depicts the percentages of improvement or worsening with respect to the results in [11].

## 7 Conclusions

We presented an evolutionary algorithm to solve in an integrated way a combination of the Order Batching and the Order Picking problems. The algorithm operates on a novel way of representing the chromosome with two genomes, allowing incorporating directly specific knowledge of the problem, yielding a more flexible treatment of the search space while at the same time providing an explicit representation of the constraints. We ran the algorithm on simulated instances of different sizes. We found that the algorithm improved in general the results presented in [11]. Future work involves addressing the problem in a cross-dock platform and under a multi-objective perspective.

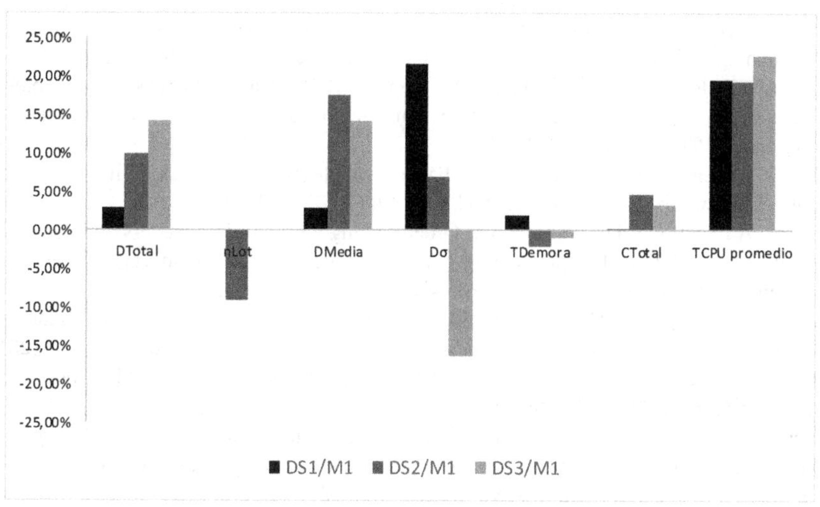


**Fig. 3.** Percentage of improvements over the results in [11].

## References


1. De Koster, R., Van der Poort, E.S., Wolters, M.: Efficient order batching methods in warehouses. Int. J. Prod. Res. **37**(7), 1479–1504 (1999)
2. De Koster, R., Le-Duc, T., Roodbergen, K.J.: Design and control of warehouse order picking: a literature review. Eur. J. Oper. Res. **182**(2), 481–501 (2007)
3. Miguel, F., Frutos, M., Tohmé, F., Rossit, D.A.: A memetic algorithm for the integral OBP/OPP problem in a logistics distribution center. Uncertain Supply Chain Manag. **7**(2019), 203–214 (2019)
4. Chen, M.C., Wu, H.P.: An association-based clustering approach to order batching considering customer demand patterns. Omega **33**(4), 333–343 (2005)
5. Henn, S., Koch, S., Wäscher, G.: Order batching in order picking warehouses: a survey of solution approaches. In: Manzini, R. (ed.) Warehousing in the Global Supply Chain, pp. 105–137. Springer, London (2012). https://doi.org/10.1007/978-1-4471-2274-6_6
6. Henn, S., Schmid, V.: Metaheuristics for order batching and sequencing in manual order picking systems. Comput. Ind. Eng. **66**(2), 338–351 (2013)
7. Henn, S., Wäscher, G.: Tabu search heuristics for the order batching problem in manual order picking systems. Eur. J. Oper. Res. **222**(3), 484–494 (2012)
8. Henn, S., Koch, S., Doerner, K., Strauss, C., Wäscher, G.: Metaheuristics for the order batching problem in manual order picking systems. Bus. Res. **3**(1), 82–105 (2010)
9. Hwang, H., Kim, D.: Order-batching heuristics based on cluster analysis in a low-level picker-to-part warehousing system. Eur. J. Oper. Res. **43**(17), 3657–3670 (2005)
10. Rana, K.: Order-picking in narrow-aisle warehouse. Int. J. Phys. Distrib. Logist. **20**(2), 9–15 (1991)
11. Tsai, C.Y., Liou, J.J., Huang, T.M.: Using a multiple-GA method to solve the batch picking problem: considering travel distance and order due time. Int. J. Prod. Res. **46**(22), 6533–6555 (2008)
12. Grosse, E.H., Glock, C.H.: The effect of worker learning on manual order picking processes. Int. J. Prod. Econ. **170**(C), 882–890 (2015)

13. Lam, C.H., Choy, K.L., Ho, G.T., Lee, C.K.: An order-picking operations system for managing the batching activities in a warehouse. Int. J. Syst. Sci. **45**(6), 1283–1295 (2014)
14. Van Gils, T., Ramaekers, K., Braekers, K., Depaire, B., Carisa, A.: Increasing order picking efficiency by integrating storage, batching, zone picking, and routing policy decisions. Int. J. Prod. Econ. **197**, 243–261 (2018)
15. Ho, Y.C., Tseng, Y.Y.: A study on order-batching methods of order-picking in a distribution centre with two cross-aisles. Int. J. Prod. Res. **44**(17), 3391–3471 (2006)
16. Scholz, A., Schubert, D., Wäscher, G.: Order picking with multiple pickers and due dates - Simultaneous solution of order batching, batch assignment and sequencing, and picker routing problems. Eur. J. Oper. Res. **263**(2), 461–478 (2017)
17. Ardjmand, E., Shakeri, H., Singh, M., Bajgiran, O.S.: Minimizing order picking makespan with multiple pickers in a wave picking warehouse. Int. J. Prod. Econ. **206**, 169–183 (2018)
18. Öztürkoğlu, Ö., Hoser, D.: A discrete cross aisle design model for order-picking warehouses. Eur. J. Oper. Res. **275**(2), 411–430 (2018)
19. Tappia, E., Roy, D., Melacini, M., De Koster, R.: Integrated storage-order picking systems: Technology, performance models, and design insights. Eur. J. Oper. **274**(3), 947–965 (2018)
20. Lu, W., McFarlane, D., Giannikas, V., Zhang, Q.: An algorithm for dynamic order-picking in warehouse operations. Eur. J. Oper. Res. **248**(1), 107–122 (2016)
21. Petersen, C.G.: An evaluation of order picking routeing policies. Int. J. Oper. Prod. Manag. **17**(11), 1098–1111 (1997)
22. Žulj, I., Glock, C.H., Grosse, E.H., Schneider, M.: Picker routing and storage-assignment strategies for precedence-constrained order picking. Comput. Ind. Eng. **123**, 338–347 (2018)
23. Theys, C., Bräysy, O., Dullaert, W., Raa, B.: Using a TSP heuristic for routing order pickers in warehouses. Eur. J. Oper. Res. **200**, 755–763 (2010)
24. Wetzel, A.: Evaluation of the Effectiveness of Genetic Algorithms in Combinatorial Optimization. University of Pittsburgh, Pittsburgh (1983)
25. Whitley, L.D., Starkweather, T., Fuquay, D.: Scheduling problems and traveling salesmen: the genetic edge recombination operator. In: Proceedings of the 3rd International Conference on Genetic Algorithms. Morgan Kaufmann Publishers Inc., San Francisco, pp. 133–140 (1989)
26. Fogel, D.B.: An evolutionary approach to the traveling salesman problem. Biol. Cybern. **60**(2), 139–144 (1988)

# Simulated Annealing Algorithm for In-Plant Milk-Run System

Islam Altin and Aydin Sipahioglu(✉)

Department of Industrial Engineering, Eskisehir Osmangazi University, Eskişehir, Turkey
{ialtin,asipahi}@ogu.edu.tr

**Abstract.** Milk-run, a cyclic material delivering system, aims to increase the efficiency of transportation and supply chain based on lean logistics perspective. There are two kinds of milk-run systems as supplier and in-plant milk-run system in the literature. In-plant milk-run system that has growing appeals with Industry 4.0 concept, is applied to manage the process of delivering materials from warehouse to assembly stations in plants. This system can be implemented using Autonomous Vehicles (AV), which provide automated materials handling. However, a challenging problem arises in determining milk-run routes and periods for each AV, simultaneously. Besides, this problem becomes even harder in the presence of assembly stations with buffer stock constraint and requiring more than one commodity (multi-commodity). Since this problem is quite difficult to handle with exact solution methods, this problem is tackled here by using Simulated Annealing algorithm. To evaluate the performance of the proposed algorithm, we carry out experiments using a range of test problems. The computational results indicate that the suggested algorithm is efficient to obtain both milk-run routes and periods for each AV in reasonable computation times.

**Keywords:** Autonomous Vehicles · In-plant milk-run system · Buffer stock area · Multi-commodity · Simulated Annealing algorithm

## 1 Introduction

Over the past decades, competition among companies around the world has become increasing. To survive in this competition environment, material handling system is one of the most important issue that should be well managed for companies. Some studies have revealed that it can account for 30–75% of the total cost and a well operated material handling system can reduce plant's operating cost by 15–30% [1]. Moreover, efficient material handling system ensures on time delivery, which is a key factor in the implementation of a lean logistics. It is a logistics application in lean manufacturing environment and provides the delivery of the products at the right time to the right place, effectively [2]. The implementation of lean logistics provides some benefits such as balancing production line, reduction of stock levels and elimination of delays in the logistics process [3].

© Springer Nature Switzerland AG 2021
D. A. Rossit et al. (Eds.): ICPR-Americas 2020, CCIS 1407, pp. 187–201, 2021.
https://doi.org/10.1007/978-3-030-76307-7_15

Lean logistics activities consist of 3 parts such as in-bound, out-bound and in-plant logistics [4]. While in-plant logistics deals with logistics in the factory, in-bound and out-bound logistics deal with obtaining raw materials and delivering goods to the customer, respectively [5]. These three types of logistics affect each other implicitly and one problem in any of them signify that there will be a problem in the others. For this reason, the logistics process should be carried out in a good way at all levels, simultaneously.

There is a correlation between a just in time plan and a good logistics strategy [6]. For in-plant logistics just in time material supply is a vital issue, because early material supply causes inventory holding cost and late material supply causes stopping assembly lines due to the parts shortage [7]. Milk-Run system, cyclic goods taking, is able to reduce these kind of delays in manufacturing especially in assembly lines [8]. Indeed, milk-run system can also be considered as a special kind of Vehicle Routing Problem with Time Windows [9]. Since it is actually a lean logistics method, it is possible to be implemented for all three types of lean logistics. There are two different milk-run systems as supplier milk-run and in-plant milk-run system in the literature [4]. With the supplier milk-run system, goods are collected from external suppliers and delivered to a customer by following a predefined route [10, 11]. On the other hand, in-plant milk-run system helps to deliver materials from warehouse to assembly lines in plants in a cyclic manner. Moreover, implementation of this system to the factory, mass production is made, is appropriate because many different materials should be delivered to the assembly stations at certain periods [12]. Unlike supplier milk-run system, it is not affected by external factors such as weather conditions. Accordingly, schedules of in-plant milk-run system are more reliable and robust [4]. The objective of in-plant milk-run system can be defined as minimizing the total inventory holding and transportation cost to ensure no parts shortage will occur in assembly stations.

In plant Milk-Run systems has some certain advantages listed below [13];

- Improved performance of the supply chain and logistics because of effective transportation,
- Lower inventory and reduced maintenance costs,
- More effective use of buffer stock area,
- More appropriate and reliable delivery times,
- Increased capital turnovers,
- Increased flexibility in supplying the parts,
- Smooth and well-established logistic operations.

Choosing the right material handling equipment is essential to improve facility utilization, increase efficiency of material flow, reduce waste in plant and provide effective utilization of manpower [5, 14]. There is growing interest using automated material handling vehicles especially with the Industry 4.0 concept. In this sense, in-plant milk-run system can be applied using Autonomous Vehicles (AV). Contrary to Automated Guided Vehicles (AGV), AV navigates autonomously in the environment without any limitations on the paths. Therefore, it is more convenient to use AV in flexible logistics system which should be used in just in time production.

In-plant milk-run system is applied to manage process of in-plant logistics. Operating this system effectively is important not only for minimizing material handling cost, but also for ensuring continuity of production process. Demands of the assembly stations have to be satisfied from warehouse using AV by taking into account paths and time scheduling without causing any parts shortage. Therefore, it is crucial to obtain well designed milk-run routes and periods for each AV. In the meantime, milk-run period affects delivery quantities of the assembly stations and this influences milk-run route construction process because vehicle capacity is fixed [15]. So, the problem of obtaining milk-run period and route should be handled together. In addition, milk-run routes and periods should be determined considering demands of the assembly stations, capacity of buffer stock area, the number of vehicles and their capacities.

This study is set out to explore the design of in-plant milk-run system. Milk-run routes and periods must be obtained simultaneously to operate system, effectively. Although they are common issues, the inherent difficulties make this problem tough. These are the presence of AV capacity, limited buffer stock area and assembly stations requiring multi-commodity goods. Accordingly, related problem can be explained as obtaining milk-run routes and periods for AVs with the minimum total transportation cost to satisfy the demand of assembly stations in a cyclic manner.

Our research aims at developing effective solution method for obtaining milk-run routes and periods simultaneously for AVs in-plant. With this aim in mind, a novel method which is based on Simulated Annealing algorithm has been designed to solve this challenging problem in reasonable computation times. The structure of the generic Simulated Annealing algorithm is enriched by various neighbor search procedures used in probabilistic manner. To evaluate the performance of the proposed method, we carry out experiments using generated test problems. The key contribution of this work is solution representation that provides all information about milk-run routes, periods and it enables exploring the solution space effectively. Another contribution is that proposed approach gets high quality solutions at reasonable computation times even for large-scale problems.

The rest of the papers is organized as follows. The relevant literature is reviewed in Sect. 2. Section 3 describes the related problem and gives detailed information about the proposed method. Test problems and computational results are presented in Sect. 4. Finally, Sect. 5 concludes the paper and provides future researches.

## 2 Literature Review

There is an increased attention to in-plant milk-run system area. So, a significant variety of designs of this system is found in the literature. Here we will discuss the literature which is relevant to our approach.

The problem that has arisen in obtaining milk-run routes and periods in-plant has been handled by researchers in different ways. Some researchers suggested exact solution methods, others developed heuristic methods. Besides, good overviews of earlier works in milk-run logistics and in-plant milk-run system is presented by [11, 12].

First of all, studies proposing the exact solution method has been examined. Akilioglu et al. [16] proposed a mixed integer mathematical model to obtain route and cycle time for in-plant milk-run system. The objectives of this model is to minimize total inventory and distribution costs. The results of the mathematical model are evaluated by using a simulation model. Kilic et al. [5] classified the milk-run distribution system and developed a mathematical model for each case. The objective function of proposed mathematical models is to minimize the number of vehicles and the total travelled distance in-plant milk-run system. The most important finding in this study is that using multiple routed milk-run trains are more advantageous than one routed milk-run trains. Similar work has also been pursued by Kilic and Durmusoglu [17] in which they focus on periodic material delivery in lean production environment. They proposed a mixed integer linear programming model to minimize transportation costs. Volling et al. [18] developed a combined inventory-transportation system with multi-product, multi-vehicle and stochastic demand. They proposed an exact decomposition approach to handle the related problem and reported that substantial gains are achieved by this approach. Emde and Gendreau [19] explored scheduling tow-train with predefined routes to feed parts to automotive assembly lines. They developed a mixed-integer mathematical model to minimize in-process inventory cost. A more efficient implementation for in-plant milk-run system has been proposed relatively recently by Satoglu and Sipahioglu [7]. Because, they proposed two different assignment based mathematical model for just in time material supply system of the assembly lines. One of the models doesn't guarantee obtaining the global optimum solution, but this model performs better performance with regard to computation time. Mao et al. [20] introduced "Milk-run routing problem with progress lane (MRPPL)" to the literature. This problem is about to collecting automobile parts by integrating the progress-lane into the corresponding vehicle routing problem. They proposed a mixed integer linear programming model for small scale problems. Buyukozkan et al. [21] focused on in-plant milk-run system application in white goods industry. To maintain the assembly lines' operations, they proposed a mathematical model for single-vehicle case milk-run system with predetermined periods. Buyukozkan and Satoglu [22] also proposed a mathematical model for multi-vehicle case milk-run system with predetermined cycle times. Sipahioglu and Altin [23] dealt with in-plant milk-run system with predetermined periods. They developed a new mixed integer mathematical model to obtain milk-run routes and periods for AGVs. This model allows split deliveries for stations and increase vehicle capacity by adding trailers. Besides, the proposed model has less number of constraints comparing the others and quite short computational time to obtain the optimal solution.

Studies proposing the heuristic methods has been reported as follows. Golz et al. [24] explored just in time part supply system with predefined routes to minimize the required number of shuttle drivers. They developed a heuristic solution procedure consisting of two stage. Gyulai et al. [9] proposed an initial solution generation heuristics and a local search method to solve the vehicle routing problem to obtain routes for in-plant milk-run planning problem. Kilic and Durmusoglu [17] tackled periodic material delivery in lean production environment with a predefined route in equal cycle times. They proposed a heuristic method since the optimal solution could not be obtained when scale of the problem gets bigger. Fathi et al. [25] discussed the assembly line part feeding problem with predefined tours. They suggested a modified particle swarm optimization algorithm to solve two sub-problems which are tour scheduling and tow-train loading problems. Emde and Gendreau [19] also offered a decomposition heuristic to solve instances of realistic size in acceptable computation time for scheduling tow-train with predefined routes. Mao et al. [20] also proposed genetic algorithm to obtain trip routes to collect automobile parts within a reasonable computation time for large size instances. Buyukozkan et al. [21] focused on in-plant milk-run system with predetermined period, multi product and limited buffer stock area. They proposed matheuristic algorithm that solve the proposed single vehicle milk-run model iteratively for multi vehicle case. Buyukozkan and Satoglu [22] proposed artificial bee colony algorithm to solve large scale in-plant milk-run instances with predetermined cycle time, multi product, multi vehicle and limited buffer stock area.

It is clear that, previous works are aimed at obtaining milk-run routes and periods for vehicles to minimize total cost of in-plant logistics. It is understood from these studies that the related problem can be extended in terms of the problem size and computation times. On the other hand, a solution approach that enables obtaining milk-run routes and periods simultaneously for large scale problems will contribute to the literature. In this sense, we proposed Simulated Annealing algorithm for in-plant milk-run system with multi commodity, multi vehicle and limited buffer stock area.

## 3 Problem Definition and Proposed Method

Companies take care not to stop assembly stations due to the parts shortage and make an effort to ensure continuity of production in plant. Therefore, it is vital to satisfy demands of the assembly stations on time. Since different operations are carried out at each assembly stations, the material types and quantities required by the stations may differ from each other. Accordingly, assembly stations must be fed periodically by AVs to minimize total transportation costs. In this feeding process, routes and periods for AVs should be obtained in accordance with the following conditions and assumptions:

- Each route starts and ends at the depot.
- Each AV is operated at most only one route.
- Each station is visited only once.
- AVs are homogenous.
- Each assembly stations have only delivery demands.
- Demands of each station are known but varies with regard to milk-run period.
- Distances between assembly stations are known and used as a distance matrix.
- Unloading time at each station is limited to 0.5 min.
- Speed of each vehicle is 10 m per minute.
- There are limited and known buffer stock area for each station.
- Stations require different types of goods.

Taking into account all of these, it is necessary to determine milk-run routes and periods for each AGV in order to operate in-plant milk-run system. However, a challenging problem arises in determining milk-run routes and periods simultaneously. This problem is difficult to handle by exact solution method in reasonable computation times. Therefore, we have developed Simulated Annealing (SA) algorithm.

SA algorithm was proposed by Kirkpatrick et al. [26] and Černý [27] as a single solution based metaheuristic approach. SA is a stochastic neighborhood search method that imitates the physical annealing process where a solid is slowly cooled until the minimum energy state is achieved. Therefore, not only improving solutions are always accepted, but also non-improving (inferior) solutions can be accepted by acceptance probability of the Metropolis criterion. This acceptance probability function is the main mechanism of SA algorithm and prevents getting stuck into local optima [28]. Thus, this algorithm provides obtaining high quality solution in a reasonable computation time. Because of its simplicity and efficiency, SA algorithm has widely been applied to solve most of the combinatorial optimization problems.

In order to apply metaheuristic algorithm effectively, an appropriate solution representation structure should be used. It should be easy to generate new and feasible solutions from a current solution. Proposed solution representation is shown in Fig. 1. It consists of positive and zero numbers. Positive numbers represent delivery customers and zero numbers indicate vehicles. So, this solution representation length is related to the number of customers and vehicles.

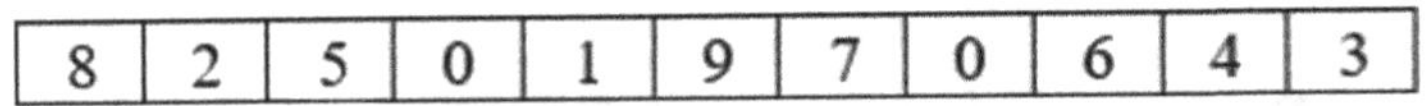

| 8 | 2 | 5 | 0 | 1 | 9 | 7 | 0 | 6 | 4 | 3 |
|---|---|---|---|---|---|---|---|---|---|---|

**Fig. 1.** An example of solution representation.

The main advantage of suggested method is the suggested solution representation. This representation provides all information about milk-run routes and periods that are vital to operate in-plant milk-run system, efficiently. Referring to Fig. 1, it is clear that there are 9 customers and 3 vehicles. Milk-run routes for each vehicle are as follows; first route is 0-8-2-5-0, second route is 0-1-9-7-0 and third route is 0-6-4-3-0. On the other hand, milk-run periods are calculated taking into account trip time of route and unloading time at each station.

SA algorithm was modified to achieve high quality solutions by using different neighbor search operators. Four different search operators have been used such as swap, insert, 3-change and reverse. These operators have been used probabilistically with regard to the scale of the problem. For small scale problems (number of customer $\leq 45$), only swap, insert and 3-change operators have been used with equal probability. On the other hand, for medium (number of customer $\epsilon$ [55, 95]) and large scale problems (number of customer $\geq 150$), all operators have been used with equal probability. Moreover, the number of iteration has been used as a stopping criteria and geometric cooling schedule has been used in this algorithm.

These neighbor search operators are robust and ensure obtaining milk-run routes and periods in any case. For example, route and period information is obtained from the customer numbers that are among zero values in the neighbor solution obtained by swap operator. On the other hand, if there is no customer number among the zero values, then it is understood that one of the vehicles is not used. So, the presented method is capable of generating feasible neighbor solution at each iteration when capacity of AVs and buffer stock areas is ignored.

Capacity of AVs and buffer stock areas is allowed to be exceeded in order to take advantage of the information about different solutions. A penalty function has been proposed to prevent the capacity to be exceeded in the final obtained solution and it also makes exploring the solution space, extensively. Pseudo code of the developed algorithm is given below.

Pseudo Code of Proposed Simulated Annealing Algorithm

**Definitions:**

s: solution
$s_0$: initial solution
$s'$: neighbor solution
$f(s_0)$: objective function value of the initial solution
$f(s')$: objective function value of the neighbor solution
$f(s_{best})$: objective function value of the best obtained solution
$s_{best}$: the best obtained solution
α: cooling rate
$T_0$: initial temperature
T: current temperature
P: penalty coefficient
e: Excess capacity value

**Preliminary Step:**

Generation of the initial solution (s0), calculate objective function (f(s0)
$s_{best} := s_0;\ f(s_{best}) := f(s_0);\ s := s_0;\ f(s) := f(s_0)$
Determine the initial temperature and cooling rate
$T := T_0;\ \propto = 0{,}99$
Determine neighbor search operators to be used based on the problem size.
Determine the number of iteration.
Determine stopping criteria for the algorithm.

```
Repeat
     Repeat
          Generate s′ neighbor solution randomly with one
of the swap, insert, 3-change and reverse operators
          Calculate f(s′) with penalty function
          f(s′):= f(s′) + (p * e);
          Δ= f(s′) − f(s);
          If Δ≤ 0;
               Accept neighbor solution (s:= s′);
          Else;
               Generate random number u = (0,1);
               If u < e^(−Δ/T),accept neighbor solution (s:= s′);
          If f(s′) < f(s_best);
               The best obtained solution (s_best:= s′)
     Until (the number of iteration)
     Update temperature T:= T *∝
Until (Stopping Criteria Satisfied)
Return the best obtained solution and objective function
value s_best, f(s_best)
```

The success of the algorithm is discussed in following section on generated test problems.

## 4 Computational Results

Here we evaluate the performance of our approach by using test instances. Firstly, environment features and parameter values of the problem are explained in detail over a small size test instance. Afterwards, this problem was solved for different scenarios and the obtained results are presented. Finally, parameter values used in generating test problems, the proposed algorithm parameters and the obtained results are mentioned.

A visual representation of generated test instance having 15 assembly stations and 1 warehouse is shown in Fig. 2. Different types of materials are available in the warehouse and homogenous Autonomous Vehicles operated to transport these materials to assembly stations are ready for use. Besides, vehicle roads in the facility are shown with dashed lines. It is important to note that material transfer is carried out without exceeding vehicle capacities and buffer stock capacities of assembly stations.

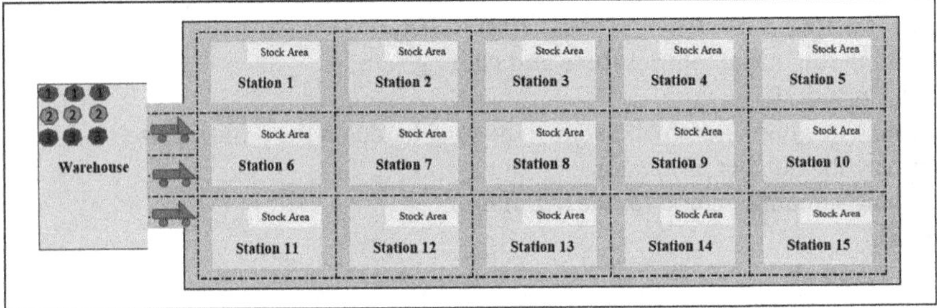


**Fig. 2.** Representation of facility layout.

There are 10 homogenous AVs available in this instance. Capacity of AVs is expressed on volume based. Therefore, the volumes of the materials to be transported should not be larger than the volume of the vehicle capacity. In addition, there is a fixed usage cost of AVs and unit material handling cost per distance covered. Capacity, fixed usage cost and unit material handling cost of these AVs are denoted in Table 1.

**Table 1.** Autonomous vehicle properties.

| Number of AVs | Capacity of AVs $\left(cm^3\right)$ | Usage cost of AVs ($) | Unit material handling cost $(\$/meter)$ |
|---|---|---|---|
| 10 | 1000 | 100 | 2 |

Assembly stations demand three different types of materials in this instance. Inventory holding cost that will occur if these materials are kept in buffer stock area and volume values of these materials are presented in Table 2. Additionally, the total inventory holding cost for each assembly station is calculated by multiplying half the amount of materials in the buffer stock area by the unit inventory holding cost.

**Table 2.** Material properties.

| Material Type | Unit inventory holding cost $\left(\$/cm^3\right)$ | Volume of a box of material $\left(cm^3\right)$ |
|---|---|---|
| Material 1 | 2 | 5 |
| Material 2 | 5 | 10 |
| Material 3 | 5 | 2 |

In this system, the demands of assembly stations must be satisfied on time and these stations must continue to operate by efficient transportation method. However, each assembly stations have a certain number of material demands in different periods based on their material consumption rates. Expressing station requests in different periods with this manner will cause complication. For this reason, 10 min' material demands of the assembly stations are calculated based on the original requests and standardization is provided in the feeding process with this way. Demands of assembly station for 10 min and capacity of buffer stock area for each station are shown in Table 3. It can be understood that the 3rd station requests 4 boxes of material-1, 7 boxes of material-2 and 3 boxes of material-3 every 10 min. These amounts are known but varies with regard to the obtained milk-run period. To illustrate, if the AV serving 3rd station has a period of 15 min, so 3rd station will demand 6 boxes of material-1, 10.5 boxes of material-2 and 4.5 boxes of material-3 every 15 min.

The proposed algorithm was run for the test problem (scenario-1) whose characteristic is given above. In addition, 2 different scenarios have been created since it is worth testing how well the proposed algorithm performs in different conditions. These scenarios were generated by using different capacity values for AVs and buffer stock areas. Capacity of AV and capacity of buffer stock area were determined as 500 ($cm^3$), 375 ($cm^3$) in scenario-2 and as 250 ($cm^3$), 150 ($cm^3$) in scenario-3. So, it will be observed how the proposed algorithm will handle tight capacity constraints. The proposed SA algorithm has run 5 times for this problem and the best obtained results have denoted in Table 4. The parameters of the algorithm (initial temperature, final temperature, number of iteration at each temperature and cooling rate) were specified as $(T_0, T_f, I, \propto) = (75, 0.1, 100, 0.99)$.

Table 4 provides the computational results of the related test problem in terms of 3 scenarios. In order to minimize total transportation costs in each scenario, milk-run routes and periods for AVs were obtained simultaneously. Column 2 and column 3 gives information about the number of assembly stations and the number of AVs operated in each scenario. The milk-run routes obtained for AVs are demonstrated in column 4. In this column, while the warehouse is denoted by 0, the assembly stations are shown by their numbers. The $5^{th}$ and $6^{th}$ columns show the travel times and the total unloading times for the relevant route, respectively. In the 7th column, milk-run periods obtained for AVs are specified. Each milk-run period is achieved by summing travel time and unloading time of the relevant route. The obtained total cost values for each scenario are shown in column 8 and the last column contains computation times which are given in seconds.

**Table 3.** Assembly station properties.

| Assembly Station | Capacity of buffer stock area $(cm^3)$ | Demand (box/10 min.) | | |
|---|---|---|---|---|
| | | Material 1 | Material 2 | Material 3 |
| 1 | 750 | 3 | 3 | 2 |
| 2 | 750 | 9 | 6 | 2 |
| 3 | 750 | 4 | 7 | 3 |
| 4 | 750 | 4 | 1 | 2 |
| 5 | 750 | 3 | 7 | 3 |
| 6 | 750 | 7 | 8 | 0 |
| 7 | 750 | 7 | 0 | 2 |
| 8 | 750 | 4 | 0 | 0 |
| 9 | 750 | 2 | 1 | 1 |
| 10 | 750 | 5 | 5 | 1 |
| 11 | 750 | 3 | 2 | 6 |
| 12 | 750 | 7 | 3 | 7 |
| 13 | 750 | 8 | 3 | 7 |
| 14 | 750 | 4 | 4 | 0 |
| 15 | 750 | 7 | 0 | 1 |

From Table 4, it can be understood that there are 3 AVs to satisfy demands of 15 assembly stations in cyclic manner for scenario-1. First AV supplies material 10, 12, 3, 13, 15, 4, 9th stations every 21.4 min, second AV supplies material 2, 8, 1st stations every 16.3 min and third AV supplies material 6, 11, 7, 5, 14th stations every 21.5 min. Thus, considering all cost items, the total cost was obtained as 1940.5 for scenario-1.

There are several points to note concerning the computational results. While 3 AVs were used in scenario-1, 4 AVs in scenario-2 and 7 AVs in scenario-3 were used. Besides, when these three scenarios are compared, the minimum total cost value was obtained in scenario-1, on the other hand the maximum total cost value was obtained in scenario-3. These implications represent the impact of tight capacity constraints. That is to say, as the capacity of AVs and buffer stock areas is decreased, the number of AV needed and the total cost values increase. Moreover, it is observed that the duration of milk-run periods is shortened with tight capacity constraints. In addition, it is seen that there is no significant difference between 3 different scenarios with regard to computation times.

Different scale of test problems was generated to demonstrate the effectiveness of the developed approach. 15 different test problems were generated and classified as small, medium and large scale. All the parameter values of these test problems were expressed in Table 5. Besides, the parameter values used in the SA algorithm are shown in Table 6.

**Table 4.** Results of the test problem.

| Scenarios | # of stations | # of AVs | Routes | Travel times (min.) | Unloading times (min.) | Periods (min.) | Total cost ($) | Computation times (second) |
|---|---|---|---|---|---|---|---|---|
| 1 | 15 | 3 | Route-1 (0-10-12-3-13-15-4-9-0) | 17.9 | 3.5 | 21.4 | 1940.5 | 11.92 |
| | | | Route-2 (0-2-8-1-0) | 14.8 | 1.5 | 16.3 | | |
| | | | Route-3 (0-6-11-7-5-14-0) | 19 | 2.5 | 21.5 | | |
| 2 | 15 | 4 | Route-1 (0-1-8-2-9-0) | 11.8 | 2 | 13.8 | 2098.6 | 12.02 |
| | | | Route-2 (0-5-13-4-15-0) | 18.1 | 2 | 20.1 | | |
| | | | Route-3 (0-10-6-11-7-0) | 13.3 | 2 | 15.3 | | |
| | | | Route-4 (0-3-12-14-0) | 16.6 | 1.5 | 18.1 | | |
| 3 | 15 | 7 | Route-1 (0-13-15-0) | 13.8 | 1 | 14.8 | 2943 | 12.04 |
| | | | Route-2 (0-3-10-0) | 11 | 1 | 12 | | |
| | | | Route-3 (0-8-2-9-0) | 11.8 | 1.5 | 13.3 | | |
| | | | Route-4 (0-4-12-0) | 17.4 | 1 | 18.4 | | |
| | | | Route-5 (0-1-11-7-0) | 16.8 | 1.5 | 18.3 | | |
| | | | Route-6 (0-5-14-0) | 11.9 | 1 | 12.9 | | |
| | | | Route-7 (0-6-0) | 8.2 | 0.5 | 8.7 | | |

The proposed algorithm was coded in Python 3.7 and run on an Intel Core i5 1.8 GHz PC with 8 GB memory. This algorithm has run 5 times for each test problem and the best obtained computational results have been presented in Table 7.

Numerical results of the different scale instances are summarized in Table 7 in terms of the number of assembly stations, the number of AVs operated, total cost and computation times. The first two columns of Table 7 contain information about test problems, such as instance name and the number of assembly stations. Third column shows the number of AVs used in test problems. Next column indicates the obtained total cost value and the last column includes computation times given in seconds. In these test problems, different milk-run routes and periods for AVs were obtained simultaneously in order to minimize total cost for in-plant milk-run system. So, the demands of the assembly stations have been satisfied by AVs in cyclic manner without violating capacity constraint for AVs and buffer stock areas. There is a significant positive correlation between problem size and the number of AVs and computation times needed. In other words, as the size of the problem increases, the number of AVs used and the computation times increases. However, even the L-15 problem which includes 500 assembly stations, 50 AVs and 20 different types of materials was solved in about 1 h by our proposed algorithm. Therefore, it is seen that the proposed algorithm is able to produce good solutions for even large scale problems at reasonable computation times.

**Table 5.** Test problem parameter values.

| Parameters | Small scale problems | Medium scale problems | Large scale problems |
|---|---|---|---|
| # of AVs | 10 | 15 | 50 |
| # of Assembly stations | [10, 45] | [55, 95] | [150, 500] |
| Capacity of AVs $\left(cm^3\right)$ | 1000 | 5000 | 40000 |
| Capacity of buffer stock area $\left(cm^3\right)$ | 750 | 1500 | 10000 |
| Usage cost of AVs ($) | 100 | 300 | 500 |
| Unit material handling cost $(\$/meter)$ | 2 | 3 | 5 |
| Distance matrix | uniform (5, 100) | uniform (5, 100) | uniform (5, 100) |
| # of Material types | uniform (2, 5) | uniform (5, 10) | uniform (10, 20) |
| Material demands of stations (box/10 min.) | uniform (0, 10) | uniform (0, 10) | uniform (0, 10) |
| Volume of a box of material $\left(cm^3\right)$ | uniform (2, 20) | uniform (2, 20) | uniform (2, 20) |
| Inventory holding cost $\left(\$/cm^3\right)$ | uniform (1, 5) | uniform (1, 5) | uniform (1, 5) |

**Table 6.** SA parameter values.

| | Small scale problems | Medium scale problems | Large scale problems |
|---|---|---|---|
| Initial temperature $(T_0)$ | 75 | 100 | 200 |
| Final temperature $(T_f)$ | 0.1 | 0.1 | 0.1 |
| Number of iteration $(I)$ | 100 | 150 | 200 |
| Cooling rate $(\propto)$ | 0.99 | 0.99 | 0.99 |

**Table 7.** Numerical results of the proposed algorithm on the test problems.

| Test problem | # of stations | # of AVs | Total cost ($) | Computation times (second) |
|---|---|---|---|---|
| S-1 | 10 | 3 | 1831 | 10 |
| S-2 | 15 | 3 | 1940.53 | 12.5 |
| S-3 | 25 | 5 | 3091.60 | 12.6 |
| S-4 | 35 | 10 | 5463.46 | 25.7 |
| S-5 | 45 | 9 | 5583.93 | 23.5 |
| M-6 | 55 | 10 | 12509.91 | 84.2 |
| M-7 | 65 | 12 | 15674.95 | 114.8 |
| M-8 | 75 | 12 | 15729.10 | 107.4 |
| M-9 | 85 | 15 | 21589.68 | 160.2 |
| M-10 | 95 | 15 | 26836 | 181.6 |
| L-11 | 150 | 28 | 64697.27 | 1043 |
| L-12 | 200 | 35 | 89325.96 | 1445.9 |
| L-13 | 300 | 50 | 120402.67 | 1583.7 |
| L-14 | 400 | 50 | 196526.37 | 2601.9 |
| L-15 | 500 | 50 | 287402.60 | 3603.2 |

## 5 Conclusions

This study is set out to explore in-plant milk-run system with regard to material feeding process. The main problem encountered here is obtaining milk-run routes and periods simultaneously for AVs with the minimum total cost to satisfy the demand of assembly stations in a cyclic manner. Besides, this problem is quite difficult taking into account capacity of AV, limited buffer stock area and assembly stations demanding different types of materials. Simulated Annealing algorithm has been proposed to solve this challenging problem. Different neighbor search operators and penalty function have been used the structure of the SA algorithm. In addition, a well-designed solution representation, provides all information about milk-run routes and periods as well as it enables exploring the solution space effectively, has been proposed for SA algorithm. Computational experiment has been carried out for a range of the generated test problems. Computational results show that the proposed approach gets different milk-run routes and periods for AVs simultaneously at reasonable computation times even for large-scale problems. The proposed algorithm can handle even the L-15 test problem including 500 assembly stations, 50 AVs and 20 different types of materials in about 1 h. Thus, it is seen that the proposed algorithm is a successful approach for in-plant milk-run system with multi commodity, multi vehicle and limited buffer stock area.

The key contributions of this work are the solution representation and obtaining high quality results at reasonable computation times. To the best of our knowledge, this is the first study to propose a metaheuristic algorithm for in-plant milk-run system with multi commodity, multi vehicle, limited buffer stock area and not using predetermined route and period.

In future work, it may be useful to study on a metaheuristic approach that allows split deliveries for assembly stations in-plant milk-run system and different metaheuristics can be used.

## References


1. Sule, D.R.: Manufacturing Facilities: Location, Planning and Design, 2nd edn. PWS Publishing Company, Boston (1994)
2. Patel, M.B.: Optimization approach of vehicle routing by a milk-run material supply system. Int. J. Sci. Res. Dev. **1**(6), 1357–1360 (2013)
3. Wronka, A.: Lean logistics. J. Posit. Manag. **7**(2), 55–63 (2016)
4. Baudin, M.: Lean Logistics: The Nuts and Bolts of Delivering Materials and Goods. Productivity Press, New York (2005)
5. Kilic, H.S., Durmusoglu, M.B., Baskak, M.: Classification and modeling for in-plant milk-run distribution systems. Int. J. Adv. Manuf. Technol. **62**(9–12), 1135–1146 (2012)
6. Ji-li, K., Guo-zhu, J., Cui-ying, G., A new mathematical model of vehicle routing problem based on milk-run. In: International Conference on Management Science and Engineering 20th Annual Conference, Harbin, China (2013)
7. Satoglu, S.I., Sipahioglu, A.: An assignment based modelling approach for the inventory routing problem of material supply systems of the assembly lines. Sigma: J. Eng. Nat. Sci./Mühendislik ve Fen Bilimleri Dergisi **36**(1), 161–177 (2018)
8. You, Z., Jiao, Y.: Development and application of milk-run distribution systems in the express industry based on saving algorithm. Math. Probl. Eng. 1–6 (2014)
9. Gyulai, D., Pfeiffer, A., Sobottka, T., Váncza, J.: Milkrun vehicle routing approach for shop-floor logistics. Procedia CIRP **7**, 127–132 (2013)
10. Brar, G.S., Saini, G.: Milk run logistics: literature review and directions. In: World Congress on Engineering, London, United Kingdom (2011)
11. Eroglu, D.Y., Rafele, C., Cagliano, A.C., Sevilay, M.S., Ippolito, M.: Simultaneous routing and loading method for milk-run using hybrid genetic search algorithm. In: XII International Logistics and Supply Chain Congress, Istanbul, Turkey (2014)
12. Alnahhal, M., Ridwan, A., Noche, B.: In-plant milk run decision problems. In: International Conference on Logistics Operations Management, pp. 85–92 (2014)
13. Sadjadi, S.J., Jafari, M., Amini, T.: A new mathematical modeling and a genetic algorithm search for milk run problem (an auto industry supply chain case study). Int. J. Adv. Manuf. Technol. **44**(1–2), 194–200 (2009)
14. Chan, F.T.S., Ip, R.W.L., Lau, H.: Integration of expert system with analytic hierarchy process for the design of material handling equipment selection system. J. Mater. Process. Technol. **116**(2–3), 137–145 (2001)
15. Satoglu, S.I., Sahin, I.E.: Design of a just-in-time periodic material supply system for the assembly lines and an application in electronics industry. Int. J. Adv. Manuf. Technol. **65**(1–4), 319–332 (2013)
16. Akillioglu, H., et al.: Dizel Enjektör Üretimi Yapan Bir Şirket İçin Fabrika İçi Çekme Esaslı Tekrarlı Dağıtım Sistemi Tasarımı. Endüstri Mühendisliği Dergisi **17**(3), 2–15 (2006)
17. Kilic, H.S., Durmusoglu, M.B.: A mathematical model and a heuristic approach for periodic material delivery in lean production environment. Int. J. Adv. Manuf. Technol. **69**(5–8), 977–992 (2013)

18. Volling, T., Grunewald, M., Spengler, T.S.: An integrated inventory-transportation system with periodic pick-ups and leveled replenishment. Bus. Res. **6**(2), 173–194 (2013)
19. Emde, S., Gendreau, M.: Scheduling in-house transport vehicles to feed parts to automotive assembly lines. Eur. J. Oper. Res. **260**(1), 255–267 (2017)
20. Mao, Z., Huang, D., Fang, K., Wang, C., Lu, D.: Milk-run routing problem with progress-lane in the collection of automobile parts. Ann. Oper. Res. **282**(1–2), 1–28 (2019)
21. Buyukozkan, K., Bal, A., Oksuz, M.K., Kapukaya, E.N., Satoglu, S.I.: A mathematical model and a matheuristic for in-plant milk-run systems design and application in white goods industry. In: Calisir, F., Cevikcan, E., Camgoz Akdag, H. (eds.) Industrial Engineering in the Big Data Era. LNMIE, pp. 99–112. Springer, Cham (2019). https://doi.org/10.1007/978-3-030-03317-0_9
22. Buyukozkan, K., Satoglu, S.I.: A mathematical model and an artificial bee colony algorithm for in-plant milk-run design. In: Calisir, F., Korhan, O. (eds.) GJCIE 2019. LNMIE, pp. 106–118. Springer, Cham (2020). https://doi.org/10.1007/978-3-030-42416-9_11
23. Sipahioglu, A., Altin, I.: A mathematical model for in-plant milk-run routing. Pamukkale Üniversitesi Mühendislik Bilimleri Dergisi **25**(9), 1050–1055 (2019)
24. Golz, J., Gujjula, R., Günther, H.O., Rinderer, S., Ziegler, M.: Part feeding at high-variant mixed-model assembly lines. Flex. Serv. Manuf. J. **24**(2), 119–141 (2012)
25. Fathi, M., Rodríguez, V., Fontes, D.B., Alvarez, M.J.: A modified particle swarm optimisation algorithm to solve the part feeding problem at assembly lines. Int. J. Prod. Res. **54**(3), 878–893 (2016)
26. Kirkpatrick, S., Gelatt, C.D., Vecchi, M.P.: Optimization by simulated annealing. Science **220**(1983), 671–680 (1983)
27. Cerny, V.: Thermodynamical approach to the travelling salesman problem: an efficient simulation algorithm. J. Optim. Theory Appl. **45**, 41–51 (1985)
28. Talbi, E.G.: Metaheuristics: From Design to Implementation, vol. 74. John Wiley & Sons, Hoboken (2009)

# A Column Generation Based Algorithm for Solving the Log Transportation Problem

Maximiliano R. Bordón(✉), Jorge M. Montagna, and Gabriela Corsano

Instuto de Desarrollo y Diseño (INGAR), Avellaneda 3657, Santa Fe, Argentina
{mbordon,mmontagna,gcorsano}@santafe-conicet.gov.ar

**Abstract.** In this work, a solution approach based on column generation for the log transportation problem in the Argentine forest industry is presented. This problem involves making decisions that covers aspects related to the size of the truck fleet, different types of delivered raw materials, different regional bases where trucks begin and end the routes, legal and contractual restrictions, among others, generating a problem very difficult to solve. This work is focused on the determination of the routes to be performed by the truck fleet to supply the plants with the available raw materials in the forest sites, at minimum cost. A representative case study of the Argentine forestry sector is presented and the corresponding results are detailed. The proposed approach allows solving problems with more than 500 full truck loads to transport per day (approximately 13,500 tons of raw materials) in less than 15 min of execution and with a negligible optimality gap.

**Keywords:** Transportation planning · Forest industry · Column generation

## 1 Introduction

According to the Ministry of Agroindustry [1], the industrial forest region of northeast Argentina (involving the provinces of Entre Ríos, Corrientes and Misiones) encompasses 766 production plants, which represents 37% of the country's forestry companies. Of this total, 42% corresponds to micro-companies (annual production level less than 940 m$^3$), 39% corresponds to small companies (between 940 and 4,720 m$^3$), 17% corresponds to medium companies (between 4,720 and 23,585 m$^3$), and the remaining 2% corresponds to large companies (annual production level greater than 23,585 m$^3$). Regarding the level of employment, this region generates direct employment for 12,332 people, 50% of whom are in the province of Misiones. The province of Misiones is the main consumer and producer of logs: 61% of the raw material (logs) was obtained in this province (more than 2 million tons per year) in the last census year [2]. The log extraction level in Misiones was 5,914,877 m$^3$ (4,731,902 tons) during the last census year, which is equivalent to 172,069 full truckloads to transport per year (550 full truckloads per day, approximately). Since log transportation is a very expensive activity, significant savings can be obtained through efficient transportation planning.

In this work, the distribution of logs from the harvest areas to the plants is addressed. This problem involves complex decisions, where the truck fleet size and the raw materials

© Springer Nature Switzerland AG 2021
D. A. Rossit et al. (Eds.): ICPR-Americas 2020, CCIS 1407, pp. 202–216, 2021.
https://doi.org/10.1007/978-3-030-76307-7_16

supply are the typical ones. There are several aspects to be considered, such as legal and contractual restrictions (time spent on the road, number of allowed daily trips, among others), that make the problem complex and difficult to be solved. The log transportation problem tackled in this work involves decisions related to: supply allocation (assignment of raw material from harvest areas to plants), truck routing (assignment of trips to trucks and the order in which they are made), and the size of the truck fleet (the number of trucks that are used to satisfy plants requirements). Many times scheduling decisions (determination of specific times for each truck activity) are also considered, but this is beyond the scope of this work. Taking into account the involved decisions, integrated formulations are required, which enlarges the model size and hinder the resolution. In this way, developing a tool capable of solving the problem efficiently and in a reasonable computing time is relevant for this industry.

However, there are few works presented in the literature that jointly address the allocation and routing problems. In fact, regardless of the methodology adopted, they resort to decomposition approaches in order to simplify problem resolution. Trips are assumed to be known in advance or are generated prior to the trucks routes determination; that is, first, the supply of each harvest area to each plant is determined (allocation problem) and, subsequently, the routes of each vehicle are generated (routing problem) and scheduled (scheduling problem) taking into account the previously selected flows. Many times, allocation problem is solved from a tactical perspective, assuming that the solution will be effectively implemented through appropriate routing and scheduling formulations [3–5]. However, from a strictly mathematical point of view, this decomposition approach does not guarantee to reach the optimal solution of the problem, although, from a productive perspective it allows to achieve appropriate solutions. In this sense, metaheuristics approaches have been also used to address some of the aforementioned problems [6, 7].

In a previous work, Bordón et al. [8] propose a mixed integer linear programming (MILP) model to solve a simplified version of the aforesaid problem, assuming that each plant demands a single type of raw material. This assumption is a strong simplification that significantly limits the application of the proposed approach, taking into account that different raw materials are employed and transported in this industry. Unlike such work, in this paper is assumed that different raw materials are required by each plant taking into account that diverse specific production requirements must be daily met. In this way, the supply of specific raw materials is another decision variable to be considered. These assumptions significantly increase the model complexity and the number of model variables.

In addition, for real-practice problems of this industry, the model size is large, with numerous trips and involved vehicles, and, then, computational performance worsens. Therefore, an adequate solution approach must be developed in order to achieve reasonable performance for the real problems of this industry. So, a Column Generation (CG) based approach is proposed in order to obtain an integer feasible (near optimal) solution in a short computing time. The works of Palmgrem et al. [9] and Rix et al. [10] are outstanding antecedents in the use of this methodology to solve transportation problems in the forest industry, although they solve different problems to the one posed in this article (i.e., allocation is not taken into account but scheduling is considered).

Summarizing, the focus of this article is on the operational planning, specifically, on the simultaneous optimization of allocation and routing problems in order to deliver logs and forest raw materials from harvest areas to production plants, at minimum cost. The capabilities of the proposed methodologies are shown through an example from the Argentine forest industry.

In the following section, the addressed problem is stated. In Sects. 3 and 4 the proposed mathematical model and the CG-based approach are presented, respectively. In Sect. 5 a case study is analyzed and, finally, in the last section, conclusions are exposed.

## 2 Problem Statement

The problem addressed in this work considers a set $F$ of harvest areas. At each harvest area $f$, there is available a number of full-truckloads of raw material $m$, $supply_{f,m}$, $m \in M$, used to supply a set of plants $I$, which demand a certain number of full-truckloads, $demand_{i,m}$, of raw material $m$.

To transport the raw materials (logs) from harvest areas to plants, a set of trucks $C$ is available. Each truck $c$, $c \in C$, begins and finishes its route in a regional base $p$, $p \in P$. In addition, each truck $c$ can perform a limited number of trips, and transport only one type of raw material $m$ at a time. Also, each truck $c$ has a maximum number of working hours, *maxt*.

The route representation proposed by Bordón et al. [8] is used in this work. Accordingly, each route is composed of a series of trips $v$, $v \in V$, where a trip $v$ is a sequence

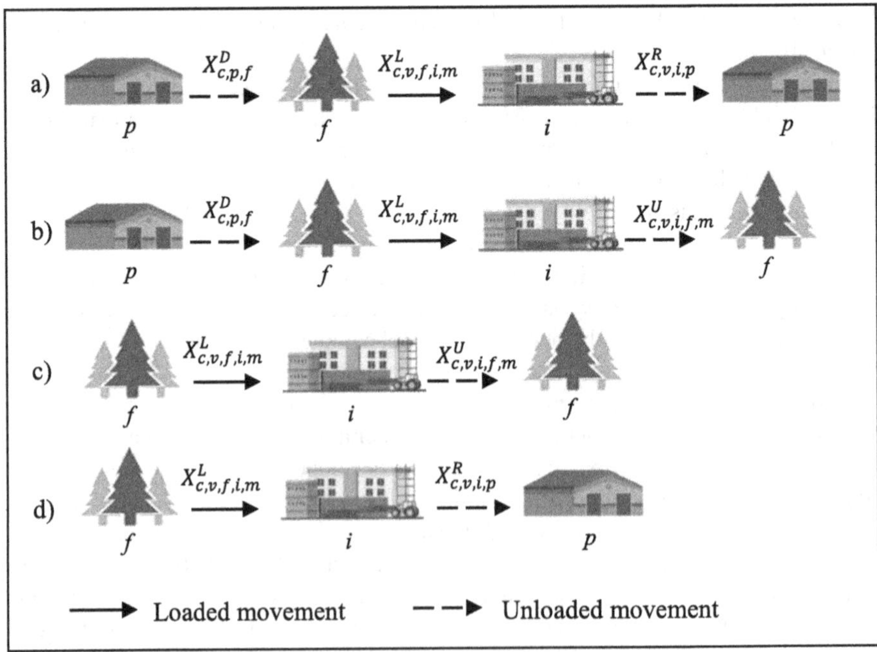


**Fig. 1.** Trips composition

of different movements. As can be seen in Fig. 1, the first two types of trips are made up by three movements, two unloaded (dashed lines) and another loaded (solid lines), meanwhile the two remaining are conformed by two movements, one unloaded and another loaded. Therefore, according to the above trip definition, the proposed approach assembles the routes through trips composition which are simultaneously assigned to the truck in the overall model, using binary variables, one for each type of movement.

Thus, the objective of the present work is the minimization of the total transportation costs, subject to demand fulfillment and legal and contractual restrictions.

## 3 Mathematical Model

In this section the proposed mathematical model (P1) is presented. Note that the parameters are written in lower case, while the variables are written in upper case.

$$\begin{aligned} \min Z = & \sum_{p\in P}\sum_{c\in C_p}\sum_{f\in F} cd_{p,f}\,dpf_{p,f}X^{D}_{c,p,f} \\ & + \sum_{c\in C}\sum_{v\in V}\sum_{f\in F}\sum_{i\in I}\sum_{m\in M_i} cl_{f,i}\,dfi_{f,i}X^{L}_{c,v,f,i,m} \\ & + \sum_{c\in C}\sum_{v\in V}\sum_{f\in F}\sum_{i\in I}\sum_{m\in M_i} cu_{i,f}\,dif_{i,f}X^{U}_{c,v,i,f,m} \\ & + \sum_{p\in P}\sum_{c\in C_p}\sum_{v\in V}\sum_{i\in I} cr_{i,p}\,dip_{i,p}X^{R}_{c,v,i,p} + \sum_{p\in P}\sum_{c\in C_p}\sum_{f\in F} cfixX^{D}_{c,p,f} \end{aligned} \tag{1}$$

Subject to:

$$\sum_{c\in C}\sum_{v\in V}\sum_{i\in I} X^{L}_{c,v,f,i,m} \le supply_{f,m} \quad f\in F, m\in M_i \tag{2}$$

$$\sum_{c\in C}\sum_{v\in V}\sum_{f\in F} X^{L}_{c,v,f,i,m} \ge demand_{i,m} \quad i\in I, m\in M_i \tag{3}$$

$$\begin{aligned} \sum_{c}\sum_{v}\sum_{i}\sum_{m\in M_i} X^{L}_{c,v,f,i,m} = & \sum_{c\in C}\sum_{p\in C_p} X^{D}_{c,p,f} \\ & + \sum_{c\in C}\sum_{v\in V}\sum_{i\in I}\sum_{m\in M_i} X^{U}_{c,v,i,f,m} \quad f\in F \end{aligned} \tag{4}$$

$$\begin{aligned} \sum_{c\in C}\sum_{v\in V}\sum_{f\in F}\sum_{m\in M_i} X^{L}_{c,v,f,i,m} = & \sum_{p\in P}\sum_{c\in C_p}\sum_{v\in V} X^{R}_{c,v,i,p} \\ & + \sum_{c\in C}\sum_{v\in V}\sum_{f\in F}\sum_{m\in M_i} X^{U}_{c,v,i,f,m} i\in I \end{aligned} \tag{5}$$

$$\sum_{f\in F} X^{D}_{c,p,f} \le 1 \quad p\in P, c\in C_p \tag{6}$$

$$X^{D}_{c,p,f} = \sum_{i\in I}\sum_{m\in M_i} X^{L}_{c,v,f,i,m} p\in P, c\in C_p,\ f\in F, v=1 \tag{7}$$

$$\sum_{f\in F} X^{D}_{c,p,f} \le \sum_{v\in V}\sum_{f\in F}\sum_{i\in I}\sum_{m\in M_i} X^{L}_{c,v,f,i,m} \quad p\in P, c\in C_p \tag{8}$$

$$\sum_{f\in F} X^{D}_{c,p,f} = \sum_{v\in V}\sum_{i\in I} X^{R}_{c,v,i,p} \quad p\in P, c\in C_p \tag{9}$$

$$\sum_{f\in F}\sum_{i\in I}\sum_{m\in M_i} X^L_{c,v,f,i,m} \leq \sum_{f\in F} X^D_{c,p,f} \quad p\in P, c\in C_p, v\in V \tag{10}$$

$$\sum_{i\in I}\sum_{f\in F}\sum_{m\in M_i} X^U_{c,v,i,f,m} \leq \sum_{f\in F} X^D_{c,p,f} \quad p\in P, c\in C_p, v\in V \tag{11}$$

$$\sum_{f\in F}\sum_{m\in M_i} X^L_{c,v,f,i,m} = X^R_{c,v,i,p} + \sum_{f\in F}\sum_{m\in M_i} X^U_{c,v,i,f,m} \quad p\in P, c\in C_p, v\in V, i\in I \tag{12}$$

$$\sum_{i\in I}\sum_{f\in F}\sum_{i'=i}\sum_{f'\in F}\left(X^L_{c,v,f,i,m} - X^U_{c,v,i',f',m}\right) \geq 0 \quad c\in C, v\in V, m\in M_i \tag{13}$$

$$\sum_{i\in I}\sum_{m\in M_i} X^U_{c,v-1,i,f,m} \leq \sum_{i\in I}\sum_{m\in M_i} X^L_{c,v,f,i,m} \quad c\in C, f\in F, v>1 \tag{14}$$

$$X^L_{c,v+1,f,i,m} \leq \sum_{f'\in F}\sum_{i'\in I}\sum_{m'\in M_i} X^L_{c,v,f',i',m'} \quad c\in C, f\in F, i\in I, m\in M_i, v<|V| \tag{15}$$

$$\sum_{f\in F}\sum_{i\in I}\sum_{m\in M_i} X^L_{c,v+1,f,i,m} \leq \sum_{f\in F}\sum_{i\in I}\sum_{m\in M_i} X^L_{c,v,f,i,m} \quad c\in C, v<|V| \tag{16}$$

$$\begin{aligned} maxt\sum_{f\in F} X^D_{c,p,f} \geq & \sum_{f\in F}\left(\frac{dpf_{p,f}}{vd_{p,f}}\right)X^D_{c,p,f} \\ & + \sum_{v\in V}\sum_{f\in F}\sum_{i\in I}\sum_{m\in M_i}\left(\frac{dfi_{f,i}}{vl_{f,i}} + tload_f + tunload_i\right)X^L_{c,v,f,i,m} \\ & + \sum_{v\in V}\sum_{i\in I}\sum_{f\in F}\sum_{m\in M_i}\left(\frac{dif_{i,f}}{vu_{i,f}}\right)X^U_{c,v,i,f,m} \\ & + \sum_{v\in V}\sum_{i\in I}\left(\frac{dip_{i,p}}{vr_{i,p}}\right)X^R_{c,v,i,p} \quad p\in P, c\in C_p \end{aligned} \tag{17}$$

$$X^L_{c,v,f,i,m}, X^U_{c,v,i,f,m}, X^R_{c,v,i,p}, X^D_{c,p,f} \in \{0,1\} \tag{18}$$

The aim of the proposed model is to minimize total transportation costs ($Z$), including costs per traveled kilometers (with and without load) and fixed costs per used trucks. Equation (1) defines this objective function. The first four terms correspond to the costs for the different movements that constitute a trip while the last one represents the fixed costs per use of trucks. Parameters $cd_{p,f}$, $cl_{f,i}$, $cu_{i,f}$ and $cr_{i,p}$ represent the costs per kilometer traveled (in \$/km) between the corresponding nodes, $cfix$ is the fixed cost per use of truck (in \$) and the parameters $dpf_{p,f}$, $dfi_{f,i}$, $dif_{i,f}$ and $dip_{i,p}$ define the distances (in km) between the corresponding nodes.

Constraints (2) and (3) establish the maximum supply for each harvest area and the demand to be satisfied for each plant, respectively. Constraints (4) and (5) establish the node balances, that is, the number of trips that arrive at each harvest area/plant must coincide with the number of trips that departs from those nodes.

Restrictions (6) state that a truck can arrive at most once to a harvest area coming from its regional base, while restrictions (7) state that if a truck departs from the regional base, it must perform a loaded movement to any plant in its first trip.

Restrictions (8) state that if a truck is used, then it must perform at least one loaded movement. Constraints (9) force each truck to return to the regional base from which it started its route, while restrictions (10) and (11) establish that if a truck is not used then it must not have any loaded movement assigned.

In constraints (12) it is established that, after completing a loaded movement, the truck can either return to the regional base to finish the route or go to a harvest area to perform a new loaded movement.

Restrictions (13) guarantee that if a loaded movement is not performed, the associated unloaded movement must not be performed either. In addition, when a route is made up of more than one trip, before making an unloaded movement to return to the regional base (on the last trip) the corresponding loaded movement must be completed (Eq. (14)).

To avoid alternative solutions, the trips are created in ascending order, that is, if the trip $v$ is not performed, neither does the trip $v + 1$ (constraints (15) and (16)).

Finally, in restrictions (17) the maximum time of use of a truck is defined (the working day is limited by $maxt$ hours). Parameters $vd_{p,f}$, $vl_{f,i}$, $vu_{i,f}$ and $vr_{i,p}$ represent the average speed between the nodes (in km/h) while $tload_f$ and $tunload_i$ define the average loading (in harvest area) and unloading (in plant) times (in h), respective. Constraints (18) establish the nature of the involved variables.

## 4 Solution Approach

Due to combinatorial nature of this type of formulation, the (P1) model for large-size problems cannot be optimally solved in a reasonable computing time. Therefore, in this section a solution approach is proposed in order to reduce the solution times. A closer look on the structure of the problem presented in the previous section shows that constraints (2) to (5) are coupling the trucks, while the remaining restrictions are dealing with each truck separately. This structure strongly suggests that the problem could be decomposed into a Master Problem (MP) with the linking constraints and a Sub-Problem (SP) per truck. A CG based approach appears to be an appropriate method to tackle the problem.

The idea behind CG is that many linear programs are too large to consider explicitly all the variables. Since most of the variables are non-basic and take value zero in the optimal solution, only a subset of variables needs to be considered when solving the problem. CG takes advantage of this idea to generate only variables which have the potential to improve the objective function, that is, to find variables with negative reduced costs (for a minimization problem). In other words, in this methodology the original problem is decomposed into a MP and a SP: the first decides which variable (or column) is used while the second generates such variables (or columns). There is an extensive literature about CG; for further details, the reader is recommended the works by Desaulniers et al. [11] and Lübbecke and Desrosiers [12].

Thus, the original formulation (P1) is decomposed into a MP, in which only the constraints that links the trucks are considered (constraints (2) to (5)), and a SP per truck, including the rest of the restrictions (constraints (6) to (17)). The decision variables in the MP are related to used route (that is, which column is selected, $Y_r$), while in the SP the variables are related to the conformation of the routes (that is, which column is generated). According to the formulation proposed in this work, each generated column contains the information about the trips involved in the route (combination of movements from regional base to harvest area, between harvest area and plants, and from plants to regional base). The solution approach used in this work is presented in Fig. 2.

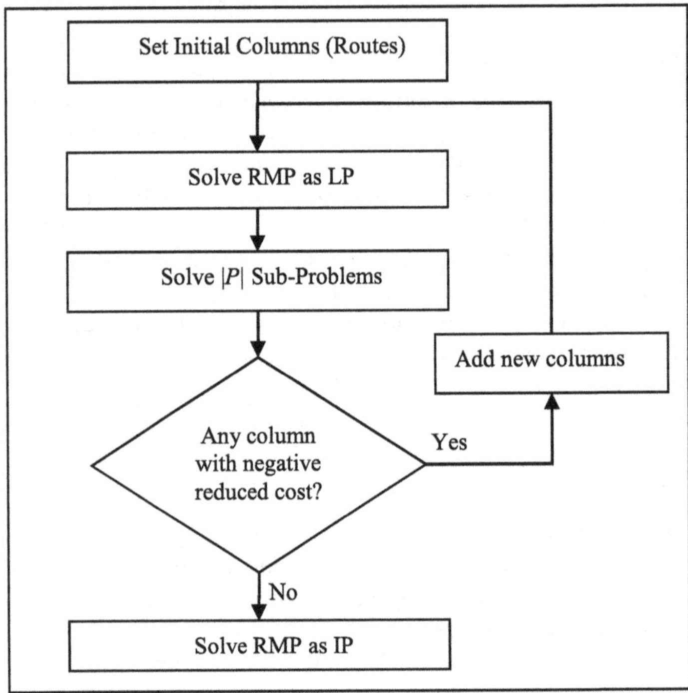


**Fig. 2.** Column generation-based approach

The algorithm presented in Fig. 2 performs the following: first, the initial columns (initial routes) are generated. After that, the relaxation of the Restricted Master Problem (RMP), i.e., the MP with a limited number of columns or variables is solved to determine which columns (routes) are used and the corresponding dual information is obtained. With that information (the values of the dual variables are fixed as parameters), the SP is solved for each regional base $p$ to obtain new columns. If one or more columns with negative reduced costs are obtained, they are added to the RMP and the process is repeated. Otherwise, the RMP is solved as IP (with the already generated columns) to obtain a feasible integer solution to the problem.

The MP model is the following:

$$\min Z_{MP} = \sum\nolimits_{r \in R^+} croute_r Y_r \tag{19}$$

Subject to:

$$\sum_{r\in R^+}\sum_{p\in P}\sum_{v\in V}\sum_{i\in I}A^{XL}_{p,v,f,i,m,r}Y_r \le supply_{f,m} f\in F, m\in M_i \tag{20}$$

$$\sum_{r\in R^+}\sum_{p\in P}\sum_{v\in V}\sum_{f\in F}A^{XL}_{p,v,f,i,m,r}Y_r \ge demand_{i,m} \quad i\in I, m\in M_i \tag{21}$$

$$\begin{aligned}\sum_{r\in R^+}\sum_{p\in P}\sum_{v\in V}\sum_{i\in I}A^{XL}_{p,v,f,i,m,r}\ Y_r = \sum_{r\in R^+}\sum_{p\in P}A^{XD}_{p,f,r}Y_r\\ +\sum_{r\in R^+}\sum_{p\in P}\sum_{v\in V}\sum_{i\in I}\sum_{m\in M_i}A^{XU}_{p,v,i,f,m,r}\ Y_r f\in F\end{aligned} \tag{22}$$

$$\begin{aligned}\sum_{r\in R^+}\sum_{p\in P}\sum_{v\in V}\sum_{f\in F}\sum_{m\in M_i}A^{XL}_{p,v,f,i,m,r}Y_r = \sum_{r\in R^+}\sum_{p\in P}\sum_{v\in V}A^{XR}_{p,v,i,r}Y_r\\ +\sum_{r\in R^+}\sum_{p\in P}\sum_{f\in F}\sum_{m\in M_i}\sum_{v\in V}A^{XU}_{p,v,i,f,m,r}Y_r \quad i\in I\end{aligned} \tag{23}$$

$$\sum_{r\in R^+}\sum_{f\in F}A^{XD}_{p,f,r}Y_r \le ntrucks_p \quad p\in P \tag{24}$$

$$Y_r\in\{0,1\} \tag{25}$$

Regarding the objective function, it consists of the minimization of the routing cost $Z_{MP}$ (Eq. (19)), where $croute_r$ represents the cost of the feasible route $r$. Constraints (20) to (23) correspond to Eqs. (2) to (5) of the original formulation, respectively. Restriction (24) must be incorporated into the MP model to consider the size of the truck fleet belonging to the regional base $p$, that is, the number of routes beginning at regional base $p$ must be less or equal than the number of available trucks in that regional base ($ntrucks_p$). $R^+$ is the set of all feasible routes and the parameters $A^{XD}_{p,f,r}$, $A^{XL}_{p,v,f,i,m,r}$, $A^{XU}_{p,v,i,f,m,r}$ and $A^{XR}_{p,v,i,r}$ contain the information about the composition of each route $r$.

As was previously mentioned, since it is not practical to consider all feasible routes, the CG procedure works with a subset of them ($R^-\subseteq R^+$), where $R^-$ is the set of the already generated routes. Thus, the RMP model is given by constraints (19) to (24), relaxing the binary condition of $Y_r$.

Since the truck fleet is homogeneous, it is not necessary to solve $|C|$ sub-problems (that is, a sub-problem per vehicle). It suffices to solve one SP per regional base considered, i.e. $|P|$ sub-problems, because the possible routes to be performed by each truck belonging to the regional base $p$ are the same. Thus, the SP model is the following:

$$\begin{aligned}\min\ Z^p_{SP} = \sum_{f\in F}cd_{p,f}dpf_{p,f}X^{DSP}_f + \sum_{v\in V}\sum_{f\in F}\sum_{i\in I}\sum_{m\in M_i}cl_{f,i}dfi_{f,i}X^{LSP}_{v,f,i,m}\\ +\sum_{v\in V}\sum_{f\in F}\sum_{i\in I}\sum_{m\in M_i}cu_{i,f}dif_{i,f}X^{USP}_{v,i,f,m} + \sum_{v\in V}\sum_{i\in I}cr_{i,p}dip_{i,p}X^{RSP}_{v,i}\\ +\sum_{f\in F}cfixX^{DSP}_f - \sum_{v\in V}\sum_{f\in F}\sum_{i\in I}\sum_{m\in M_i}\pi^1_{f,m}X^{LSP}_{v,f,i,m}\\ -\sum_{v\in V}\sum_{f\in F}\sum_{i\in I}\sum_{m\in M_i}\pi^2_{i,m}X^{LSP}_{v,f,i,m}\\ -\sum_{v\in V}\sum_{f\in F}\sum_{i\in I}\sum_{m\in M_i}\pi^3_f\left(X^{DSP}_f + X^{USP}_{v,i,f,m} - X^{LSP}_{v,f,i,m}\right)\end{aligned}$$

$$-\sum_{v\in V}\sum_{f\in F}\sum_{i\in I}\sum_{m\in M_i}\pi_i^4\left(X_{v,f,i,m}^{LSP}-X_{v,i,f,m}^{USP}-X_{v,i}^{RSP}\right)-\pi^0 \tag{26}$$

Subject to:

$$\sum_{f\in F}X_f^{DSP}\leq 1 \tag{27}$$

$$X_f^{DSP}=\sum_{i\in I}\sum_{m\in M_i}X_{v,f,i,m}^{LSP}\quad f\in F, v=1 \tag{28}$$

$$\sum_{f\in F}X_f^{DSP}\leq\sum_{v\in V}\sum_{f\in F}\sum_{i\in I}\sum_{m\in M_i}X_{v,f,i,m}^{LSP} \tag{29}$$

$$\sum_{f\in F}X_f^{DSP}=\sum_{v\in V}\sum_{i\in I}X_{v,i}^{RSP} \tag{30}$$

$$\sum_{f\in F}\sum_{i\in I}\sum_{m\in M_i}X_{v,f,i,m}^{LSP}\leq\sum_{f\in F}X_f^{DSP}\quad v\in V \tag{31}$$

$$\sum_{i\in I}\sum_{f\in F}\sum_{m\in M_i}X_{v,i,f,m}^{USP}\leq\sum_{f\in F}X_f^{DSP}\quad v\in V \tag{32}$$

$$\sum_{f\in F}\sum_{m\in M_i}X_{v,f,i,m}^{LSP}=X_{v,i}^{RSP}+\sum_{f\in F}\sum_{m\in M_i}X_{v,i,f,m}^{USP}\quad v\in V, i\in I \tag{33}$$

$$\sum_{i\in I}\sum_{f\in F}\sum_{i'=i}\sum_{f'\in F}\left(X_{v,f,i,m}^{LSP}-X_{v,i',f',m}^{USP}\right)\geq 0\quad v\in V, m\in M_i \tag{34}$$

$$\sum_{i\in I}\sum_{m\in M_i}X_{v-1,i,f,m}^{USP}\leq\sum_{i\in I}\sum_{m\in M_i}X_{v,f,i,m}^{LSP}\quad f\in F, v>1 \tag{35}$$

$$X_{v+1,f,i,m}^{LSP}\leq\sum_{f'\in F}\sum_{i'\in I}\sum_{m'\in M_i}X_{v,f',i',m'}^{LSP}\quad f\in F, i\in I, m\in M_i, v<|V| \tag{36}$$

$$\sum_{f\in F}\sum_{i\in I}\sum_{m\in M_i}X_{v+1,f,i,m}^{LSP}\leq\sum_{f\in F}\sum_{i\in I}\sum_{m\in M_i}X_{v,f,i,m}^{LSP} v<|V| \tag{37}$$

$$\begin{aligned} maxt\sum_{f\in F}X_f^{DSP} &\geq\sum_{f\in F}\left(\frac{dpf_{p,f}}{vd_{p,f}}\right)X_f^{DSP}\\ &+\sum_{v\in V}\sum_{f\in F}\sum_{i\in I}\sum_{m\in M_i}\left(\frac{dfi_{f,i}}{vl_{f,i}}+tload_f+tunload_i\right)X_{v,f,i,m}^{LSP}\\ &+\sum_{v\in V}\sum_{i\in I}\sum_{f\in F}\sum_{m\in M_i}\left(\frac{dif_{i,f}}{vu_{i,f}}\right)X_{v,i,f,m}^{USP}+\sum_{v\in V}\sum_{i\in I}\left(\frac{dip_{i,p}}{vr_{i,p}}\right)X_{v,i}^{R} \end{aligned} \tag{38}$$

$$X_{v,f,i,m}^{LSP}, X_{v,i,f,m}^{USP}, X_{v,i}^{RSP}, X_f^{DSP}\in\{0,1\} \tag{39}$$

Constraint (26) performs the pricing (it determines the reduced costs), i.e., this objective function evaluates if the generated route could potentially improve the objective function of the RMP model by incorporating it into the $R^-$ set. The dual variables correspond to Eqs. (20) to (24) and are given by $\pi_{f,m}^1$, $\pi_{i,m}^2$, $\pi_f^3$, $\pi_i^4$ and $\pi^0$, respectively. The aim of the |P| sub-problems is to determine the route with the most negative reduce cost. Constraints (27) to (38) are related to restrictions (6) to (17) of the (P1) model; in turn, since the SP model is solve by regional base $p$ (regardless of the truck $c$ for which the route is created) the variables $X^{D}_{c,p,f}$, $X^{L}_{c,v,f,i,m}$, $X^{U}_{c,v,i,f,m}$ and $X^{R}_{c,v,i,p}$ must be rewritten as follows:

$$X_{f}^{DSP} = \sum\nolimits_{r\in R+} \sum\nolimits_{p\in P} A^{XD}_{p,f,r} \quad Y_r f \in F \tag{40}$$

$$X_{v,f,i,m}^{LSP} = \sum\nolimits_{r\in R} + \sum\nolimits_{p\in P} A^{XL}_{p,v,f,i,m,r} Y_r \; f \in F, v \in V, i \in I, m \in M_i \tag{41}$$

$$X_{v,i,f,m}^{USP} = \sum\nolimits_{r\in R+} \sum\nolimits_{p\in P} A^{XU}_{p,v,i,f,m,r} Y_r \quad f \in F, v \in V, i \in I, m \in M_i \tag{42}$$

$$X_{v,i}^{RSP} = \sum\nolimits_{r\in R+} \sum\nolimits_{p\in P} A^{XR}_{p,v,i,r} Y_r \quad v \in V, i \in I \tag{43}$$

A set of initial routes is required to proceed with the column generation algorithm. All unitary length routes that can be made (regardless of the actual number of available trucks) are generated so as to achieve the first dual solution and thus initialize the CG algorithm. There are several heuristic techniques for route generation. In this work, a linear programming model (a classical transportation problem) is used to obtain the minimum supply distances traveled with load. Subsequently, this solution is converted into routes of unitary length (that is, routes with a single trip).

## 5 Illustrative Example

To test both the mathematical programing model (P1) and the CG based algorithm, an example is developed. The considered supply chain involves 5 plants that demand 5 types of raw materials, which can be obtained from 15 harvest areas. There are 750 available trucks (spread across 12 regional bases) for hauling a total of 750 full-truckloads demanded by the plants. In Table 1 the used parameters are presented. In addition each truck can make at most 3 loaded trips beyond considering its maximum available working time (*maxt*).

**Table 1.** Model parameters.

| Parameter | Value | Parameter | Value | Parameter | Value |
|---|---|---|---|---|---|
| $cd_{p,f}$ | 0.8 $/km | $vd_{p,f}$ | 65 km/h | $tload_f$ | 0.5 h |
| $cl_{f,i}$ | 1.2 $/km | $vl_{f,i}$ | 55 km/h | $tunload_i$ | 0.5 h |
| $cu_{i,f}$ | 0.8 $/km | $vu_{i,f}$ | 65 km/h | *cfix* | 30 $ |
| $cr_{i,p}$ | 0.8 $/km | $vr_{i,p}$ | 65 km/h | *maxt* | 10 h |

**Table 2.** Supply/demand of raw materials (in full-truckloads).

| | $m_1$ | $m_2$ | $m_3$ | $m_4$ | $m_5$ |
|---|---|---|---|---|---|
| $f_1$ | 14 | 21 | 11 | 7 | 15 |
| $f_2$ | 11 | 7 | 3 | 10 | 10 |
| $f_3$ | 1 | 15 | 12 | 12 | 11 |
| $f_4$ | 11 | 1 | 10 | 13 | 3 |
| $f_5$ | 16 | 9 | 18 | 12 | 21 |
| $f_6$ | 17 | 11 | 11 | 9 | 11 |
| $f_7$ | 13 | 11 | 17 | 5 | 1 |
| $f_8$ | 16 | 6 | 3 | 13 | 21 |
| $f_9$ | 6 | 8 | 15 | 1 | 15 |
| $f_{10}$ | 15 | 21 | 5 | 13 | 6 |
| $f_{11}$ | 8 | 4 | 10 | 21 | 10 |
| $f_{12}$ | 11 | 21 | 11 | 11 | 10 |
| $f_{13}$ | 3 | 11 | 21 | 14 | 6 |
| $f_{14}$ | 11 | 15 | 11 | 16 | 15 |
| $f_{15}$ | 17 | 9 | 13 | 13 | 15 |
| $i_1$ | 35 | 50 | – | – | 40 |
| $i_2$ | 50 | 85 | 40 | – | – |
| $i_3$ | 35 | – | 50 | 15 | – |
| $i_4$ | 30 | 15 | 30 | 60 | 50 |
| $i_5$ | – | – | 30 | 75 | 60 |

In Table 2 supply and demand of raw materials are detailed. The distances between the nodes are presented in Table 3.

All the models are implemented and solved in GAMS [13] 24.7.3 version, using CPLEX 12.6.3 solver in an Intel(R) Core(TM) i7-4790, 3.60 GHz.

The (P1) model involves 978,750 binary variables and 448,601 equations. This formulation cannot be solved in one hour of execution using the selected solver. However, if the resolution time limit in (P1) is removed, the optimal solution (0% optimality gap) is found in 70,285 s (near 20 h). The optimal solution has an objective value equal to $60,940 and 250 trucks are used to satisfy the demand. Each truck completes 3 loaded movements in their respective route.

By applying the CG based approach, the solution is reached in 122.95 s (resolution time of the column generation phase: 121.42 s; resolution time of the final RMP: 1.53 s). The final RMP involves 9,820 binary variables (columns) and the objective function is equal to $60,964, while 250 trucks are used (also making 3 trips each). As it can be noted, the proposed solution approach reaches a solution with a difference of 0.039% in

**Table 3.** Distances between the nodes (in kilometers).

| | $p_1$ | $p_2$ | $p_3$ | $p_4$ | $p_5$ | $p_6$ | $p_7$ | $p_8$ | $p_9$ | $p_{10}$ | $p_{11}$ | $p_{12}$ | $i_1$ | $i_2$ | $i_3$ | $i_4$ | $i_5$ |
|---|---|---|---|---|---|---|---|---|---|---|---|---|---|---|---|---|---|
| $f_1$ | 32 | 70 | 70 | 120 | 100 | 100 | 82 | 58 | 136 | 164 | 158 | 45 | 71 | 50 | 120 | 114 | 108 |
| $f_2$ | 61 | 114 | 20 | 150 | 61 | 85 | 22 | 78 | 82 | 108 | 98 | 81 | 106 | 40 | 32 | 92 | 60 |
| $f_3$ | 30 | 85 | 14 | 125 | 54 | 70 | 22 | 50 | 86 | 114 | 106 | 50 | 78 | 14 | 60 | 81 | 58 |
| $f_4$ | 50 | 100 | 14 | 136 | 50 | 73 | 10 | 64 | 76 | 103 | 94 | 67 | 92 | 32 | 40 | 81 | 51 |
| $f_5$ | 91 | 130 | 54 | 156 | 57 | 85 | 40 | 95 | 61 | 81 | 71 | 102 | 121 | 73 | 10 | 86 | 50 |
| $f_6$ | 45 | 81 | 30 | 114 | 32 | 51 | 14 | 45 | 64 | 92 | 85 | 51 | 72 | 36 | 54 | 60 | 36 |
| $f_7$ | 36 | 40 | 76 | 90 | 85 | 78 | 81 | 36 | 121 | 148 | 143 | 22 | 41 | 58 | 122 | 92 | 95 |
| $f_8$ | 45 | 54 | 54 | 86 | 32 | 32 | 42 | 20 | 67 | 94 | 89 | 32 | 45 | 50 | 81 | 45 | 41 |
| $f_9$ | 85 | 90 | 73 | 103 | 14 | 32 | 51 | 63 | 22 | 50 | 45 | 76 | 80 | 81 | 67 | 28 | 10 |
| $f_{10}$ | 73 | 51 | 82 | 64 | 36 | 10 | 67 | 36 | 60 | 82 | 81 | 50 | 41 | 80 | 99 | 22 | 45 |
| $f_{11}$ | 113 | 112 | 98 | 114 | 42 | 51 | 76 | 89 | 10 | 22 | 20 | 103 | 102 | 108 | 81 | 40 | 36 |
| $f_{12}$ | 108 | 95 | 102 | 92 | 41 | 36 | 81 | 78 | 28 | 40 | 41 | 92 | 85 | 108 | 95 | 22 | 40 |
| $f_{13}$ | 91 | 36 | 112 | 28 | 72 | 45 | 100 | 51 | 92 | 110 | 110 | 60 | 32 | 104 | 135 | 51 | 81 |
| $f_{14}$ | 110 | 64 | 124 | 40 | 72 | 45 | 108 | 71 | 81 | 92 | 95 | 82 | 58 | 120 | 136 | 42 | 78 |
| $f_{15}$ | 60 | 30 | 81 | 58 | 51 | 32 | 71 | 20 | 81 | 104 | 102 | 32 | 20 | 73 | 108 | 45 | 61 |
| $i_1$ | 63 | 10 | 92 | 51 | 71 | 51 | 86 | 28 | 100 | 124 | 122 | 32 | 71 | 50 | 120 | 114 | 108 |
| $i_2$ | 22 | 86 | 20 | 130 | 67 | 81 | 36 | 54 | 100 | 128 | 120 | 50 | 106 | 40 | 32 | 92 | 60 |
| $i_3$ | 90 | 134 | 51 | 163 | 64 | 92 | 41 | 98 | 71 | 91 | 81 | 104 | 78 | 14 | 60 | 81 | 58 |
| $i_4$ | 89 | 73 | 90 | 76 | 32 | 14 | 71 | 57 | 41 | 61 | 60 | 71 | 92 | 32 | 40 | 81 | 51 |
| $i_5$ | 78 | 91 | 63 | 108 | 10 | 36 | 41 | 61 | 28 | 57 | 50 | 73 | 121 | 73 | 10 | 86 | 50 |

the objective function given by the (P1) model, but in a considerable shorter resolution time.

It is interesting to note that the solution obtained from the last solved RMP can be used as initial starting point to solve the original formulation (P1) to global optimality. According to this, using a time limit of 15 min, then the objective function obtained is equal to \$60,948 (0.009% optimality gap). While, if the time limit is omitted, the global optimal solution starting at the RMP solution is reached in 17,234 s, which represents a reduction of 75% with respect to the resolution of the (P1) model without starting point.

Table 4 shows the resolution times for five instances (the case study presented here correspond to the scenario 5). The characteristics of scenarios 1 to 4 are as follows:

- regional bases: 12, 8, 5 and 12, respectively
- plants: 5, 4, 4 and 5, respectively
- types of raw materials: 2, 4, 8 and 8, respectively
- harvest areas: 15, 10, 10 and 15, respectively
- total demand: 150, 250, 150 and 500, respectively.

**Table 4.** Resolution times for all the scenarios.

| Scenario | Gen.Time (s) | F.RMP (s) | F.RMP gap (%) | Trucks CG | P1+IP (s) | P1+IP gap (%) | Trucks P1[a] | P1-IP (s) | P1-IP gap (%) | Trucks P1[b] |
|---|---|---|---|---|---|---|---|---|---|---|
| 1 | 26.16 | 0.43 | 0 | 50 | 17.24 | 0 | 50 | 18.01 | 0 | 50 |
| 2 | 39.29 | 900 | 0.032 | 84 | 900 | 0.099 | 84 | 900 | 0.1 | 84 |
| 3 | 30.98 | 1.47 | 0 | 52 | 16.91 | 0 | 50 | 21.17 | 0 | 50 |
| 4 | 126.81 | 900 | 0.046 | 168 | 900 | 0.147 | 168 | – | – | – |
| 5 | 121.42 | 1.53 | 0 | 250 | 900 | 0.009 | 250 | – | – | – |

In this table, "Gen.Time (s)" corresponds to the time needed to generate all the columns (feasible routes); "F.RMP (s)" is the resolution time of the final RMP; "F.RMP gap (%)", "P1 + IP (s)" and "P1 + IP gap (%)" represent the optimality gap of the CG-based approach, the (P1) model with initial point and the (P1) model without initial point, respectively; "Trucks CG", "Trucks P1[a]" and "Trucks P1[b]" represents the number of used trucks in the CG-based approach, in the (P1) model with initial point and in the (P1) model without initial point, respectively; and finally, "P1 + IP (s)" and "P1-IP (s)" contain the time needed to solve the (P1) with and without initial point, respectively. It is worth mentioning that the time required to generate the starting point is not included in "P1 + IP (s)". From the showed results, it can be derived that both formulations (the (P1) model and the CG-based approach) are effective to solve small/medium-size problems. Clearly, for large size problems, the CG-based approach is an efficient methodology to take into account.

In Table 5 the objective functions of both CG-based approach and (P1) model with the initial point given by CG are compared to the LP relaxation of (P1) for the five scenarios. There is no time limit in the resolution of the LP relaxation of (P1). A time limit of 15 min is used in the final RMP and the same time is set for the (P1) model with the starting point defined by the CG-based approach. It is well known that the objective function of LP relaxation represents a lower bound for the objective function of the (P1) model, and therefore the quality of the solution can be assessed.

**Table 5.** Comparison between the objectives functions.

| Scenario | LP relaxation of P1 ($) | CG-based approach ($) | Optimality gap (%) | P1 with initial point ($) | Optimality gap (%) |
|---|---|---|---|---|---|
| 1 | 10,501.73 | 10,542.8 | 0.39 | 10,517.2 | 0.147 |
| 2 | 24,362 | 24,394 | 0.131 | 24,384.4 | 0.092 |
| 3 | 12,733.73 | 12,912.8 | 1.387 | 12,734.8 | 0.008 |
| 4 | 39,813.33 | 39,895.2 | 0.205 | 39,872 | 0.147 |
| 5 | 60,935.46 | 60,964 | 0.047 | 60,942 | 0.011 |

The obtained results are closer to the LP relaxation of (P1), as can be seen in Table 5, which implies that the proposed approach can be applied in real size problems and guarantee a very good solution in less than 15 min of computing time.

## 6 Final Remarks

Transportation planning in forest industry is a critical issue taking into account the high incidence of transportation costs in the profitability of the firms, given the large transported volumes and traveled distances. In the Argentine context, firms generally make decisions regarding the distribution of logs based on their experience. Therefore, the use of a planning tool to make these decisions could lead to significant improvements in the obtained results and the performance of the truck fleet.

In this work, a mathematical programming model is presented. Unlike the previous published works in the literature, the (P1) model simultaneously address product allocation and routing decisions, taking into account different raw materials and several depots or regional bases. In addition, the (P1) model allows determining the size of the truck fleet in such a way to guarantee the supply at minimum cost. The (P1) model is a generalization of the model developed in [8], in which the plants demand a single type of raw material. This generalization allows a better representation of the real problem. The developed mathematical model is robust enough and allows solving problems adjusted to the reality of the Argentine forest industry. In any case, beyond the analyzed example and the obtained results, the formulation could be improved to achieve a more detailed problem description, such as different types of trucks.

From the computational performance point of view, the (P1) model represents a highly combinatorial problem. Therefore, resorting to a CG based approach to solve the model in a reasonable computing time is a crucial aspect, especially if it is intended to have a planning tool that supports logistic planners on a day-to-day basis. An example was analyzed and solved with both solution methods. The CG based approach is able to find a good solution in a short computing time. Since the solution times of the CG based approach depends on the time taken by the column generation phase (the solution times of the sub-problems), it could be improved taking into account acceleration techniques in that phase to achieved a reduction in the total computing time; but this problem is out of the scope of this work and it will be considered in future works.

Finally, it should be noted that, although the CG based approach is not exact (reaching the optimal solution is not guaranteed), it allows obtaining solutions with negligible optimality gap, i.e., solutions very close to the optimal one. This tradeoff between solution quality and solution times makes this approach a viable option for its implementation in a real environment.

## References

1. National Census of Sawmills, Ministry of Agroindustry. https://www.magyp.gob.ar/sitio/areas/ss_desarrollo_foresto_industrial/censos_inventario/_archivos/censo/000000_Provincia%20de%20Misiones%20(Marzo%202018).pdf. Accessed 06 Jul 2020

2. Forestry Report, Ministry of Agroindustry. https://www.agroindustria.gob.ar/sitio/areas/ss_desarrollo_foresto_industrial/estadisticas/_archivos//000000_Sector%20Forestal/000000_Informes/170000_2017%20-%20Sector%20Forestal.pdf. Accessed 06 Jul 2020
3. Gronalt, M., Hirsch, P.: Log-truck scheduling with a tabu search strategy. In: Doerner, K.F., Gendreau, M., Greistorfer, P., Gutjahr, W., Hartl, R.F., Reimann, M. (eds.) Metaheuristics. Operations Research/Computer Science Interfaces Series, pp. 65–88. Springer, Boston (2007)
4. Rix, G., Rousseau, L.-M., Pesant, G.: Solving a multi-period log-truck scheduling. In: 34th Council on Forest Engineering, Quebec, Canada, pp. 1–10. https://www.cirrelt.ca/cofe2011/proceedings/29-rix.pdf. Accessed 20 Feb 2020
5. El Hachemi, N., El Hallaoui, I., Gendreau, M., Rousseau, L.-M.: Flow-based integer linear programs to solve the weekly log-truck scheduling problem. Ann. Oper. Res. **232**, 87–97 (2014)
6. Derigs, U., Pullman, M., Vogel, U., Oberscheider, M., Gronalt, M., Hirsch, P.: Multi-level neighborhood search for solving full truckload routing problems arising in timber transportation. Electron. Not. Discrete Math. **39**, 281–288 (2012)
7. Lin, P., Contreras, M., Dai, R., Zhang, J.: A multilevel ACO approach for solving forest transportation planning problems with environmental constraints. Swarm Evol. Comput. **28**, 78–87 (2016)
8. Bordón, M.R., Montagna, J.M., Corsano, G.: An exact mathematical formulation for the optimal log transportation. For. Policy Econ. **95**, 115–122 (2018)
9. Palmgren, M., Ronnqvist, M., Varbrand, P.: A near-exact method for solving the log-truck scheduling problem. Int. Trans. Oper. Res. **11**, 447–464 (2004)
10. Rix, G., Rousseau, L.-M., Pesant, G.: A column generation algorithm for tactical timber transportation planning. J. Oper. Res. Soc. **66**, 278–287 (2014)
11. Desaulniers, G., Desrosiers, J., Solomon, M.: Column Generation. Springer, Boston (2005)
12. Lübbecke, M., Desrosiers, J.: Selected topics in column generation. Oper. Res. **53**(6), 1007–1023 (2005)
13. Rosenthal, R.E.: A GAMS tutorial. https://www.gams.com/31/docs/index.html. Accessed 06 Jul 2020

# A Branch-and-Bound Method to Minimize the Total Flow Time in a Permutation Flow Shop with Blocking and Setup Times

João Vítor Silva Robazzi[1,2], Marcelo Seido Nagano[1], and Mauricio Iwama Takano[3(✉)]

[1] School of Engineering of São Carlos, University of São Paulo, Av. Trabalhador São-Carlense, 400, São Carlos, SP, Brazil

[2] Federal Institute of Education, Science and Technology of São Paulo, R. Américo Ambrósio 269, Sertãozinho, SP, Brazil

[3] Federal University of Technology – Paraná, Avenida Alberto Carazzai, 1640, Cornélio Procópio, PR, Brazil

takano@utfpr.edu.br

**Abstract.** This paper presents an improvement of the branch-and-bound algorithm for the blocking-in-process and setup times permutation flow shop problem with total flow time criterion, which is known to be NP-Hard for $m \geq 2$. For that, a new machine-based lower bound which exploits the machine idleness and the occurrence of blocking is proposed. Computational experiments were performed using a database that contains 27 classes of problems, varying the number of jobs (n) and the number of machines (m). Results show that the algorithm can handle most of the $n < 20$ problems in less than one hour. Therefore, this work can support the scheduling of many applications in manufacturing systems with limited buffers and setup times.

**Keywords:** Scheduling · Permutation flow shop · Blocking · Setup · Total flow time · Branch-and-bound

## 1 Introduction

In the permutation flow shop scheduling problem, $n$ jobs must be processed on each of the machines in the same order. It is also assumed that every job has the same processing sequence on every machine, each machine can process at most a single job and no interruptions are allowed. The processing time of each job on each machine is given. The total flow time objective measures the time the jobs stay inside the system, and its minimization results in cost reductions related to initial inventory and in-process maintenance costs. For $m \geq 2$, the flow shop with total flow time criterion problem is NP-hard [1].

Many papers considered the problem with unlimited buffers [2] or considering the setup time embedded in the processing time of the job [3–5]. In this paper, the setup time is separated from the processing time, allowing a machine to be prepared to initiate

© Springer Nature Switzerland AG 2021
D. A. Rossit et al. (Eds.): ICPR-Americas 2020, CCIS 1407, pp. 217–232, 2021.
https://doi.org/10.1007/978-3-030-76307-7_17

the processing of a job before the previous machine has finished processing this job. This ensures better flexibility for scheduling, allowing for a better use of time, hence a possible reduction in the makespan. Furthermore, the setup time depends on the sequence of jobs, as well as the machine. In other words, there is a different setup time for each pair of jobs in each machine.

In an environment with blocking, there are limited buffers in between the machines. In this paper, a zero buffer constraint is considered, i.e., if machine $k$ finishes the processing of job $j$ and machine $k + 1$ is not able to receive the job (because it is still processing job $j - 1$ or is still being set up), the job remains on machine $k$, blocking it. In this case machine $k$ is unable to receive the next job of the sequence. The absence of buffers can cause the occurrence of blocking, which is when a job completed on a specific machine blocks it until the following one is ready for processing. Another reason for the occurrence of blocking lies in the production technology itself [6]. For instance, in robotic cells, a machine may be blocked while waiting for the robot arm to pick up the finished job and move it to the next stage [7].The flow shop problem with zero buffer can be used to model any flow shop problem with limited buffer because a unit capacity buffer can be represented by a machine with zero processing time for all jobs [3]. Figure 1 shows an example of the problem in a Gantt chart.

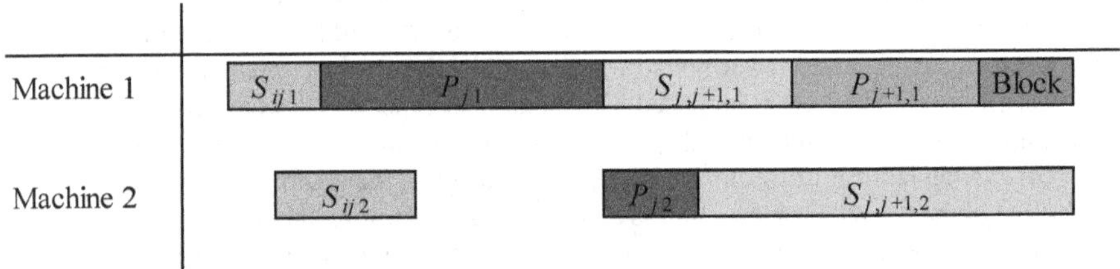


**Fig. 1.** Example of a blocking flow shop scheduling problem with setup times

In Fig. 1 is shown the moment when job $i$ just finished being processed by both Machine 1 and 2. Thus, both machines begin by being setup to start processing job $j$ ($S_{ij1}$ and $S_{ij2}$). It can be noted that Machine 2 must wait for Machine 1 to finish processing job $j$ before it can start processing it, so it stays idle for a while. Right after finishing processing job $j$ each machine can start setting up for the next job ($S_{j,j+1,1}$ and $S_{j,j+1,2}$). In this case, notice that Machine 1 finished processing job $j + 1$ before Machine 2 had finished the setup process. Therefore, a *block* occurs, preventing Machine 1 from starting to setup for the next job.

Because of its complexity, large-size blocking flow shop problems are usually solved by heuristics. [8] are among the first authors that explored the blocking environment. Their work consisted of reducing the 2-machine flow shop with zero buffer and makespan minimization to a special case of the Traveling Salesman Problem, which can be solved in polynomial time using the Gilmore-Gomory algorithm [9]. [3] developed the Profile Fitting constructive heuristic, which fits an unscheduled job that provides the least blocking and machine idle time in front of a partial sequence.

[10] developed a two-machine case branch-and-bound to minimize the mean completion time and a three-machine case for makespan on a flow shop environment. [11]

extended this machine based lower bound to m machines when the objective is minimizing the sum of completion times, or total flow time when all release dates are zero. [12] developed a machine based lower bound with the preemptive relaxation, which dominated the formulation of [11] in terms of value. [13] proposed a new total flow time lower bound for flow shop that dominated both previous formulations. The results point out that the algorithm can easily handle $n \leq 15$ and can solve most of $15 < n \leq 20$ problems.

A reduced number of papers explores the blocking flow shop solved by a branch-and-bound algorithm. [14] presented a branch-and-bound algorithm for flow shops with zero buffer. The authors developed a departure time based lower bound that was used for the total tardiness and makespan objectives. [15] outperformed the lower bound for flow shop with blocking in order to minimize makespan proposed by [14] in 85.2% of the cases. Their lower bound calculated an underestimate departure time of the last job in non-partial sequence. [16] developed a lower bound for flow shop with zero buffer constraint with the total flow time criterion. The algorithm solved $n \leq 18$ and $m \leq 10$ problems and 17 instances of Taillards benchmark [17] in less than 20 min. [18] proposed an initial upper bound (UB) to reduce the computational time of the branch-and-bound algorithm. The initial UB was provided by a constructive heuristic, and a significant reduction of the computational time was noticed.

Just a few papers addressed the blocking with setup times in flow shop. [19] were one of the first to explore this problem. They adapted the lower bound presented by [10]. [20] was the first paper to address the use of a branch-and-bound algorithm for the problem. A lower bound was developed for the makespan in a flow shop with blocking and sequence dependent setup times, addressing two structural properties of the problem: an upper bound for the machine idle and a lower bound for the machine blocking time.

In this article a lower bound for the branch-and-bound algorithm for blocking with setup time permutation flow shop problems with total flow time criterion is presented, in order to optimally solve small and medium size problems. In Sect. 2 the problem is defined, and some basic calculations are shown. The branch-and-bound algorithm and the lower bound development are presented in Sect. 3. The computational tests are showed in Sect. 4 and the conclusions are presented in Sect. 5.

## 2 Problem Definition

The flow shop problem with blocking and setup can be defined by the following set of equations. Let $\sigma = \{1, 2, \ldots, i, j, \ldots, n\}$ be an arbitrary sequence of jobs, $k = \{1, 2, \ldots, m\}$ be the sequence of available machines, $i$ be the job that directly precedes job $j$ in the sequence, $P_{jk}$ be the processing time of the $j$-th job in the sequence on machine $k$, $S_{ijk}$ be the setup time of machine $k$ between the $i$-th and the $j$-th job in the sequence, $S_{01k}$ be the setup time of machine $k$ before processing the first job in the sequence, $R_{jk}$ be the completion time of the setup of machine $k$ to the $j$-th job in the sequence and $D_{jk}$ be the departure time of the $j$-th job in the sequence on machine $k$. Equations 1 to 5 show how the departure times are calculated for a given sequence:

$$R_{1k} = S_{01k} \; \forall 1 \leq k \leq m \tag{1}$$

$$D_{j1} = \max\left(R_{j2}, R_{j1} + P_{j1}\right)\ \forall 1 \leq j \leq n \tag{2}$$

$$D_{jk} = \max\left(R_{j,k+1}, D_{j,k-1} + P_{jk}\right)\ \forall 1 \leq j \leq n;\ 2 \leq k \leq m - 1 \tag{3}$$

$$D_{jm} = D_{j,m-1} + P_{jm}\ \ \forall 1 \leq j \leq n \tag{4}$$

$$R_{jk} = D_{ik} + S_{ijk}\ \ \forall 2 \leq j \leq n;\ 1 \leq k \leq m \tag{5}$$

Initially, the setup completion times of the machines for the first job in the sequence are calculated by Eq. 1. Then, the departure times of the first job in all machines are calculated by Eqs. 2, 3, and 4. Then, Eq. 5 is used to calculate the setup completion time of all machines for the following job. Equations 2, 3, and 4 are used again to calculate the departure times of the following job in all machines. The total flow time is the sum of all departure times of all jobs on the last machine. Therefore, Eq. 6 is used to calculate the total flow time.

$$TFT = \sum\nolimits_{j=1}^{n} D_{jm} \tag{6}$$

## 3 The Branch-and-Bound Algorithm

The branch-and-bound algorithm branches an original high complexity problem in two or more sub-problems, called nodes, that are easier to solve. This process can be done recursively until a feasible solution is obtained. The nodes are defined by an exclusive sub-sequence which is called Partial Sequence (PS). The set of jobs not included in PS is called Non-Partial Sequence (NPS).

When a node is branched, one or more partial sequences (nodes) are created by adding up a new job from |NPS| to the partial sequence associated with the branched node. The search tree is created using the depth first rule, in which the node with more jobs in the PS is branched. The ties are broken by the minimum lower bound. For each node, a lower bound of the total flow time is computed and this node is fathomed if the incumbent solution is smaller than its lower bound. As the focus of this article is to evaluate the tightness of the proposed lower bound, the initial incumbent solution is set to a very high value and is updated every time a better feasible solution is reached.

### 3.1 Proposed Lower Bound

The effectiveness of a branch-and-bound algorithm is heavily related to the lower bounds used. Such bounds are more useful if their computational complexities are small and if they are tight. The proposed lower bound assumes a relaxation in which one machine can process one job at a time and the others can handle the whole set of n jobs at a time. Since the SPT rule is optimal for total flow time criterion on a single machine

[21], it is possible to obtain a lower bound for the flow shop problem. The lower bound considering a partial sequence σ of size s is computed as follows:

$$LB = TFT(\sigma) + \max_{1 \le k \le m} LB_k \tag{7}$$

$$LB_k = \sum\nolimits_{t=s+1}^{n} \left( ED_{tk} + \sum\nolimits_{r=k+1}^{m} p_{tr} \right) \tag{8}$$

In which $ED_{tk}$ is the underestimate of the departure time of a job in position $t > s$ on machine k, as shown in Eq. 9. Where $s$ is the last job in σ and $p_{[j],k}$ is the (s-j)-th smaller processing time on machine k.

$$ED_{tk} = max \begin{pmatrix} max\left( D_{sk} + \sum\nolimits_{j=s+1}^{t-1} p_{[j]k}; D_{s,k+1} + \sum\nolimits_{j=s+1}^{t-2} p_{[j],k+1} + Setup_{k+1}(t-1) \right) \\ + Setup_k(t); D_{s,k-1} + \sum\nolimits_{j=s+1}^{t} p_{[j],k-1} \end{pmatrix} + p_{[t]k} \tag{9}$$

For $k = 1$ and $k = m$, the formulation excludes the $k-1$ and $k+1$ terms, respectively.

## 4 Computational Results

This section describes the computational tests done in order to evaluate the effectiveness of the proposed lower bound (LB) when applied to a branch-and-bound algorithm. The experimentation used the problem instances database from [14], in which the processing times were randomly generated in a uniform distribution ranging from 1 to 99. The database consists of 27 different problem sizes, with 20 unique problems each, totaling 540 instances.

The database provided by [15] does not include setup times. Therefore, in order to analyze the influence of the setup times in the algorithm, four different setup time databases were generated for these tests. To create these databases, the values were uniformly distributed between:

- Database 1: 01 and 09;
- Database 2: 01 and 49;
- Database 3: 01 and 99;
- Database 4: 01 and 124;

The setup data ranges were defined to approximately: 10%; 50%; 100%; and 150% of the processing time. By doing so, it is possible to evaluate the influence of the setup time on the methods. If the results in problems where the setup time is 10% of the processing time and where the setup time is 150% of the processing time are similar, then it can be stated that the value of the setup time does not influence in which is the best method.

The processing time database was combined with each of the setup time database. Thus, a total of 2160 problems were analyzed.

The experimentation codes were written in C and ran on an Intel Core i7-8700 K 3.7 GHz, 16 GB RAM DDR4 3000 MHz using Microsoft Visual Studio Community 2017. The computational times were obtained through the function clock(). A time limit of 3600 s was set for the execution of the algorithm.

As no other lower bound for the $F_m|block, S_{ijk}|\sum C_{jm}$ was found in the literature, the tests were performed by comparing three variations of the proposed lower bound (LB). The purpose of the tests is to evaluate which variations of the LB solve the problems in a viable computational time. The three variations are:

- $LB_1$: $S_{i[1]k}$ and $S_{[1]jk}$ were defined as zero and were not included in the searches;
- $LB_2$: $Setup_k(j)$ was defined as zero and was not included in the searches;
- $LB_3$: LB was fully calculated.

To compare the performance of the lower bounds, the relative percentage deviation was calculated for the average number of created nodes and computational times, using the following equation:

$$RPD_{Variation} = \frac{\sigma_{Variation} - \sigma^*}{\sigma^*} * 100 \tag{10}$$

In Eq. 10, $\sigma_{Variation}$ is the mean value of either the computational time or the number of created nodes obtained with the LB variation analyzed and $\sigma^*$ is the best value obtained among all the LB variations.

Each problem was solved using all variations of the LB, and all algorithms were run in the same conditions. Tables 1, 2, 3, and 4 show the average CPU time spent, the average number of created nodes, and the number of unsolved problems considering database 1, 2, 3, and 4, respectively, for the setup times.

The RPD was calculated for all databases separately, and the results are shown in Table 5.

From Tables 1, 2, 3, and 4 it was possible to notice that the bigger the setup times, the smallest is the number of nodes for all variations of the LB. Thus, it can be noted that this factor is not significant for the comparisons. From Table 5 it was possible to observe that for all different databases the lower bound $LB_3$ was the one that most reduced the number of nodes. On the other hand, $LB_2$ solved more problems in less computational time. The results of $LB_1$ proved to be inefficient for all databases. As the most important measure for the B&B algorithm is computational time, it can be considered that the search of $Setup_k(j)$ is impracticable for the problem studied in this paper.

**Table 1.** Comparison of the variations of LB – Database 1

| Size | | Ave. node count | | | Ave. CPU time (ms) | | | Number of unsolved | | |
|---|---|---|---|---|---|---|---|---|---|---|
| n | m | $LB_1$ | $LB_2$ | $LB_3$ | $LB_1$ | $LB_2$ | $LB_3$ | $LB_1$ | $LB_2$ | $LB_3$ |
| 10 | 2 | 581 | 576 | **530** | 0.01 | **0.008** | 0.01 | **0** | **0** | **0** |
| 10 | 3 | 8760 | 8510 | **8390** | 0.071 | **0.06** | 0.069 | **0** | **0** | **0** |
| 10 | 4 | 44300 | 40400 | **40200** | 0.293 | **0.223** | 0.28 | **0** | **0** | **0** |
| 10 | 5 | 382000 | **381000** | 382000 | 1.904 | **1.722** | 1.944 | **0** | **0** | **0** |
| 10 | 7 | 696000 | **661000** | 672000 | 3.28 | **3.185** | 3.398 | **0** | **0** | **0** |
| 10 | 10 | **1560000** | 1580000 | 1600000 | **7.891** | 8.873 | 8.768 | **0** | **0** | **0** |
| 12 | 2 | 2170 | 2180 | **1930** | 0.061 | **0.043** | 0.058 | **0** | **0** | **0** |
| 12 | 3 | 99500 | 82700 | **81300** | 0.84 | **0.575** | 0.689 | **0** | **0** | **0** |
| 12 | 4 | 5510000 | 5260000 | **5210000** | 26.32 | **26.221** | 28.437 | **0** | **0** | **0** |
| 12 | 5 | 13500000 | **13200000** | 13300000 | 66.189 | **63.332** | 67.047 | **0** | **0** | **0** |
| 12 | 7 | 41600000 | 41500000 | **40900000** | 223.579 | **205.25** | 224.6 | **0** | **0** | **0** |
| 12 | 10 | 90900000 | **77800000** | 79400000 | 539.634 | **437.121** | 490.666 | **0** | **0** | **0** |
| 14 | 2 | 16900 | 18100 | **15100** | 0.644 | **0.456** | 0.578 | **0** | **0** | **0** |
| 14 | 3 | 8360000 | **2840000** | 8130000 | 57.063 | **21.574** | 57.114 | **0** | **0** | **0** |
| 14 | 4 | **73100000** | 111000000 | 109000000 | **427.835** | 550.287 | 595.402 | **1** | 2 | 2 |
| 14 | 5 | 205000000 | **193000000** | **193000000** | 1191.026 | **1076.282** | 1129.617 | 3 | **2** | 3 |
| 14 | 7 | 299000000 | 307000000 | **289000000** | 2101.254 | **1976.018** | 2098.374 | 8 | **7** | 8 |
| 14 | 10 | 589000000 | 614000000 | **575000000** | 3510.736 | **3495.732** | 3511.206 | **19** | **19** | **19** |

*(continued)*

**Table 1.** *(continued)*

| Size | | Ave. node count | | | Ave. CPU time (ms) | | | Number of unsolved | | |
|---|---|---|---|---|---|---|---|---|---|---|
| n | m | $LB_1$ | $LB_2$ | $LB_3$ | $LB_1$ | $LB_2$ | $LB_3$ | $LB_1$ | $LB_2$ | $LB_3$ |
| 16 | 2 | 104000 | 117000 | **92200** | 5.135 | **3.696** | 4.657 | **0** | **0** | **0** |
| 16 | 3 | 3460000 | 3530000 | **3180000** | 129.559 | **89.907** | 119.861 | **0** | **0** | **0** |
| 16 | 4 | 165000000 | **141000000** | 160000000 | 1105.745 | **878.71** | 1065.442 | 4 | **3** | 4 |
| 18 | 2 | 843000 | 1030000 | **750000** | 55.86 | **40.276** | 51.584 | **0** | **0** | **0** |
| 18 | 3 | **27100000** | 37800000 | 34600000 | 840.463 | **711.829** | 873.607 | **0** | 1 | 1 |
| 18 | 4 | **204000000** | 270000000 | 225000000 | 2410.37 | **2313.624** | 2359.329 | **9** | 10 | **9** |
| 20 | 2 | 6180000 | 7690000 | **5380000** | 495.834 | **381.78** | 452.126 | **0** | **0** | **0** |
| 20 | 3 | 72400000 | 62400000 | **51000000** | 2653.069 | **2306.895** | 2488.527 | 12 | **9** | 10 |
| 20 | 4 | **256000000** | 281000000 | 260000000 | **3600.001** | **3600.001** | **3600.001** | **20** | **20** | **20** |
| Average | | 76439526.33 | 80479313.56 | **76138653.70** | 720.54 | **673.84** | 712.35 | 2.81 | **2.70** | 2.81 |

**Table 2.** Comparison of the variations of LB – Database 2

| Size | | Ave. node count | | | Ave. CPU time (ms) | | | Number of unsolved | | |
|---|---|---|---|---|---|---|---|---|---|---|
| n | m | $LB_1$ | $LB_2$ | $LB_3$ | $LB_1$ | $LB_2$ | $LB_3$ | $LB_1$ | $LB_2$ | $LB_3$ |
| 10 | 2 | 1420 | 1310 | **1100** | 0.028 | **0.018** | 0.02 | **0** | **0** | **0** |
| 10 | 3 | 20800 | **19600** | 24400 | 0.191 | **0.152** | 0.187 | **0** | **0** | **0** |
| 10 | 4 | 72100 | 48800 | **45100** | 0.544 | **0.387** | 0.432 | **0** | **0** | **0** |
| 10 | 5 | **206000** | 242000 | 214000 | 1.382 | 1.355 | **1.333** | **0** | **0** | **0** |
| 10 | 7 | 519000 | **496000** | 513000 | 3.305 | **2.867** | 3.095 | **0** | **0** | **0** |
| 10 | 10 | 1310000 | **1200000** | 1210000 | **6.753** | 6.917 | 7.048 | **0** | **0** | **0** |
| 12 | 2 | 12400 | 12900 | **9590** | **0.183** | 0.248 | 0.277 | **0** | **0** | **0** |
| 12 | 3 | **703000** | 821000 | 800000 | 5.097 | 4.963 | **4.076** | **0** | **0** | **0** |
| 12 | 4 | 4090000 | **2430000** | 2790000 | 29.105 | **17.793** | 21.131 | **0** | **0** | **0** |
| 12 | 5 | **5250000** | 6080000 | 6140000 | 39.324 | **37.569** | 41.504 | **0** | **0** | **0** |
| 12 | 7 | **24600000** | 25500000 | 25400000 | 147.206 | **142.323** | 153.871 | **0** | **0** | **0** |
| 12 | 10 | 64900000 | 51100000 | **50300000** | 447.156 | **335.646** | 359.675 | **0** | **0** | **0** |
| 14 | 2 | 69600 | 74100 | **54900** | 2.657 | **1.658** | 2.177 | **0** | **0** | **0** |
| 14 | 3 | 8220000 | 6020000 | **5760000** | 97.614 | **70.584** | 72.341 | **0** | **0** | **0** |
| 14 | 4 | 66200000 | **32600000** | 34200000 | 550.658 | **300.324** | 369.081 | 1 | **0** | **0** |
| 14 | 5 | 178000000 | **135000000** | 161000000 | 1312.986 | **985.334** | 1193.124 | 4 | **2** | 4 |
| 14 | 7 | **302000000** | 394000000 | 315000000 | 2661.016 | 2828.815 | **2654.844** | **11** | **11** | **11** |
| 14 | 10 | 484000000 | 484000000 | **447000000** | 3365.358 | **3238.832** | 3308.02 | 18 | **16** | 17 |

(*continued*)

**Table 2.** (*continued*)

| Size | | Ave. node count | | | Ave. CPU time (ms) | | | Number of unsolved | | |
|---|---|---|---|---|---|---|---|---|---|---|
| n | m | $LB_1$ | $LB_2$ | $LB_3$ | $LB_1$ | $LB_2$ | $LB_3$ | $LB_1$ | $LB_2$ | $LB_3$ |
| 16 | 2 | 520000 | 595000 | **407000** | 27.855 | **19.407** | 22.54 | **0** | **0** | **0** |
| 16 | 3 | 42300000 | 33900000 | **31400000** | 724.443 | **539.128** | 646.615 | 2 | **1** | **1** |
| 16 | 4 | 262000000 | 198000000 | **192000000** | 2491.518 | **1731.481** | 1995.478 | 9 | **7** | **7** |
| 18 | 2 | 4530000 | 5420000 | **3550000** | 314.373 | **221.093** | 251.807 | **0** | **0** | **0** |
| 18 | 3 | 122000000 | 174000000 | **113000000** | 3171.886 | **2801.018** | 2947.936 | 15 | **12** | 14 |
| 18 | 4 | 120000000 | 150000000 | **99800000** | 3576.416 | **3407.678** | 3501.395 | 19 | **16** | 18 |
| 20 | 2 | 21800000 | 30700000 | **18600000** | 2053.918 | **1574.208** | 1736.984 | 3 | **2** | **2** |
| 20 | 3 | 59100000 | 86000000 | **57200000** | **3600.001** | **3600.001** | **3600.001** | **20** | **20** | **20** |
| 20 | 4 | 297000000 | 280000000 | **275000000** | **3600.001** | **3600.001** | **3600.001** | **20** | **20** | **20** |
| Average | | 76645345.19 | 77713359.63 | **68200707.04** | 1045.59 | **943.33** | 981.30 | 4.52 | **3.96** | 4.22 |

**Table 3.** Comparison of the variations of LB – Database 3

| Size | | Ave. node count | | | Ave. CPU time (ms) | | | Number of unsolved | | |
|---|---|---|---|---|---|---|---|---|---|---|
| n | m | $LB_1$ | $LB_2$ | $LB_3$ | $LB_1$ | $LB_2$ | $LB_3$ | $LB_1$ | $LB_2$ | $LB_3$ |
| 10 | 2 | 2350 | 2050 | **1580** | 0.039 | 0.032 | **0.03** | **0** | **0** | **0** |
| 10 | 3 | 25300 | 17200 | **17100** | 0.274 | **0.177** | 0.205 | **0** | **0** | **0** |
| 10 | 4 | 49500 | 49300 | **38400** | 0.542 | **0.421** | 0.446 | **0** | **0** | **0** |
| 10 | 5 | 153000 | 124000 | **118000** | 1.284 | **0.973** | 1.068 | **0** | **0** | **0** |
| 10 | 7 | 405000 | 296000 | **293000** | 2.272 | **2.124** | 2.431 | **0** | **0** | **0** |
| 10 | 10 | **698000** | 808000 | 771000 | **4.728** | 5.416 | 5.95 | **0** | **0** | **0** |
| 12 | 2 | 12000 | 11500 | **8660** | 0.317 | **0.218** | 0.26 | **0** | **0** | **0** |
| 12 | 3 | 255000 | **200000** | 250000 | 3.771 | **2.31** | 2.953 | **0** | **0** | **0** |
| 12 | 4 | **603000** | 748000 | 639000 | 10.668 | **8.469** | 9.621 | **0** | **0** | **0** |
| 12 | 5 | 3140000 | **2330000** | 2360000 | 35.531 | **22.606** | 27.407 | **0** | **0** | **0** |
| 12 | 7 | 11300000 | 11100000 | **10700000** | 97.695 | **82.675** | 90.788 | **0** | **0** | **0** |
| 12 | 10 | 33200000 | **20600000** | 22100000 | 266.644 | **171.832** | 218.601 | **0** | **0** | **0** |
| 14 | 2 | 76200 | 76200 | **55600** | 2.665 | **1.983** | 2.378 | **0** | **0** | **0** |
| 14 | 3 | **1480000** | 2290000 | 1590000 | **36.153** | 37.268 | 40.787 | **0** | **0** | **0** |
| 14 | 4 | 24100000 | **10500000** | 16500000 | 328.047 | **162.731** | 239.095 | **0** | **0** | **0** |
| 14 | 5 | 104000000 | **46500000** | 65900000 | 1052.792 | **554.925** | 795.59 | 2 | **0** | 1 |
| 14 | 7 | **216000000** | 223000000 | 219000000 | 2088.795 | **1897.037** | 2051.695 | **7** | **7** | **7** |
| 14 | 10 | 365000000 | 358000000 | **337000000** | 3335.304 | **3050.372** | 3238.649 | 15 | **14** | **14** |

(*continued*)

**Table 3.** (*continued*)

| Size | | Ave. node count | | | Ave. CPU time (ms) | | | Number of unsolved | | |
|---|---|---|---|---|---|---|---|---|---|---|
| n | m | $LB_1$ | $LB_2$ | $LB_3$ | $LB_1$ | $LB_2$ | $LB_3$ | $LB_1$ | $LB_2$ | $LB_3$ |
| 16 | 2 | 1270000 | 1060000 | **728000** | 57.811 | **33.615** | 37.202 | **0** | **0** | **0** |
| 16 | 3 | 31500000 | 33000000 | **28100000** | 619.392 | **455.64** | 527.823 | **1** | **1** | **1** |
| 16 | 4 | 104000000 | 110000000 | **92800000** | 2197.687 | **1757.055** | 1922.422 | **6** | **6** | 7 |
| 18 | 2 | 3480000 | 4190000 | **2640000** | 253.645 | **167.221** | 196.373 | **0** | **0** | **0** |
| 18 | 3 | 61500000 | 89000000 | **58100000** | 3342.785 | **2933.593** | 3050.296 | 14 | 11 | **10** |
| 18 | 4 | **140000000** | 183000000 | **140000000** | 3600.001 | **3546.032** | 3592.311 | 20 | **19** | **19** |
| 20 | 2 | 21600000 | 30500000 | **18600000** | 2077.88 | **1618.644** | 1799.302 | 6 | **1** | 2 |
| 20 | 3 | **62200000** | 93400000 | 64100000 | **3600.001** | **3600.001** | **3600.001** | **20** | **20** | **20** |
| 20 | 4 | 144000000 | 147000000 | **125000000** | **3600.001** | **3600.001** | **3600.001** | **20** | **20** | **20** |
| Average | | 49261087.04 | 50659342.59 | **44718901.48** | 985.80 | **878.27** | 927.91 | 4.11 | **3.67** | 3.74 |

**Table 4.** Comparison of the variations of LB – Database 4

| Size | | Ave. node count | | | Ave. CPU time (ms) | | | Number of unsolved | | |
|---|---|---|---|---|---|---|---|---|---|---|
| n | m | $LB_1$ | $LB_2$ | $LB_3$ | $LB_1$ | $LB_2$ | $LB_3$ | $LB_1$ | $LB_2$ | $LB_3$ |
| 10 | 2 | 1860 | 1560 | **1200** | 0.033 | **0.022** | 0.024 | **0** | **0** | **0** |
| 10 | 3 | 13200 | **11600** | 17900 | 0.174 | **0.156** | 0.187 | **0** | **0** | **0** |
| 10 | 4 | 51900 | 52600 | **40100** | 0.524 | **0.417** | 0.419 | **0** | **0** | **0** |
| 10 | 5 | 111000 | 96500 | **91900** | 1.108 | **0.834** | 0.94 | **0** | **0** | **0** |
| 10 | 7 | 305000 | **242000** | 249000 | 2.849 | **2.016** | 2.415 | **0** | **0** | **0** |
| 10 | 10 | 691000 | **526000** | 554000 | 4.369 | **4.197** | 4.892 | **0** | **0** | **0** |
| 12 | 2 | 17000 | 16500 | **11900** | **0.242** | 0.318 | 0.341 | **0** | **0** | **0** |
| 12 | 3 | 168000 | 135000 | **121000** | 3.159 | **2.036** | 2.497 | **0** | **0** | **0** |
| 12 | 4 | 859000 | 646000 | **593000** | 10.668 | **5.908** | 8.88 | **0** | **0** | **0** |
| 12 | 5 | 1060000 | **944000** | 1260000 | 16.945 | **9.8** | 17.235 | **0** | **0** | **0** |
| 12 | 7 | 6850000 | **6480000** | 6530000 | 70.271 | **57.843** | 67.667 | **0** | **0** | **0** |
| 12 | 10 | 21100000 | **17900000** | 18400000 | 222.954 | **164.728** | 196.393 | **0** | **0** | **0** |
| 14 | 2 | 90300 | 95800 | **65900** | 1.999 | **1.874** | 2.691 | **0** | **0** | **0** |
| 14 | 3 | 4120000 | 3060000 | **2000000** | 62.388 | 43.66 | **41.695** | **0** | **0** | **0** |
| 14 | 4 | **26700000** | 43300000 | 31200000 | 361.938 | 401.366 | **358.702** | **0** | **0** | **0** |
| 14 | 5 | 16800000 | **14400000** | **14400000** | 376.944 | **255.084** | 337.281 | **0** | **0** | **0** |
| 14 | 7 | **208000000** | 234000000 | 215000000 | 2362.521 | **2197.001** | 2275.122 | **8** | 9 | **8** |
| 14 | 10 | 260000000 | 259000000 | **236000000** | 3452.525 | **3085.414** | 3310.143 | 16 | **11** | 14 |

*(continued)*

**Table 4.** (*continued*)

| Size | | Ave. node count | | | Ave. CPU time (ms) | | | Number of unsolved | | |
|---|---|---|---|---|---|---|---|---|---|---|
| n | m | $LB_1$ | $LB_2$ | $LB_3$ | $LB_1$ | $LB_2$ | $LB_3$ | $LB_1$ | $LB_2$ | $LB_3$ |
| 16 | 2 | 464000 | 560000 | **365000** | 25.5 | **18.455** | 20.377 | **0** | **0** | **0** |
| 16 | 3 | 10900000 | 10900000 | **7580000** | 435.885 | **279.153** | 317.886 | **0** | **0** | **0** |
| 16 | 4 | 81400000 | **73200000** | 74600000 | 2695.172 | **1872.869** | 2437.614 | 9 | **4** | 6 |
| 18 | 2 | 3680000 | 4570000 | **2730000** | 265.789 | **178.755** | 203.287 | **0** | **0** | **0** |
| 18 | 3 | 45400000 | 67100000 | **44600000** | 2864.808 | **2692.193** | 2823.943 | 13 | **9** | 12 |
| 18 | 4 | 132000000 | **121000000** | 130000000 | 3531.263 | **3492.868** | 3507.656 | **19** | **19** | **19** |
| 20 | 2 | 14700000 | 22700000 | **12300000** | 1417.233 | **1157.732** | 1159.092 | **1** | **1** | **1** |
| 20 | 3 | 64900000 | 93100000 | **62300000** | **3600.001** | **3600.001** | **3600.001** | **20** | **20** | **20** |
| 20 | 4 | 140000000 | 149000000 | **100000000** | **3600.001** | **3600.001** | **3600.001** | **20** | **20** | **20** |
| Average | | 38532676.30 | 41593983.70 | **35592996.30** | 940.27 | **856.47** | 899.90 | 3.93 | **3.44** | 3.70 |

**Table 5.** RPD of the variations of the LB for all databases

| Database | Ave. node count | | | Ave. CPU time (ms) | | | Number of unsolved | | |
|---|---|---|---|---|---|---|---|---|---|
| | $LB_1$ | $LB_2$ | $LB_3$ | $LB_1$ | $LB_2$ | $LB_3$ | $LB_1$ | $LB_2$ | $LB_3$ |
| Database 1 | **0.00** | 0.06 | **0.00** | 0.07 | **0.00** | 0.06 | 0.04 | **0.00** | 0.04 |
| Database 2 | 0.12 | 0.14 | **0.00** | 0.11 | **0.00** | 0.04 | 0.14 | **0.00** | 0.07 |
| Database 3 | 0.10 | 0.13 | **0.00** | 0.12 | **0.00** | 0.06 | 0.12 | **0.00** | 0.02 |
| Database 4 | 0.08 | 0.17 | **0.00** | 0.10 | **0.00** | 0.05 | 0.14 | **0.00** | 0.08 |
| Average | 0.08 | 0.13 | **0.00** | 0.10 | **0.00** | 0.05 | 0.11 | **0.00** | 0.05 |

## 5 Conclusion

This paper considers a blocking permutation flow shop scheduling problem with the total flow time criterion, which is known to be NP-hard for $m \geq 2$. A machine based lower bound that exploits the occurrence of blocking and idle time (LB) is proposed and applied to the branch-and-bound algorithm. Four setup time databases were generated, each with different values range. The tests show that this range value did not affect the LB performance.

As no other lower bound was found in the literature, three different ways of applying the proposed LB were studied and compared. $LB_1$ got the worst computational time and was the variation that solved less problems within 3600 s, also got the second worst number of nodes. So, it can be concluded that including both $S_{i[1]k}$ and $S_{[1]jk}$ in the search is important for the LB.

Overall $LB_3$ got the smallest number of nodes and $LB_2$ got the smallest computational time and number of unsolved problems. This can mean that including $Setup_k(j)$ in the search is not interesting for the LB.

For future works, the development of a dominance rule for pruning a higher number of nodes and the use of efficient heuristics as initial upper bounds are suggested. These suggestions aim at improving the algorithm, while applied along with $LB_2$. Furthermore, other exploration rules may be tested, such as the best bound rule, or a hybrid technique. Considering the efficiency of the proposed LB, another suggestion is to apply the rules used for incorporating the blocking constraints in this LB into a general case with a limited buffer greater than or equal to zero.

## References

1. Garey, M.R., Johnson, D.S., Sethi, R.: The complexity of flowshop and jobshop scheduling. Math. Oper. Res. **1**(2), 117–129 (1976)
2. Nawaz, M., Enscore, E.E., Ham, I.: A heuristic algorithm for the m-machine, n-job flow-shop sequencing problem. Omega **11**(1), 91–95 (1983)
3. Mccormick, S., Pinedo, M.J. Shenker, S., Wolf, B.: Sequencing in an assembly line with blocking to minimize cycle time. Oper. Res. **37**(6), 925–935 (1989)

4. Pan, Q.K., Wang, L.: Effective heuristics for the blocking flowshop scheduling problem with makespan minimization. Omega **40**(2), 218–229 (2012)
5. Ronconi, D.P.: A note on constructive heuristics for the flowshop problem with blocking. Int. J. Prod. Econ. **87**(1), 39–48 (2004)
6. Hall, N.G., Sriskandarajah, C.: A survey of machine scheduling problems with blocking and no-wait in process. Oper. Res. **44**(3), 510–525 (1996)
7. Sethi, S.P., Sriskandarajah, C., Sorger, G., Blazewicz, J., Kubiak, W.: Sequencing of parts and robot moves in a robotic cell. Int. J. Flex. Manufact. Syst. **4**(3), 331–358 (1992)
8. Reddi, S., Ramamoorthy, C.: On the flow-shop sequencing problem with no wait in process. J. Oper. Res. Soc. **23**(3), 323–331 (1972)
9. Gilmore, P.C., Gomory, R.E.: Sequencing a one state-variable machine: A solvable case of the traveling salesman problem. Oper. Res. **12**(5), 655–679 (1964)
10. Ignall, E., Schrage, L.: Application of the branch and bound technique to some flow-shop scheduling problems. Oper. Res. **13**(3), 400–412 (1965)
11. Bansal, S.P.: Minimizing the sum of completion times of n jobs over m machines in a flowshop: a branch and bound approach. A I I E Trans. **9**(3), 306–311 (1977). https://doi.org/10.1080/05695557708975160
12. Ahmadi, R.H., Bagchi, U.: Improved lower bounds for minimizing the sum of completion times of n jobs over m machines in a flow shop. Eur. J. Oper. Res. **44**(3), 331–336 (1990). https://doi.org/10.1016/0377-2217(90)90244-6
13. Chung, C.S., Flynn, J., Kirca, O.: A branch and bound algorithm to minimize the total flow time for m-machine permutation flowshop problems. Int. J. Prod. Econ. **79**(3), 185–196 (2002). https://doi.org/10.1016/S0925-5273(02)00234-7
14. Ronconi, D.P., Armentano, V.A.: Lower bounding schemes for flowshops with blocking in-process. J. Oper. Res. Soc. **52**(11), 1289–1297 (2001). https://doi.org/10.1057/palgrave.jors.2601220
15. Ronconi, D.P.: A branch-and-bound algorithm to minimize the makespan in a flowshop with blocking. Ann. Oper. Res. **138**(1), 53–65 (2005). https://doi.org/10.1007/s10479-005-2444-3
16. Moslehi, G., Khorasanian, D.: Optimizing blocking flow shop scheduling problem with total completion time criterion. Comput. Oper. Res. **40**(7), 1874–1883 (2013). https://doi.org/10.1016/j.cor.2013.02.003
17. Taillard, E.: Benchmarks for basic scheduling problems. Eur. J. Oper. Res. **64**(2), 278–285 (1993)
18. Sanches, F.B., Takano, M.I., Nagano, M.S.: Evaluation of heuristics for a branch and bound algorithm to minimize the makespan in a flowshop with blocking. Acta Scientiarum Technol. **38**(3), 321 (2016)
19. Rios-Mercado, R.Z., Bard, J.F.: A branch-and-bound algorithm for flowshop scheduling with setup times. IIE Trans. Sched. Logist. **31**(8), 721–731 (1999)
20. Takano, M.I., Nagano, M.S.: A branch-and-bound method to minimize the makespan in a permutation flow shop with blocking and setup times. Cogent Eng. **4**(1) (2017). https://doi.org/10.1080/23311916.2017.1389638
21. Pinedo, M.L.: Scheduling: Theory, Algorithms, and Systems, 3rd edn. Springer Publishing Company, New York (2008)

# Production Planning in a Seru Production System, Considering Heterogeneity to Balance Production Times and Minimize Energy Consumption

Sofía M. Escobar Forero and Ciro A. Amaya Guio(✉)

Department of Industrial Engineering, Universidad de Los Andes, Bogota, Colombia
{sm.escobar,ca.amaya}@uniandes.edu.co

**Abstract.** With the rise of Industry 4.0 an era of real-time global connectivity between customers and the processes that transform consumer goods is imminent. Faced with the need to satisfy the current dynamic demand, the Japanese technology industry has developed an innovative production system called seru, which is distinguished by its satisfactory results in economic and environmental terms. This study poses a production planning problem in a seru production system considering heterogeneous workers and the energy consumption of production. To solve it, a multi-objective mathematical model is proposed in order to balance the production time between workers (intra-seru) and between serus (inter-seru) and minimize the total energy consumption. For larger instances, a metaheuristic algorithm is developed based on the evolutionary algorithm NSGA-II. Finally, the effectiveness of the proposed genetic algorithm is evaluated by comparing the area under the curve obtained with the solutions of the pareto frontier that it generates and the area under the curve from the pareto frontier of the multi-objective linear optimization model created via $\varepsilon$ - constraint.

**Keywords:** Seru production system · Production planning · Line balancing · Heterogeneity · Energy consumption · $CO_2$ Emissions

## 1 Introduction

The seru production system (SPS) is recognized for its economic and environmental benefits (double E) [2]. Due to the growing concern about environmental problems, studies related to sustainable production systems have acquired greater relevance. However, the SPS has not been widely studied in this area. According to [2], only those companies that lead a transformation towards environmental efficiency will be competitive. On the contrary, those that do not modify their traditional production processes will incur in high costs for waste,

© Springer Nature Switzerland AG 2021
D. A. Rossit et al. (Eds.): ICPR-Americas 2020, CCIS 1407, pp. 233–247, 2021.
https://doi.org/10.1007/978-3-030-76307-7_18

fines for violation of environmental legislation and high costs for energy consumption.

According to a 2020 report by UNEP (United Nations Environment Program), the record for global concentrations of carbon dioxide has been reached despite the COVID-19 crisis. Through the drilling of an ice core it was determined that in the last 800,000 years the current concentration of $CO_2$ in the atmosphere had never been reached, 416.21 ppm. According to the World Energy Outlook in 2018 industrial activities generated 24% of global $CO_2$ emissions.

Anticipating the growing need to develop manufacturing techniques to stay competitive, technology company Sony implemented SPS in 1992. Later, other companies in the technology industry such as Canon adapted this production system [4]. Both companies experienced economic and environmental benefits that have caught the attention of academics, regulators and companies [2]. Among the benefits there is a decrease in assembly time, in the number of workers required, in the space used for production, in the cost of machinery, in energy consumption and $CO_2$ emissions [7].

To ensure the effective implementation of SPS. [5] proposed three problems that should be studied: 1) determine the correct moment to implement this new production system, 2) determine a suitable allocation plan of workers and products and 3) develop a practical production planning system. This study focuses on the production planning in a SPS through the assignment of workers-serus, batches-serus and workers-tasks. The problem considers heterogeneous workers, that is, they are trained to perform a specific set of tasks. Additionally, it considers the proficiency of each worker to perform a particular task. According to [5] the aforementioned allocation problem can be associated with a traditional allocation problem with two levels, the first level being the allocation of workers to serus and the second level being the allocation of products to serus. These levels generate an additional worker-task assignment problem.

The characteristics of the SPS allow it to respond quickly to the fluctuating demand of Industry 4.0. This is defined as a type of manufacturing cell differentiated by two main characteristics, it is reconfigurable and includes additional operations to the production of the product such as packaging and quality processes. Its use is justified by three global advantages: increased productivity, quality and flexibility. Additionally, this system provides a *trade off* between efficiency (lean manufacturing) and responsiveness (agile manufacturing) [4].

In recent years, multiple investigations have been carried out on the seru production system given its results in real cases of the Japanese technology industry. [2] proposed a bi-objective mathematical to minimize $CO_2$ emissions from production considering the energy consumption to produce a unit of a product and decrease the makespan. Additionally, they create a NSGA-II based algorithm to solve larger instances, as proposed by [3]. On the other hand, [10] simultaneously addressed minimizing the total cost of production and energy consumption in an assembly line. Additionally, it considered the skills set of each of the workers to make a more efficient assignment. For small instances, they presented the mathematical formulation in order to generate an optimal pareto frontier through the $\varepsilon$

constraint method. They also design three metaheuristic solutions using an NSGA-II, MOSA (multi-objective simulated annealing) and a PT-EC SFR (processing time and energy consumption sorted-first rule).

Additionally, because this Japanese production system recognizes workforce as the main resource [8], researchers have shown interest in the heterogeneity of workers to carry out the allocation problem. [1] proposed a mathematical model that balances the production time between serus (inter-seru) and between workers (intra-seru) considering the heterogeneity and proficiency of each worker. Using the weighted sum method, they give equal weight to both objective functions of the exact model. For larger instances they used a NSGA-II based algorithm. [11] have investigated the problem of assigning and training workers. To do this, they proposed a multi-objective mathematical model that aims to minimize the total cost composed of the cost of training and the cost of balancing the processing times of the workers of the same seru. Additionally, they design a two-phase heuristic called the SAIG algorithm, based on the concepts of the SA (simulated annealing) metaheuristics and the IG (iterated greedy) algorithms. [12] investigated the same problem and presented the same objective functions, under a scenario where an assembly line is reconfigured into multiple serus. To solve the assignment problems between task-worker and worker-seru, they developed a three-stage heuristic algorithm. [13], proposed a bi-objective mathematical model that minimizes the makespan and the workload imbalance considering workers transferring between serus. Additionally, created a NSGA-II based algorithm given the non-polynomial (NP-hard) nature of the worker assignment problem.

## 2 Problem Description

The production planning problem is established in a seru production system considering heterogeneous workers. To generate a solution, the batches size must be determined for each type of product, the allocation of batches and workers to the serus and the assignment of tasks to workers according to their skills and proficiency. To carry out the assignment, it must be considered that each worker is trained to carry out a set of tasks $subS \subseteq S$. Additionally, each worker has a proficiency that determines their ability to perform a specific task. Each of the products has a deterministic demand that must be satisfied. To do so, it is possible to distribute the demand of a specific product in as many batches as serus exist, allowing a flexible allocation.

## 3 Mathematical Formulation

### 3.1 Model Assumptions

As assumptions of the model, we have: **1.** The demand for the products is known and deterministic. **2.** The available time of each worker is known. **3.** All products can be produced in any seru. **4.** The upper bound of workers in a seru is known.

**5.** The task processing time of each product is known and deterministic. **6.** Workers can only perform tasks of their skills set. **7.** The workers' proficiency is constant. The learning curve is not considered. **8.** workers transfer between serus is not allowed. **9.** Batch transfer between serus is not allowed. **10.** Only one worker can be assigned to perform a batch's task. **11.** Each worker of a seru must perform at least one task from each batch assigned to the same seru. **12.** The travel time of workers between one workstation to another is negligible. **13.** The reconfiguration cost of the serus is negligible. **13.** The electrical power $[kW]$ of the machines used to perform the tasks is constant and known.

### 3.2 Notation

To solve small instances, the following mathematical formulation is presented. The formulation presented by [1] was taken as a basis to solve the production planning problem. They proposed a bi-objective model that aims to balance production time between workers and between serus. In the formulation presented in its model, the batches size of each type of product is known, unlike the mathematical formulation of this study where a variable is modeled to determine the batches size in order to allow a better balance of production time. Additionally, the product demand is allowed to be produced in all the available serus. Additionally, the model presented by [2] was considered to model the calculation of energy consumption depending on the machine used for each task and the time of use of this.

**Sets.** The following sets are defined: $C$: set of serus $\{1, ..., n_C\}$, $B$: set of batches $\{1, ..., n_B\}$, $S$: set of tasks $\{1, ..., n_S\}$, $P$: set of products $\{1, ..., n_P\}$, $W$: set of workers $\{1, ..., n_W\}$.

**Parameters.** To model the proficiency of each worker $w$ to perform a task $s$, the parameter $f_{w,s}$ is considered. Being $t_{p,s}$ the time required to carry out a task $s$ of a type of product $p$ in minutes, those workers with a proficiency lower than 1 finish the task in less time with respect to the standard time $t_{p,s}$. On the contrary, if the proficiency of a worker is greater than 1, it indicates that he is slower to perform the task, taking a longer time with respect to the standard time $t_{p,s}$.

To calculate the energy consumption of performing a task $s$ of a product $p$ by a worker $w$, the parameters $pot_s$, $t_{p,s}$ are required and $f_{w,s}$. The first parameter ($pot_s$) refers to the electrical power in kilowatts [kW] of the machine with which the task $s$ is carried out. The multiplication of the standard time required to perform a task $t_{p,s}$, the proficiency of the worker who performs the task $f_{w,s}$ and the electrical power of the machine used to perform it $pot_s$, allows to obtain the energy consumption of performing the task $s$ in kilowatts per minute $[kW * min]$. Additionally, the parameter $\zeta$ indicates the $CO_2$ emission factor $[gCO_2/kWmin]$. This allows determining the $CO_2$ emissions generated in production from energy consumption.

$d_p$, $M$, $N$, $G$, $E$ are also defined. The first refers to the demand of a product $p$, the second and third, represent the upper bound of workers assigned to a seru and of tasks assigned to a worker in a batch. Additionally, the parameter $G$ represents the total available time of each worker in minutes and the parameter $E$ is the upper bound of grams of $CO_2$ generated in production.

Finally, three binary parameters are defined. $x_{b,p}$ indicates whether batch $b$ belongs to product $p$; $y_{p,s}$ represents if the task $s$ is required by the product $p$; finally, the parameter $e_{w,s}$ indicates whether the worker $w$ is trained to do the task $s$.

**Decision Variables.** The binary variable $l_{b,c}$ is defined, which indicates whether the batch $b$ is assigned to the seru $c$. The binary variable $a_{w,}$ takes the value of 1 if worker $w$ is assigned to seru $c$. The binary variable $z_{w,b,s}$, represents whether the worker $w$ performs the task $s$ of batch $b$. Additionally, the binary variable $ex_b$ indicates whether batch $b$ is used or not in production. Finally, the variable $q_b$ represents the batch size of $b$.

The proposed optimization model has two objective functions. The first objective function ($FO_1$) aims to balance the production times between workers and between serus. The second objective function ($FO_2$) aims to minimize the total energy consumption of production. The first objective function is composed of two parts: the inter-seru balance (between serus) and the intra-seru balance (between workers). Equations (1) and (2) represent the variances of the production times of serus and workers, respectively. The $FO_1$ is formulated as the sum of these variances (3). Next, the mathematical formulation of the model is presented.

$$WO_1 = \frac{1}{C}\sum_{c=1}^{C}\left[\sum_{w=1}^{W}\sum_{b=1}^{B}\sum_{p=1}^{P}\sum_{s=1}^{S} x_{b,p}y_{p,s}e_{w,s}l_{b,c}a_{w,c}z_{w,b,s}q_b t_{p,s}f_{w,s} - \frac{1}{C}\sum_{c=1}^{C}\sum_{w=1}^{W}\sum_{b=1}^{B}\sum_{p=1}^{P}\sum_{s=1}^{S} x_{b,p}y_{p,s}e_{w,s}l_{b,c}a_{w,c}z_{w,b,s}q_b t_{p,s}f_{w,s}\right]^2 \quad (1)$$

$$WO_2 = \frac{1}{W}\sum_{W=1}^{w}\left[\sum_{c=1}^{C}\sum_{b=1}^{B}\sum_{p=1}^{P}\sum_{s=1}^{S} x_{b,p}y_{p,s}e_{w,s}l_{b,c}a_{w,c}z_{w,b,s}q_b t_{p,s}f_{w,s} - \frac{1}{W}\sum_{w=1}^{W}\sum_{c=1}^{C}\sum_{b=1}^{B}\sum_{p=1}^{P}\sum_{s=1}^{S} x_{b,p}y_{p,s}e_{w,s}l_{b,c}a_{w,c}z_{w,b,s}q_b t_{p,s}f_{w,s}\right]^2 \quad (2)$$

Minimize

$$FO_1 = WO_1 + WO_2 \quad (3)$$

$$FO_2 = \sum_{c=1}^{C}\sum_{b=1}^{B}\sum_{p=1}^{P}\sum_{s=1}^{S}\sum_{w=1}^{W} x_{b,p}y_{p,s}e_{w,s}l_{b,c}a_{w,c}z_{w,b,s}q_bt_{p,s}f_{w,s}pot_s \tag{4}$$

Subject to,

$$\sum_{c=1}^{C} l_{b,c} = ex_b \quad \forall \quad b \in B \tag{5}$$

$$\sum_{c=1}^{C} a_{w,c} = 1 \quad \forall \quad w \in W \tag{6}$$

$$z_{w,b,s} \leq e_{w,s} \quad \forall \quad w \in W, b \in B, s \in S \tag{7}$$

$$\sum_{w=1}^{W} z_{w,b,s} = \sum_{p=1}^{P} x_{b,p}y_{p,s}ex_b \quad \forall \quad b \in B, s \in S \tag{8}$$

$$\sum_{w=1}^{W} a_{w,c} \leq N \quad \forall \quad c \in C \tag{9}$$

$$\sum_{s=1}^{S} z_{w,b,s} \leq M \quad \forall \quad b \in B, w \in W \tag{10}$$

$$\sum_{w=1}^{W} a_{w,c}e_{w,s} \geq \sum_{p=1}^{P} l_{b,c}x_{b,p}y_{p,s} \quad \forall \quad c \in C, s \in S, b \in B \tag{11}$$

$$z_{w,b,s} \leq \sum_{c=1}^{C} l_{b,c}a_{w,c} \quad \forall \quad w \in W, b \in B, s \in S \tag{12}$$

$$\sum_{p=1}^{P}\sum_{s=1}^{S} x_{b,p}y_{p,s}z_{w,b,s} \geq \sum_{c=1}^{C} l_{b,c}a_{w,c} \quad \forall \quad b \in B, w \in W \tag{13}$$

$$\sum_{b=1}^{B} q_bx_{b,p} \geq d_p \quad \forall \quad p \in P \tag{14}$$

$$ex_b \sum_{p=1}^{P} d_px_{b,p} \geq q_b \quad \forall \quad b \in B \tag{15}$$

$$ex_b \leq q_b \quad \forall \quad b \in B \tag{16}$$

$$\sum_{c=1}^{C}\sum_{b=1}^{B}\sum_{p=1}^{P}\sum_{s=1}^{S} x_{b,p}y_{p,s}e_{w,s}l_{b,c}a_{w,c}z_{w,b,s}q_bt_{p,s}f_{w,s} \leq G \quad \forall \quad w \in W \tag{17}$$

$$\sum_{c=1}^{C}\sum_{b=1}^{B}\sum_{p=1}^{P}\sum_{s=1}^{S}\sum_{w=1}^{W} x_{b,p}y_{p,s}e_{w,s}l_{b,c}a_{w,c}z_{w,b,s}q_bt_{p,s}f_{w,s}pot_s\zeta \leq E \tag{18}$$

$$l_{b,c}, a_{w,c}, z_{w,b,s}, ex_{(b)} \in \{0,1\}, \\ q_b \in z^+ \tag{19}$$

As mentioned before, from Eqs. (1) and (2) the $FO_1$ is modeled as the sum of the variances of the production times of the serus and of the workers (3). It should be clarified that the model assumes that a product cannot be processed in the next workstation until it is completely processed in the immediately preceding workstation. With respect to the $FO_2$, the total energy consumption in production is obtained from the product between the total time required to perform a task and the power in kilowatts of the machine used to carry out that task (4). The result of this objective function is the energy consumption measured in $kilowatts * minute$.

The model restrictions are found in Eqs. (5)–(19). The first restriction (5) ensures that each batch is assigned to a single seru. Similarly, constraint (6) ensures that each worker is assigned to only one seru. The third restriction guarantees that a task is performed only by workers that are trained to perform the task (7). Constraint (8) ensures that each task in each batch can only be performed by one worker, as long as the batch size is greater than zero. Restriction (9) limits the number of workers in each seru to $N$ and restriction (10) limits the number of tasks assigned to a worker in a batch to $M$. Restriction (11) guarantees that for each task in a batch assigned to a seru there is at least one worker trained to carry out that task in the same seru. Restriction (12) prohibits a task from being performed by a worker that is not in the same seru as the batch that requires that task. Restrictions (11) and (12) prevent movement of batches and workers between serus. Restriction (13) ensures that all workers perform at least one task from each batch assigned to the same seru. Constraint (14) ensures that the demand for each product is satisfied. Constraints (15) and (16) model the relationship between the variables $ex_b$ and $q_b$. The restriction (17) ensures that the total time used by a worker in production does not exceed its available time and the restriction (18) guarantees that the upper bound of grams of $CO_2$ generated in production is not exceeded.

The model presented is not linear given that there are multiplication of variables in the restrictions and in the objective functions. To obtain a linear model, the auxiliary variables $D_{c,b,w} = l_{b,c} \cdot a_{w,c}$, $H_{c,b,w,s} = D_{c,b,w} \cdot z_{w,b,s}$ and $R_{c,b,w,s} = H_{c,b,w} \cdot q_b$ are created. $D_{c,b,w}$ and $H_{c,b,w,s}$ are binary variables since they result from the multiplication of binary variables. The variable $R_{c,b,w,s}$ belongs to the set of positive integers including zero. To model these auxiliary variables the following restrictions are added to the model:

$D_{c,b,w}$ linearization:

$$D_{c,b,w} \leq l_{b,c} \forall c, b, w \tag{20}$$

$$D_{c,b,w} \leq a_{w,c} \forall c, b, w \tag{21}$$

$$D_{c,b,w} \geq l_{b,c} + a_{w,c} - 1 \forall c, b, w \tag{22}$$

$H_{c,b,w,s}$ linearization:

$$H_{c,b,w,s} \leq D_{c,b,w} \quad \forall \quad c \in C, b \in B, w \in W, s \in S \tag{23}$$

$$H_{c,b,w,s} \leq z_{w,b,s} \quad \forall \quad c \in C, b \in B, w \in W, s \in S \tag{24}$$

$$H_{c,b,w,s} \geq D_{c,b,w} + z_{w,b,s} - 1 \quad \forall \quad c \in C, b \in B, w \in W, s \in S \tag{25}$$

$R_{c,b,w,s}$ linearization:

$$R_{c,b,w,s} \geq q_b - (1 - H_{c,b,w,s}) \sum_{p=1}^{P} d_p x_{b,p} \quad \forall \quad c \in C, b \in B, w \in W, s \in S \tag{26}$$

$$R_{c,b,w,s} \leq q_b \quad \forall \quad c \in C, b \in B, w \in W, s \in S \tag{27}$$

$$R_{c,b,w,s} \leq H_{c,b,w,s} \sum_{p=1}^{P} d_p x_{b,p} \quad \forall \quad c \in C, b \in B, w \in W, s \in S \tag{28}$$

The above restrictions are added to the model and the auxiliary variables are replaced in Eqs. (12), (13), (17) and (18). To linearize the model, it must also be considered that $FO_1$ has quadratic terms. According to [6] the minimization of the sum of squares contained in the variance of production times between serus and between workers, can approximate the difference between the maximum and the minimum time of production of serus and workers, respectively. However, the maximum and minimum functions are not linear. To linearize them, four additional variables are created each with an associated restriction that forces them to take the maximum or minimum value of the inter-seru or intra-seru production time. In Eqs. (29–32) the new variables of the model are defined. These constraints and auxiliary variables must also be added to the model. All these new auxiliar variables are continuous and non-negative.

$$I \geq \sum_{w=1}^{W} \sum_{b=1}^{B} \sum_{p=1}^{P} \sum_{s=1}^{S} x_{b,p} y_{p,s} e_{w,s} R_{c,b,w,s} t_{p,s} f_{w,s} \quad \forall \quad c \in C \tag{29}$$

$$J \leq \sum_{w=1}^{W} \sum_{b=1}^{B} \sum_{p=1}^{P} \sum_{s=1}^{S} x_{b,p} y_{p,s} e_{w,s} R_{c,b,w,s} t_{p,s} f_{w,s} \quad \forall \quad c \in C \tag{30}$$

$$K \geq \sum_{c=1}^{C} \sum_{b=1}^{B} \sum_{p=1}^{P} \sum_{s=1}^{S} x_{b,p} y_{p,s} e_{w,s} R_{c,b,w,s} t_{p,s} f_{w,s} \quad \forall \quad w \in W \tag{31}$$

$$U \leq \sum_{c=1}^{C} \sum_{b=1}^{B} \sum_{p=1}^{P} \sum_{s=1}^{S} x_{b,p} y_{p,s} e_{w,s} R_{c,b,w,s} t_{p,s} f_{w,s} \quad \forall \quad w \in W \tag{32}$$

Replacing the new variables $I$, $J$, $K$ and $U$ in the equations $WO_1$ and $WO_2$ we get:

$$WO_1 = \frac{1}{C}(I - J) \quad (33) \qquad WO_2 = \frac{1}{W}(K - U) \quad (34)$$

# 4 NSGA-II Based Algorithm

The following is the NSGA-II based genetic algorithm [3] developed to solve large instances of the production planning problem. To create the structure of the chromosomes, the encoding presented by [1] was taken from the layers whose alleles represent the decision variables $l_{b,c}$, $a_{w,c}$ and $z_{w,b,s}$, this means, the second and third layer.

## 4.1 Encoding

Each chromosome is made up of three layers. Said layers are made up of alleles that represent the values of the decision variables presented in the mathematical formulation of the problem. The alleles in the first layer represent the value of the decision variable $q_b$. Let $n_C$ be the number of serus available in production, $n_P$ the number of types of product demanded and $n_B$ the number of batches available, the length of the first layer is $L_1 = n_C * n_P = n_B$. This structure allows each product to be distributed in as many batches as serus are available. The parameter $x_{b,p}$ defined above, determines the type of product for each batch. In order to guarantee compliance with the demand for each type of product, the sum of the batch sizes of a type of product must be equal to the value of its demand. The batch size is allowed to be zero, therefore, the binary decision variable $ex_b$ will also take this value.

The second layer of the chromosome determines the assignment of batches and workers to the serus. The alleles of this layer can take values that are within the set of serus ($C = 1, \ldots, n_C$). Consequently, alleles allow setting the values of the decision variables $l_{b,c}$ and $a_{w,c}$. This layer is made up of two lists. The length of the first list is $n_B$ and the length of the second list is $n_W$, that is, the number of available workers. Thus, the second layer has a length $L_2 = n_B + n_W$. The positions of the genes in each list make it possible to identify which batch or worker is being referred to and the allele value of each gene determines the seru to which the corresponding batch or worker was assigned. In the case in which a batch is assigned a value of zero units in the first layer ($q_b = 0$), the position of that batch in the first list of the second layer will have an allele with the value 'None'. In order to generate better solutions, it is guaranteed that each seru has at least one worker and one assigned batch. This is achieved by assigning each allele with each seru to a gene on the first and second lists of the second layer of the chromosome. Batches and workers that have not been assigned are randomly assigned to any seru.

The last layer of the chromosome allows to determine the value of the decision variable $z_{w,b,s}$. This layer is made up of as many lists as batches. Each of the lists represent a batch ($B = 1, \ldots, n_B$). A list has as many genes as tasks required by the corresponding batch. Thus, with $S_b$ being the number of tasks in a batch $b$, the length of the third layer is $\sum_{b=1}^{B} S_b$. The position of each gene within a list represents a task from that batch. The allele of each gene determines the worker ($W = \{1, \ldots, n_W\}$) who will perform the task. In the case in which a batch is assigned a value of zero units in the first layer of the chromosome ($q_b = 0$), the list corresponding to that batch will be empty in the third layer. For the worker-task assignment, it must be ensured that if a worker is to perform the task of a batch, they must be trained to perform it. Additionally, it must be in the same seru of the batch with the task that was assigned to it.

### 4.2 Feasibility Correction

Once the initial population is created, infeasible chromosomes may appear that violate some of the constraints of the linear model. One of the infeasibility may be due to some tasks not being assigned. This infeasibility refers to the alleles of the third layer that take values of 'None'. This happens when there is no competent worker to perform a specific task in a seru. To correct this infeasibility, two scenarios are considered. In the first scenario, the infeasible batch is the only one in the seru. In this case, workers from other serus skilled to perform the problematic task and if they are not the only workers from their serus , should be selected. When one is randomly selected the second and third layer are modified. In the second case, the infeasible task does not belong to the only batch in the seru. In this scenario, the infeasibility is resolved by sending the problematic batch to another seru where the assigned workers can execute the task.

Another infeasibility is that the maximum limit of workers per seru is exceeded. Again, two particular cases are conceived to correct this infeasibility. In the first case, the seru that exceeds the maximum number of workers has a worker who is not assigned to any task in the seru. In this case, the serus that could receive an additional worker without becoming infeasible must be identified and the worker assigned to one of them randomly. The second case considers redundant workers that could be sent to another seru without affecting the third layer. Redundant workers are those workers whose useful skills in the seru are contained in the skills of the other workers of the same seru. Once a redundant worker is sent to a seru that can receive an additional worker, the second and third layers must be updated.

Additionally, it is not allowed for a worker to be assigned more than $M$ tasks from the same batch. This infeasibility is prevented in the creation of the third layer of the chromosome when, when identifying the competent workers to perform a specific task, the number of times that the alleles of the batch have taken the value of that worker in the third layer is counted. If the worker is trained to do the task but has already been assigned $M$ times to tasks of that batch, it is not taken into account to be assigned to more tasks of the same batch.

Finally, it may be occur that a worker assigned to a seru is not assigned to any task from each batch of the seru. To avoid this infeasibility, the creation of the third layer seeks to balance the load of workers so that if several workers are competent to carry out a task, the one who has assigned, so far, the least amount of tasks in the batch is selected. Even with this balancing, it is possible to obtain chromosomes that present this infeasibility. This is corrected by assigning the infeasible worker a task of another worker from the same seru in which he is competent, provided that this is not the other worker's only task.

### 4.3 Fitness y Penalty Function

Fitness is calculated from the objective functions presented in the mathematical formulation and two penalty functions. The first step to obtain fitness is to determine the values of the decision variables according to the values that the alleles of the chromosome have taken in each layer. Subsequently, it is possible to obtain the value of the objective functions.

Now, there are still some infeasible chromosomes even after correcting the infeasibilities mentioned above. This is because the mathematical formulation contains two complex constraints that involve the relationship between all the decision variables. These correspond to the restriction that establishes the maximum working time per worker, and the restriction that limits the $CO_2$ emissions generated in production. Those chromosomes that violate any of these restrictions will be subjected to both penalty functions. This prevents the illegal chromosomes from being selected to pass to the next generation, therefore, it is also not possible that they have intercrossing for the generation of descendants. The penalty functions used for the fitness calculation are created from the penalty function proposed by [14]. and adapted by [1]. Given a chromosome $i$ of the population $P_t$ we have:

$\Delta v_1$(i) = constraint violation value (17) for chromosome i.
$\Delta v_2$(i) = constraint violation value (18) for chromosome i.
$k$ = parameter that determines the severity of the penalty function.
$G$ = maximum time available per worker.
$\varepsilon$ = positive number tending to zero.
$E$ = maximum $CO_2$ emissions allowed in total production.
$\Delta v_1$max = value of the maximum violation of constraint (17) in $P_t$.
$\Delta v_2$max = value of the maximum violation of the constraint (18) in $P_t$.

From the previous parameters it is possible to construct the penalty functions associated with the restriction (17) and (18) respectively:

$$penalFit1(i) = 1 + \left(\frac{\Delta v_1(i)}{\Delta v_1 max}\right)^k$$
$$\Delta v_1(i) = max\{0,\ g(i) - G\}$$
$$\Delta v_1 max = max\{\varepsilon,\ \Delta v_1(i) \mid i \in P_t\}$$

$$penalFit2(i) = 1 + \left(\frac{\Delta v_2(i)}{\Delta v_2 max}\right)^k$$
$$\Delta v_2(i) = max\{0,\ e(i) - E\}$$
$$\Delta v_2 max = max\{\varepsilon,\ \Delta v_2(i) \mid i \in P_t\}$$

The term $\Delta v_1(i)$ of the penalty function associated with restriction (17) must be calculated as $max\{0, g(i)-G\}$, where $g(i)$ is the total violation of chromosome $i$, calculated as the sum of the chromosome workers violation times. Similarly, the term $e(i)$ of the penalty function associated with restriction (18) determines the $CO_2$ emissions of chromosome $i$ in addition to those allowed ($E$).

Finally, we obtain the objective functions adjusted by the penalty functions. In order to prevent an illegal chromosome from being part of the pareto frontier, each objective function is multiplied by both penalty functions. Thus, if a chromosome violates the maximum time available per worker but does not exceed the maximum $CO_2$ emissions, both target functions will be penalized.

$$FO_1Pen(i) = FO_1 * penalFit1(i) * penalFit2(i)$$
$$FO_2Pen(i) = FO_2 * penalFit1(i) * penalFit2(i)$$

### 4.4 Crossover

To generate the crossover, only the first layer of the chromosome is considered. The above, taking into account that successive layers are built from this layer. For the interbreeding, the parents who have had a better performance in the selection tournament are selected. To generate the crossover, a random number must be selected among the types of product demanded ($P = 1, \ldots, n_P$). Subsequently, the batches of the randomly selected product must be exchanged between the parents. In this way, the distribution of the batches of the other products is maintained. Finally, the second and third layers of the descendants are generated from the result of the first layer.

### 4.5 Mutation

The mutation only applies to the first layer of the chromosome. A random number must be selected among the types of product demanded ($P = 1, \ldots, n_P$). Subsequently, the batch size of the selected product type must be recalculated. Once the mutation is made, the second and third layers are updated and the infeasibilities that occur in the chromosome are corrected. The mutation rate ($p_m$) must be very low in order to maintain elitism in the solutions.

### 4.6 Replacement

Once the genetic operators have been defined, the iterations of the metaheuristics begin. The first iteration differs from the following and must start with the random creation of a starting population $P(0)$ of desired size $nPob$. Population initiation is carried out as previously described. Then, the selection tournament is carried out among the individuals of the initial population, obtaining the non-dominance rank and the distance to the crowd of each chromosome. This allows you to sort individuals according to their fitness and the density of nearby solutions. In this first iteration, all the individuals of the initial population go to the next generation $P(t+1)$. To double the size of the population, the chromosomes arranged according to the selection tournament are crossed, giving rise to two descendants, in this way a population of descendants of size $nPob$ called $Q(t)$ is created. The individuals of this last set are subjected to mutation with a probability $p_m$. Finally, a population of size $2nPob$ is obtained by the union $P(t) + Q(t)$.

For the following iterations, the population $R(t)$ of size $2nPob$ is submitted to the selection tournament, obtaining 1) the non-dominance ranges and 2) the distance from the crowd of each individual. The individuals are selected according to the selection tournament to pass to the next generation P(t+1) of size $nPob$. This is done by prioritizing chromosomes with a lower non-dominance rank. If when adding all the individuals of a particular front the size $nPob$ is exceeded, priority is given to the chromosomes of the front that have a greater distance from the crowd. This population $P(t+1)$ is crossed to obtain the population $Q(t)$ of size $nPob$. The population $Q(t)$ is mutated to add variability. Finally, the population $R(t)$ of size $2nPob$ is obtained and it is iterated again until the maximum number of generations is reached.

## 5 Analysis of Results

In order to calculate the efficiency of the proposed metaheuristic algorithm, a small instance is solved with the exact method and the NSGA-II, thus obtaining two comparable pareto frontiers. For the creation of the instance, the following specifications are taken into account:**1.** The number of serus and number of workers is 2. **2.** The number of batches is 4. **3.** The number of tasks/skills is 5. **4.** The demand for each of the products follows a distribution $U(10, 70)$. **5.** The processing time of the tasks follows a $U(10, 70)$ distribution. **6.** The proficiency obeys a distribution $U(0.90, 1.10)$ **7.** The maximum number of workers per seru and tasks per worker is 3. **8.** The size of the skill set per worker is a random integer between [3,5].

Figure 1 shows the solutions found through the NSGA-II. Additionally, it is possible to determine each one of the fronts obtained in the final generation. The approximate pareto frontier is the one generated by the solutions of the first non-dominated front, that is, those solutions that have a non-dominance rank equal to zero. To run the NSGA-II, the following parameters were taken into account: **1.** Mutation rate $(p_m) = 0.025$ **2.** Population size $(nPob) = 60$ **3.** Maximum generation $(maxGen) = 90$ **4.** Penalty coefficient $(k) = 2$.

Through the $\varepsilon$ - constraint method proposed by [9], the pareto frontier of the multi-objective linear optimization model is generated. Next is presented the optimal pareto frontier obtained through the multi-objective linear optimization model and the approximate pareto frontier obtained when selecting the solutions of the first non-dominated frontier of the maximum generation of the NSGA-II. Finally, to obtain a notion of the efficiency of the metaheuristic and the proposed encoding, the comparison of both borders is made by calculating the areas under the curve of each of the borders, setting the minimum and maximum value on the x axis (dotted lines). With the results of the areas, the proportion of the area of each method is calculated by making 15 runs of the algorithm to obtain more reliable measurements. An average gap value of 2,773 % is determined.

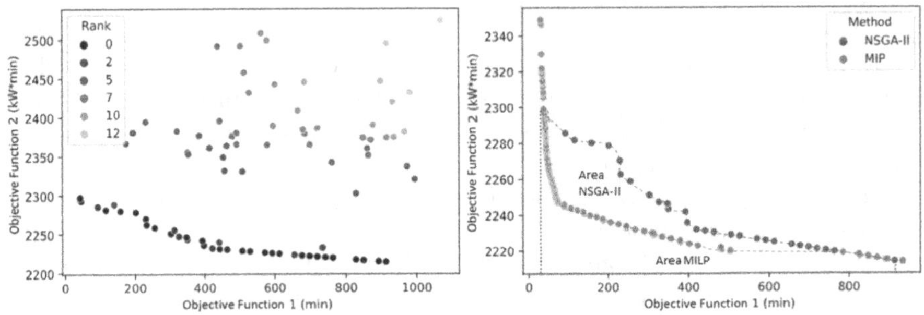


**Fig. 1.** Comparison between NSGA-II and MIPL frontiers

## 6 Conclusions and Future Work

With the accelerated increase in $CO_2$ emissions into the atmosphere and global connectivity that boosts demand, the seru production system could be the answer to maintain competitiveness given its economic and environmental benefits and its flexibility in the face of unexpected changes in demand. The concept of heterogeneous workers makes it easy to reconfigure the system in the face of unexpected changes in the production environment.

It is evident that the variables $q_b$ and $ex_b$ of the exact model allow greater flexibility to carry out the assignment of batches, workers and tasks since it is possible to produce a fraction of the demand for a type of product in each of the available serus. The algorithm based on NSGA-II demonstrates a good solution regarding the optimal boundary generated with the MILP, allowing to find good solutions to the production planning problem in a reasonable computational time.

For future studies, it is proposed to identify in which other industries, apart from the tech industry, the seru production system could be adapted and simulate it in a real case to quantify its benefits. Additionally, a more accurate study of the production time of a batch must be carried out, considering the WIP. On the other hand, given that the seru production system focuses on workforce, it is proposed to introduce qualitative measures such as the learning curve, teamwork and autonomy to determine how the performance of the system could be improved and what configurations of seru should be used in specific production environments.

## References


1. Lian, J., Liu, C., Li, W., Yin, Y.: Multi-skilled worker assignment in seru production systems considering worker heterogeneity. Comput. Ind. Eng. **118**, 366–382 (2018). https://doi.org/10.1016/j.cie.2018.02.035
2. Liu, C., Dang, F., Li, W., Lian, J., Evans, S., Yin, Y.: Production planning of multi-stage multi-option seru production systems with sustainable measures. J. Clean. Prod. **105**, 285–299 (2014). https://doi.org/10.1016/j.jclepro.2014.03.033
3. Deb, K., Pratap, A., Agarwal, S., Meyarivan, T.: A fast and elitist multiobjective genetic algorithm: NSGA-II. IEEE Trans. Evol. Comput. **6**(2), 182–197 (2002). https://doi.org/10.1109/4235.996017
4. Yin, Y., Stecke, K.E., Swink, M., Kaku, I.: Lessons from seru production on manufacturing competitively in a high cost environment. J. Oper. Manag. **49–51**, 67–76 (2017). https://doi.org/10.1016/j.jom.2017.01.003
5. Liu, C., Stecke, K.E., Lian, J., Yin, Y.: An implementation framework for seru production. Int. Trans. Oper. Res. **21**, 1–19 (2014). https://doi.org/10.1111/itor.12014
6. Bhaskar, K., Srinivasan, G.: Static and dynamic operator allocation problems in cellular manufacturing systems. Int. J. Prod. Res. **35**(12), 3467–3482 (1997). https://doi.org/10.1080/002075497194192
7. Kaku, I.: Is seru a sustainable manufacturing system? Procedia Manuf. **8**, 723–730 (2017). https://doi.org/10.1016/j.promfg.2017.02.093
8. Liu, C.G., Lian, J., Yin, Y., Li, W.J.: Seru Seisan- an innovation of the production management mode in Japan. Asian J. Technol. Innov. **18**(2), 89–113 (2010). https://doi.org/10.1080/19761597.2010.9668694
9. Haimes, Y.: On a bicriterion formulation of the problems of integrated system identification and system optimization. IEEE Trans. Syst. Man Cybern. **SMC-1**(3), 296–297 (1971). https://doi.org/10.1109/TSMC.1971.4308298
10. Liu, R., Liu, M., Chu, F., Zheng, F., Chu, C.: Eco-friendly multi-skilled worker assignment and assembly line balancing problem. Comput. Ind. Eng. **151**, 106944 (2020). https://doi.org/10.1016/j.cie.2020.106944. ISSN: 0360–8352
11. Ying, K.C., Tsai, Y.J.: Minimising total cost for training and assigning multiskilled workers in seru production systems. Int. J. Prod. Res. **55**(10), 2978–2989 (2017). https://doi.org/10.1080/00207543.2016.1277594
12. Liu, C., et al.: Training and assignment of multi-skilled workers for implementing seru production systems. Int. J. Adv. Manuf. Technol. **69**, 937–959 (2013). https://doi.org/10.1007/s00170-013-5027-5
13. Yılmaz, O.F.: Operational strategies for seru production system: a bi-objective optimisation model and solution methods. Int. J. Prod. Res. **58**(11), 3195–3219 (2020). https://doi.org/10.1080/00207543.2019.1669841
14. Gen, M., Cheng, R.: Genetic Algorithms and Engineering Optimization. Wiley, Hoboken (1999)

# Industry 4.0 and Cyber-Physical Systems

# Building High Performance Teams

Thais Carreira Pfutzenreuter[1](✉), Edson Pinheiro de Lima[1], and José Roberto Frega[2]

[1] Pontifical Catholic University, Imaculada Conceição, 1155, Curitiba 80215-901, Brazil
e.pinheiro@pucpr.br

[2] Federal University of Paraná, Av. Prefeito Lothário Meissner, 632, Curitiba 80210-170, Brazil

**Abstract.** Work design is being constantly transformed by technology and requirements for polyvalent workers piloting the work system instead of operating it. Performance no longer depends exclusively on working hours, but in multiple aspects that balance the vision of production systems results. The purpose of this paper is to investigate self-management teams' implementation and its effects on performance. A Brazilian cosmetics company case study of team development was used as a guidance for this investigation and company performance reports were the data source for descriptive statistics and multivariate analysis. Boxplot performance analysis was applied over team building stages along with One-way Variance Analysis to test average performance differences among stages' transitions. Tukey test was sequentially applied to identify its statistic differences in pairs. Results revealed team development reached the expected performance over team stages. Forming to storming, though, was the only transition with no performance average gains. Storming to norming was the highest improvement, which meets literature principles. The present research is limited to a single sociotechnical environment and performance measurement is based on average data. Therefore, specific conclusions imply further research to extend findings to other contexts. However, the present study identified empowerment best practices to achieve superior performance. The originality of this paper consists on providing consistent connection between team development theory and practice, exploring sociotechnical approach benefits through practices into performance.

**Keywords:** Work organization · Team performance · Self-managed teams · Empowerment

## 1 Introduction

Work rationalization, born in the Classical School with Scientific Administration, indorsed critical specification for task execution, and little flexibility in the work posts. Taylor's organizational model consisted of a rigid hierarchy, that it was highly dictating specialized and standardized tasks (Weisbord 2011; Batiz-Lazo 2019). The social scientist Elton Mayo was the first to reveal opposition to this work system. Hawthorne Experience, coordinated by the sociologist between 1924 and 1933, showed that Western Electric Company's productivity increased with the stimulus of social aspects in the work environment. The School of Human Relations contradicted Taylor's theory since

© Springer Nature Switzerland AG 2021
D. A. Rossit et al. (Eds.): ICPR-Americas 2020, CCIS 1407, pp. 251–264, 2021.
https://doi.org/10.1007/978-3-030-76307-7_19

it knocked down the preponderance of physiologic factors on the psychological ones (Zoller and Muldoon 2019). Toyota's Japanese approach also brought revolutions when they started to consider the intellectual prospect of working members as a critical success organizational factor (Simonetti and Marx 2010).

The models that came after Taylorism started to value the collective work, inside of activities that before were only mechanists. They tried to bring together the projection with the execution of tasks previously segmented by the classical approach (Joullié 2018). Autonomous teams, however, presented a differentiation factor: the minimization of the hierarchical role figure, stimulating teams for high performance (Marx 1997). The sociotechnical model was discovered, and subsequently, studied by Tavistock Institute researchers after observing the productivity's optimization in the English coal mines through self-management in 1949 (Salerno 1994; Moreira and Marx 2008).

Lee and Edmondson (2017) recently investigated current trends and expected benefits that motivate the development of self-managing organizations. Less-hierarchical organizations with employee empowerment initiatives stimulate brighter and faster responses in dynamic conditions. Also, Yin et al. (2018) argue that empowerment has been a mechanism to reduce traditional managing costs and proposed an economic perspective to explain how empowerment practices affect organizational performance indirectly through moderating the effect of the employee-employer exchange relationship.

The work with greater autonomy can increase productivity and work motivation providing competitive advantage for organizations (Manz and Sims 1996). The implantation of autonomous teams according to the sociotechnical model, however, involves great and challenging organizational changes (Salerno 1994). The introduction of these teams requires a new style of leadership and designates greater responsibilities to workers, and not everyone is willing to work in teams. The implementation of these teams is usually of medium to long term and involves high costs, demanding deeper organizational changes and there is not always a consensus between the hierarchies for political changes (Marx 1997; Cunningham and MacGregor 2000). Moreover, autonomy degrees must be consistent with strategic goals (Olsson and Bosch 2018).

Based on the problems related to the organizational challenges of this work model described above, the following research question is proposed for this paper: Can self-managed teams bring performance improvement? A case study is used as a guidance for this investigation. The case study's company introduced sociotechnical work teams willing to reach high performance and this paper explores its approach with performance improvement over team-building stages. Therefore, this paper presents a meaningful relationship between team empowerment theoretical principles and practice.

## 2 Literature Review

This section explores theoretical background of teams in three dimensions considered relevant to this paper: team building, empowerment and team performance measurement.

### 2.1 Team Building

Widely known as an effective team building design for the educational setting (Riebe et al. 2010; Aydin and Gumus 2016; Weber and Karman 2017), Tuckman approach

(1965) remains famous for team development model and it is still known for its four-sequential stages (Kur 1996; Largent 2016; Manges et al. 2017). Forming is the first interaction of team members, when they become familiar with each other and try to find out which behaviors are acceptable regarding their tasks, feeling confused on how to act and unsure about the benefits of team participation. There is suspicion, fear and anxiety on the onward work and the common goal supported by the formed team. At storming stage, members start to show resistance and hostility towards each other and conflicts are often caused by miscommunication, what may lead them to disbelief in the collective power. These problems begin to be solved at norming stage, when the whole team effectively learn how to work together and internal differences are overcome. The focus on tasks, interpersonal relationship and mutual identification among members are fortified. Finally, performing stage comes and the responsibility shared is intensified, together with creativity and the expected productivity. Mutual cooperation and team self-identity provide exceptional results (Tuckman 1965).

Tuckman and Jensen (1977) formally extended the original model and added a fifth stage, adjourning, to provide opportunity for acknowledgements. Since then, the model was studied, applied and successfully validated in many team development fields, remaining strongly accepted (Largent 2016; Aydin and Gumus 2016; Manges et al. 2017). An Indonesian research has recently used Tuckman model as a theoretical reference to investigate virtual team performance of Binus University e-learning student's capability of solving teamwork problems (Siregar et al. 2018).

Manges et al. (2017) successfully used the model in the healthcare environment as a guiding framework to improve safe patient care delivery. Nurse leaders' behaviors changed over Tuckman stages and were critical to build high performance teams. At the performing stage, nurse leaders no longer coordinated team actions and focused mainly on empowering members that developed shared leadership in the work design.

Kuhrmann and Munch (2016) highlights Tuckman theory's importance and argue that project managers must be familiar with the theoretical stages because they do not only start during project beginnings, but also when new members join team projects. The authors applied group dynamics with graduate students at Munich Technical University and Blekinge Institute of Technology and reinforced the premise that performance drops after changing teams. Conjointly, the dynamics revealed that time pressure, team size and missing strategies are the factors that mostly impact performance while task complexity and communication aspects affect work efficiency. Additionally, Largent (2016) reinforces Tuckman pattern that team's skill level rises over time and team enthusiasm starts high, drops and then returns to a high level. The author also emphasized the importance of team ability to recognize the team current stage because this awareness can provide knowledge to understand team progress and distinguish between normal and abnormal difficulties along team development.

On the other hand, Tuckman's model was also challenged by some researchers. Kur (1996) complemented that company teams can possibly transit from one stage to another during work execution. Miller (2003) argued that complex teams do not follow the linear team building performance suggested by the psychologist researcher. Rickards and Moger (2000) also criticized the theoretical model, claiming that not all teams go through all stages. According to the authors, storming stages can possibly never end and teams become dysfunctional, failing to pass a weak behavioral barrier before norming. Even though most teams overcome the weak barrier, fewer teams pass the strong performance one and do not achieve exceptional performance.

A similar model of team development was additionally proposed by Katzenbach and Smith (1993), projecting a performance curve that losses up to the team complete formation, integration and alignment of the members. Therefore, formation phase will undoubtedly need more time than a group run by a single leader to reach a desirable performance. According to Katzenbach and Smith (1993) it happens because a real team demands more from its members and team formation phase tends to be less effective. Formation requires difficult adaptation periods in the level of sharing experiences and distinct knowledge among members can bring disagreements and competition. Learning requires trust, as well as trust promotes learning. Therefore, learning and trust are not earned in a short period and only over time inevitable conflicts are solved.

Staniforth (1996) claims that many organizational practices and systems are more aligned to individuals, which inhibits team maturity growth and consequently teamwork performance. Castka et al. (2001) grouped successful factors of high performance teams implementation into two categories and seven subgroup categories: I) system factors: organizational impact; alignment and interaction with external entities; performance measures and defined focus II) human factors: knowledge and skills, individual needs and group culture.

## 2.2 Empowerment

High performance companies depend on internal policies and capabilities (Okoshi et al. 2019). Empowerment practices allow coworkers to gather relevant information from each other and prompt employees to work as teammates, which reduces communication and coordination costs. Unlike ineffective authority, autonomy can lead self-managed teams with consistent knowledge shared to make better decisions, contributing to organizational performance and reducing monitoring high costs (Yin et al. 2018).

Idris et al. (2018) found empirical evidence that employee empowerment strengthens job satisfaction for Malaysian capital local workers. Potnuru et al. (2019) recently identified that organizational learning culture remarkably influences the relationships of team building and empowerment on employee competencies in Indian cement manufacturing companies. Another interesting finding is the interrelationship between enabling management controls and staff empowerment and their mutual beneficial effects on performance in Australian companies (Baird et al. 2018).

Jian'an (2008) discussed how to build self-management teams through empowerment, proposing pushing and pulling power strategies to achieve the desired dynamic equilibrium between supply and demand of power along team evolution. Empowerment advances gradually with time, maturity and experience, consequently benefits are only obtained in long-term. Strategic guidance is a requirement to optimize business performance through previously established autonomy levels that meets organizational specific ambitions (Olsson and Bosch 2018). Autonomy degrees are often low during team forming stage and increases along team life cycle, exerting influence on outcomes (Hess 2018).

Even though moving away from a traditional hierarchical design is a demand, most organizations persist to inhibit and limit employee's participation and empowerment skills and shifting responsibilities are impossible if leaders at every level are not truly committed to empower their subordinates (Huusko 2006; Attaran and Nguyen 2000; Horner 1997).

According to Hess (2018) and Brower (1995), hierarchy must be limited to offer only guidelines as well as the confidence required to work execution and autonomous team members' expectations for top leader involvement must be aligned with previously established empowerment degrees consistently further reinforced to ensure perceptions that autonomy given is a real purpose and not purely symbolic.

In order to accelerate autonomy progress, empowerment practices must be adapted. Some approaches are commonly implied, such as training development plans for motivation and knowledge improvement, along with performance assessments connected to the organizational learning programs (Brower 1995; Holt et al. 2000; Potnuru et al. 2019). Employees' strengths must be intensified instead of focusing on their weaknesses, because complementary team members' skills compensate individual flaws to achieve common goals (Margulies and Kleiner 1995). In addition, clear goal statements through effective communication channels are meaningful. Targets and decisions must also be appraised by managers to ensure that commitment and efforts are focused on expected directions. Performance results, acknowledgments and rewards must be afforded not only to individuals, but also to the whole team, to reassure teammates take responsibility for each other's performance. (Brower 1995; Elmuti 1997; Conti and Kleiner 1997). In conjunction with open and frequent communication, all-inclusive recruitment and consistent resource allocation are other preconditions to be granted to autonomous team members (Hess 2018).

## 2.3 Team Performance Measurement

Teams with greater autonomy demand performance measurement indicators to appraise autonomy progress and results achieved over time (Marx 1997). In order to build high performance teams, it is exceedingly indispensable that team members are acquainted and in total consonance with the chosen performance measurement system. According to Aguinis (2013), there are some fundamentals for team performance assessment. Firstly, it is essential to ensure that teams are actually teams sharing common goals and not simply small groups with individual targets. Secondary steps are the investment in measurement structures and clear establishment of performance goals. Further, the selection of multiple appraisal methods focused on processes as well as outcomes and, finally, long-term changes are assessed.

Macbryde and Mendibil (2003) propose a four dimensional model for team performance measurement. I) team effectiveness, the dimension which process' results satisfy team stakeholders; II) team efficiency, the dimension in which internal processes support the achievement of process outcomes; III) learning and growth, which consists on monitoring team progress. Some features of this assessment involve transferable skills, documented learning, best practices, tools, methods, process improvement and team innovation potential; IV) satisfaction of the members, which implies motivation and personal fulfillment measurements, quantifying how teamwork contributes to the growth and personal well-being of each teammate.

Work designs, team composition and direct or indirect factors influencing high performance are widely explored by researches to support team cohesion theories (Ferreira et al. 2012; Wilsher 2015; Moura et al. 2019). Rezvani et al. (2019) recently applied partial least square regression analysis on construction project teams' surveys to measure emotional intelligence relationship with team performance. Emotional intelligence and trust as a mediate factor were positively related to team performance, while conflict mediate factor presented a negative relation. The study implies that relationship conflicts could be diminished, and trust can be reinforced by improving emotional intelligence to reach performance enhancement. Cha et al. (2014) also used the same statistical methodology with developed surveys based on literature to support member's psychological proximity has a critical effect on team performance.

Jaca et al. (2013) applied a teamwork effectiveness measurement structured by survey to investigate key performance factors employing Input-Mediator-Outcome methodology (Ilgen et al. 2005; Mathieu et al. 2005). Independence, autonomy, internal leadership, conflict management were some common highly rated and constantly measured in healthcare and manufacturing distinct environments.

Team performance measurements have been also consistently worked out to test work designs involving leadership styles to evaluate optimizing performance structures. Ciasullo et al. (2017) measured effectiveness, efficiency and satisfaction of members dimensions to compare a traditional top-down with a hybrid bottom-up team building approach. The authors used not only lead-time, an organizational performance indicator, but also applied external and internal surveys to quantify both, customer and employee's satisfaction. Bottom-up team development performed better across all three dimensions of this integrated performance assessment. Yang and Choi (2009) also identified that empowerment elements influence on team performance through linear regression.

Han et al. (2017) have indirectly explored some dimensions suggested by Macbryde and Mendibil (2003) in a survey research, testing the effect of shared leadership on team perceived performance. Statistical analysis performed have supported shared leadership indirectly increases performance through stimulating knowledge share, commitment to goals and activity coordination as team mediate factors. Complementary, Müller et al. (2018) have also found that shared leadership positively affects team performance on a laboratory team decision-making exercise. Statistical analysis showed not only that sharing leadership brought quality improvement, but also perceived task complexity strengthens this effect. Even with constant complexity, when recognized as harder tasks, shared leadership was intensified and fewer errors were made in work execution.

## 3 Research Design

This paper presents a case study of team building, introduced by a cosmetics Brazilian company with the purpose to reach higher performance with the development of shop-floor self-managed teams. Empowerment practices with technical and behavioral trainings were provided to gradually achieve the desired team autonomy degrees over time. This research consists on measuring quantitative team performance over team building stages and evaluate the team development project, that was segmented by five learning levels. These five progressive stages structured by the company, presented by Fig. 1, involved all factory teams and each level was planned to last the period of a year. Leadership left coordination role and gradually started to act as internal coachers and consultants, limiting to provide only generic directions along with confidence for members to accomplish autonomy. An important procedure followed by the company is that, for each stage advancement, besides the mandatory approval after trainings, auditing carried out to reassure that behavioral and professional improvement were successfully achieved to certify team breakthrough.

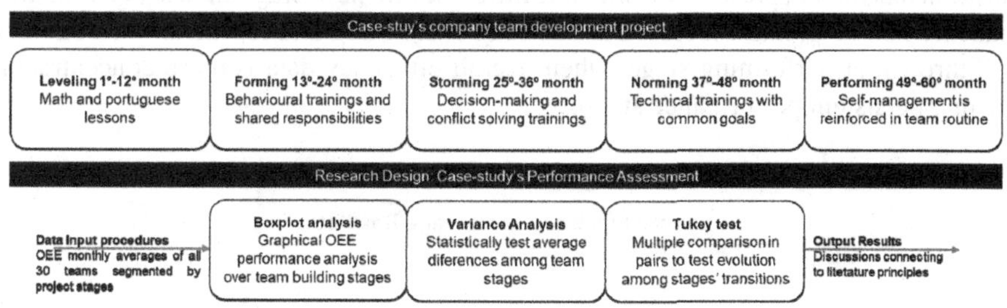


**Fig. 1.** Research design.

Quantitative performance reports provided by the company at the end of the project were the original data source for all the analyses. The main shop-floor performance indicator, overall efficiency, was chosen for the assessment. Overall line efficiency is measured through the formula Availability x Performance x Quality and team evaluation was applied by project consultants and team leaders. This indicator considers three percentage target dimensions, which only good parts are produced (100% quality), at the maximum cycle time speed (100% performance), and without equipment interruption (100% availability). Since the purpose is generic conclusions, team performance data of all 30 developed teams, were converted to monthly performance averages as a single team for all the following analyses described below.

1. Boxplot with all stages of overall efficiency's performance in a single chart was applied to graphically analyze performance over team learning stages.
2. One-way variance analysis was applied to statistically test performance average differences among team stages.
3. Tukey test was applied to test overall efficiency's multiple comparison in pairs to test performance evolution analysis between stages' transitions.

The performance assessment exhibited by Fig. 1 was applied in academic environment to evaluate performance enhancement with the new intended work design and consequently answer the research proposed question regarding the ability of self-managing teams to bring high performance.

## 4 Results

Overall efficiency exhibits wide-ranging performance evolution over the stages, presented by Fig. 2. It is perceptible forming to storming is the only particular distinguishable transition that did not present considerable difference in the average values. Performance staunchly had its greatest increase at storming to norming and repeatedly had another breakthrough at performing stage, when overall efficiency data is more concentrated, reaching persistent performance stability.

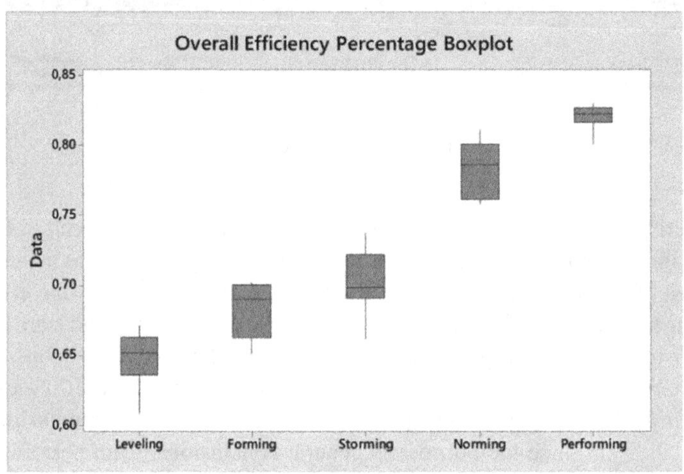


**Fig. 2.** Boxplot analysis.

Performance behavior presented in boxplot reinforces some literature principles further explored in discussion section. Boxplot method provided graphical evolution overview, summarizing each stage's enhancement along team development. A single-factor variance analysis for overall efficiency was sequentially appropriate to test significant average difference among stages. Table 1 details the hypothesis test, performed with 95% confidence interval for this analysis.

**Table 1.** Hypothesis test.

| Method | |
|---|---|
| Null hypothesis | All Means are equal |
| Alternative hypothesis | Not all means are equal |
| Significance level | α = 0,05 |
| Equal variances were assumed for the analysis. | |

| Factor Information | | |
|---|---|---|
| Factor | Levels | Values |
| Factor | 5 | Leveling; Forming; Storming; Norming; Performing |

| Analysis of Variance | | | | | |
|---|---|---|---|---|---|
| Source | DF | Adj SS | Adj MS | F-Value | P-Value |
| Factor | 4 | 0,24884 | 0,062210 | 184,25 | 0,000 |
| Error | 55 | 0,01857 | 0,000338 | | |
| Total | 59 | 0,26741 | | | |

Before conclusions, test premises were challenged. Since residues showed a large adherence to the standard normal distribution and residual variances were approximately equal, premises were ratified. Considering a 95% confidence level, H0 was rejected, since p-value < 0.05. Therefore, statistical evidence of significant average difference for at least a couple of overall efficiency's stages was found. Tukey's test was then conveniently useful to test average differences between stages in pairs, presented by Fig. 3.

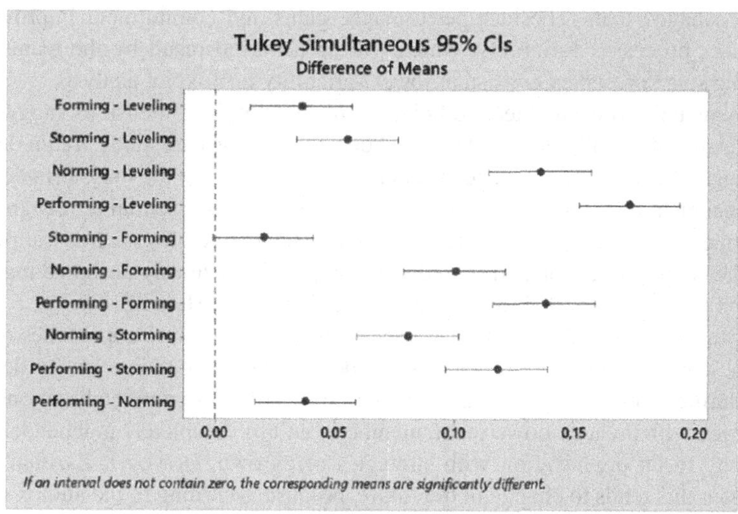


**Fig. 3.** Tukey test.

By means of the 95% confidence interval for differences between averages, Tukey's test showed that only forming to storming stage had no significant difference, as suspected upon boxplots descriptive analysis. In addition, it also implied a respective improvement for all the other transitions compared in pairs since the difference is always positive, hence, it essentially diagnoses overall efficiency gain over all other transitions, except forming-storming.

## 5 Discussion

The training development approach adopted by the company to introduce team development matches Tuckman's initial model (1965). The autonomy given was previously planned to attend organizational demands as highlighted in literature by Olsson and Bosch (2018) and autonomy was initially low (Hess 2018). Both initial Boxplot and Variance Analysis approaches imply performance improvement along team autonomy progress.

Boxplot presented consistent performance breakthrough over team development, even though some stage transitions noticeably had different improvement scales. Although there was no performance loss from forming to storming as defended by theory (Tuckman 1965; Katzenbach and Smith 1993), it was the only transition with no performance average gains since Tukey's test showed no significant average differences between both stages. Storming to norming is highlighted by Tuckman theory as the greatest performance evolution, which also coincides with the case study's pattern. The highest improvement also corresponds to Katzenbach and Smith's theory (1993), that implies a huge evolution from potential team to real team. Tukey's test analysis also implies that teams have succeeded at passing both, weak behavioral and strong performance barriers, suggested by Rickards and Moger (2000).

Salerno (1994), Manz and Sims (1996), and Simonetti and Marx (2010) emphasized that self-managed teams, besides performance gains and commitment improvement, also reduces processes variability, which was an achieved result by the teams, since performing stage presented consistent lower variability in Boxplot analysis.

Learning and growth and team efficiency are some of the dimensions suggested by Macbryde and Mendibil (2003) related and explored in this case study. Team development brought the high performance expectation achievement in the long-term defended by sociotechnical researchers (Moreira and Marx 2008). It is essential to recognize that empowerment practices applied are related to theoretical principles, such as training programs, open communication and shared leadership with previously designed individual responsibilities (Brower 1995; Margulies and Kleiner 1995; Hess 2018).

This study also implies theoretical discussion related to future approaches of organizational work. Salerno (1994) defends that despite the autonomy given for decision-making during work execution, teams cannot be regarded as completely autonomous, after all even with great empowerment, members can't be completely independent since they belong to an organization with strategies of its own. However, Zarifian (1997) believes that this tends to change in the future, because according to the author, it is not possible to keep two management systems built on different principles for too long. Self-organizations do not only reduces coordination costs (Yin et al. 2018) but remain studied because they also respond more effectively in dynamic scenarios (Lee and Edmondson 2017).

This paper supports sociotechnical approach principles, bringing evidence that the case study's company achieved its ambition to reach high performance with the new work design with empowered teams (Salerno 1994; Marx 1997). Empowerment benefits remarkably appeared in the long-term, as implied by researchers (Jian'an 2008; Olsson and Bosch 2018). This study suggests that a team building prosperity has a connection to empowerment practices that meets literature suggestions (Margulies and Kleiner 1995; Elmuti 1997; Brower 1995) such as internal alignment between managers and teams (Hess 2018) and proper strategic guidance before team development to gradually introduce autonomy (Olsson and Bosch 2018).

## 6 Conclusion

This paper strengthens the hypothesis that self-managed teams can reach high performance, since quantitative analysis showed performance evolution on the presented case study, and positively answered the research investigative proposed question. Team building academic contributors and sociotechnical researchers provided a strong theoretical background, making it possible to connect theory to practice identifying similarities and divergences shown in the discussion section. The consistency of this integration brings relevance and originality to this paper in work organizational field.

Limitations are also relevant to direct following research to cover some topics left unexplored in the present case study. For this reason, cost-exchange perspective must be a further study, to quantify managing costs reduction brought by empowerment practices defended by literature and examine the investment made on training programs comparing to team development's cost reduction. It is equally important to emphasize that we do not generalize and extend our conclusions to other realities because the present study was restricted to a single company case-study.

This paper focused on measuring team performance across team development, exploring learning and growth versus team efficiency dimensions. The unexplored dimensions could be further assessed by internal surveys to measure employee's satisfaction individually and inside their own teams, together with external surveys to evaluate customer satisfaction, as explored by team building research present in the visited literature. Complimentarily, empowerment measurement itself through external questionnaires along with internal learning and shared leadership mediate factors' influence on performance could be meaningful techniques to extend our case study's analysis.

Statistical analysis presented in research design were applied using team performance averages for general conclusions. Consequently, a further study exploring teams individually could also contribute to additional and more specific studies. It is also suggested the Myers-Briggs Type Indicator test application for team building and leadership development. Based on the ideas of analytical psychology, the test provides individual personality analysis and their behavior in different situations, considering specific preferences, values and motivations. Therefore, another complementary research involving members' personalities inside teams could be also an interesting following theme to test relations between team personality structure and performance.

## References

Aguinis, H.: Performance Management, 3rd edn. Pearson, Boston (2013)

Attaran, M., Nguyen, T.T.: Creating the right structural fit for self-directed teams. Team Perform. Manag. **6**(1/2), 25–33 (2000). https://doi.org/10.1108/13527590010731952

Aydin, I.E., Gumus, S.: Sense of classroom community and team development. Turkish Online J. Distance Educ. **17**(1), 60–77 (2016). https://doi.org/10.17718/tojde.09900

Baird, K., Su, S., Munir, R.: The relationship between the enabling use of controls, employee empowerment, and performance. Pers. Rev. **47**(1), 257–274 (2018). https://doi.org/10.1108/PR-12-2016-0324

Batiz-Lazo, B.: What is new in "a new history of management"? J. Manag. Hist. **25**(1), 114–124 (2019). https://doi.org/10.1108/JMH-07-2018-0033

Brower, M.J.: Empowering teams: what, why, and how. Empowerment Organ. **3**(1), 13–25 (1995). https://doi.org/10.1108/09684899510079780

Castka, P., Bamber, C.J., Sharp, J.M., Belohoubek, P.: Factors affecting successful implementation of high performance teams. Team Perform. Manag. **7**(7/8), 123–134 (2001). https://doi.org/10.1108/13527590110411037

Cha, M., Park, J.G., Lee, J.: Effects of team member psychological proximity on teamwork performance. Team Perform. Manag. **20**(1), 81–96 (2014). https://doi.org/10.1108/TPM-03-2013-0007

Ciasullo, M.V., Cosimato, S., Gaeta, M., Palumbo, R.: Comparing two approaches to team building: a performance measurement evaluation. Team Perform. Manag. **23**(7–8), 333–351 (2017). https://doi.org/10.1108/TPM-01-2017-0002

Conti, B., Kleiner, B.H.: How to increase teamwork in organizations. Train. Qual. **5**(1), 26–29 (1997). https://doi.org/10.1108/09684879710156496

Cunningham, J.B., MacGregor, J.: Trust and the design of work complementary constructs in satisfaction and performance. Hum. Relat. **53**(12), 1575–1591 (2000). https://doi.org/10.1177/00187267005312003

Elmuti, D.: The perceived impact of team- based management systems on organizational effectiveness. Team Perform. Manag. **3**(3), 179–192 (1997)

Ferreira, P.G.S., De Lima, E.P., Da Costa, S.E.G.: Perception of virtual team's performance: a multinational exercise. Int. J. Prod. Econ. **140**(1), 416–430 (2012). https://doi.org/10.1016/j.ijpe.2012.06.025

Han, S.J., Lee, Y., Beyerlein, M., Kolb, J.: Shared leadership in teams. Team Perform. Manag. **24**(3/4), 150–168 (2017). https://doi.org/10.1108/tpm-11-2016-0050

Hess, J.P.: Autonomous team members' expectations for top-leader involvement. Team Perform. Manag. **24**(5–6), 283–297 (2018). https://doi.org/10.1108/TPM-10-2017-0060

Holt, G.D., Love, P.E.D., Nesan, L.J.: Employee empowerment in construction: an implementation model for process improvement. Team Perform. Manag. **6**(3/4), 47–51 (2000). https://doi.org/10.1108/13527590010343007

Horner, M.: Leadership theory: past, present and future. Team Perform. Manag. **3**(4), 270–287 (1997). https://doi.org/10.1108/13527599710195402

Huusko, L.: The lack of skills: an obstacle in teamwork. Team Perform. Manag. **12**(1/2), 5–16 (2006). https://doi.org/10.1108/13527590610652756

Idris, A., See, D., Coughlan, P.: Employee empowerment and job satisfaction in urban Malaysia. J. Organ. Change Manag. **31**(3), 697–711 (2018). https://doi.org/10.1108/jocm-04-2017-0155

Ilgen, D.R., Hollenbeck, J.R., Johnson, M., Jundt, D.: Team in organizations: from input-process-output models to IMOI models. Annu. Revision Psychol. **56**, 517–543 (2005). https://doi.org/10.1146/annurev.psych.56.091103.070250

Jaca, C., Viles, E., Tanco, M., Mateo, R., Santos, J.: Teamwork effectiveness factors in healthcare and manufacturing industries. Team Perform. Manag. **19**(3), 222–236 (2013). https://doi.org/10.1108/TPM-06-2012-0017

Jian'an, C.: Research on strategies and empowerment process to achieve self-management team. In: 2008 International Conference on Wireless Communications, Networking and Mobile Computing, WiCOM 2008, pp. 1–5 (2008). https://doi.org/10.1109/WiCom.2008.1731

Joullié, J.E.: Management without theory for the twenty-first century. J. Manag. Hist. **24**(4), 377–395 (2018). https://doi.org/10.1108/JMH-05-2018-0024

Katzenbach, J.R., Smith, D.: The Wisdom of Teams: Creating the High-Performance Organization. Harvard Business School Press, Boston (1993)

Kuhrmann, M., Münch, J.: When teams go crazy: an environment to experience group dynamics in software project management courses. In: Proceedings - International Conference on Software Engineering, pp. 412–421 (2016). https://doi.org/10.1145/2889160.2889194

Kur, E.: The faces model of high performing team development. Manag. Dev. Rev. **9**(6), 25–35 (1996). https://doi.org/10.1108/09622519610151624

Largent, D.L.: Measuring and understanding team development by capturing self-assessed enthusiasm and skill levels. ACM Trans. Comput. Educ. **16**(2), 1–27 (2016). https://doi.org/10.1145/2791394

Lee, M.Y., Edmondson, A.C.: Self-managing organizations: exploring the limits of less-hierarchical organizing. Res. Organ. Behav. **37**, 35–58 (2017). https://doi.org/10.1016/j.riob.2017.10.002

MacBryde, J.C., Mendibil, K.: Designing performance measurement systems for teams: theory or practice. Manag. Decis. **14**(8), 722–733 (2003). https://doi.org/10.1108/00251740310496233

Manges, K., Scott-Cawiezell, J., Ward, M.M.: Maximizing team performance: the critical role of the nurse leader. Nurs. Forum **52**(1), 21–29 (2017). https://doi.org/10.1111/nuf.12161

Manz, C.C., Sims, H.P.: Empresas sem chefes. Makron Books, São Paulo (1996)

Margulies, J.S., Kleiner, B.H.: New designs of work groups: applications of empowerment. Empowerment Organ. **3**(2), 12–18 (1995). https://doi.org/10.1108/09684899510089284

Marx, R.: Autonomia, trabalho em grupo e estratégia empresarial. O que há de novo neste final de século?. São Paulo Em Perspectiva **11**(4), 67–75 (1997). https://produtos.seade.gov.br/produtos/spp/v11n04/v11n04_08.pdf

Mathieu, J.E., Heffner, T.S., Goodwin, G.F., Cannon-Bowers, J.A., Salas, E.: Scaling the quality of teammates' mental models: equifinality and normative comparisons. J. Organ. Behav. **26**, 37–56 (2005). https://doi.org/10.1002/job.296

Miller, D.L.: The stages of group development: a retrospective study of dynamic team processes. Can. J. Adm. Sci. **20**(2), 121–134 (2003). https://doi.org/10.1111/j.1936-4490.2003.tb00698.x

Moreira, L.F.D.C., Marx, R.: Evolução da organização do trabalho fabril: aplicação do modelo de semi-autonomia em empresa nacional de cosméticos. In: Anais. ABEPRO, Rio de Janeiro (2008)

Moura, I., Dominguez, C., Varajão, J.: Information systems project teams: factors for high performance. Team Perform. Manag. **25**(1–2), 69–83 (2019). https://doi.org/10.1108/TPM-03-2018-0022

Müller, E., Pintor, S., Wegge, J.: Shared leadership effectiveness: perceived task complexity as moderator. Team Perform. Manag. **24**(5–6), 298–315 (2018). https://doi.org/10.1108/TPM-09-2017-0048

Okoshi, C.Y., Pinheiro de Lima, E., Gouvea Da Costa, S.E.: Performance cause and effect studies: analyzing high performance manufacturing companies. Int. J. Prod. Econ. **210**, 27–41 (2019). https://doi.org/10.1016/j.ijpe.2019.01.003

Olsson, H.H., Bosch, J.: Singing the praise of empowerment: or paying the cost of chaos. In: Proceedings - 44th Euromicro Conference on Software Engineering and Advanced Applications, SEAA 2018, pp. 17–21 (2018). https://doi.org/10.1109/SEAA.2018.00012

Potnuru, R.K.G., Sahoo, C.K., Sharma, R.: Team building, employee empowerment and employee competencies: moderating role of organizational learning culture. Eur. J. Train. Dev. **43**(1–2), 39–60 (2019). https://doi.org/10.1108/EJTD-08-2018-0086

Rezvani, A., Barrett, R., Khosravi, P.: Investigating the relationships among team emotional intelligence, trust, conflict and team performance. Team Perform. Manag. **25**(1–2), 120–137 (2019). https://doi.org/10.1108/TPM-03-2018-0019

Rickards, T., Moger, S.: Creative leadership processes in a project team. Br. J. Manag. **11**(4), 273–283 (2000). https://doi.org/10.1111/1467-8551.00173

Riebe, L., Roepen, D., Santarelli, B., Marchioro, G.: Teamwork: effectively teaching an employability skill. Educ. Train. **52**(6), 528–539 (2010). https://doi.org/10.1108/00400911011068478

Salerno, M.S.: Mudança organizacional e trabalho direto em função de flexibilidade e perfomance da produção industrial. Produção **4**(1), 5–22 (1994). https://doi.org/10.1590/S0103-65131994000100001

Simonetti, P.E., Marx, R.: Estudo sobre implementação de trabalho em grupos com autonomia: pesquisa quantitativa numa amostra de empresas operando no Brasil. Production **20**(3), 347–358 (2010). https://doi.org/10.1590/S0103-65132010005000051

Siregar, C., Pane, M.M., Ruman, Y.S.: The virtual team performance in solving teamwork conflict problems. In: Proceedings of the 2018 International Conference on Distance Education and Learning, ICDEL 2018, vol. 1, pp. 1–5 (2018). https://doi.org/10.1145/3231848.3231850

Staniforth, D.: Teamworking, or individual working in a team? Team Perform. Manag. **2**(3), 37–41 (1996). https://doi.org/10.1108/13527599610126256

Tuckman, B.W.: Developmental sequence in small groups. Psychol. Bull. **63**(6), 384–399 (1965). https://doi.org/10.1037/h0022100

Tuckman, B.W., Jensen, M.A.C.: Stages of small-group development revisited. Group Organ. Manag. **2**(4), 419–427 (1977). https://doi.org/10.1177/105960117700200404

Weber, M.D., Karman, T.A.: Student group approach to teaching using Tuckman model of group development. Adv. Physiol. Educ. **261**(6), S12 (2017). https://doi.org/10.1152/advances.1991.261.6.s12

Weisbord, M.: Taylor, McGregor and me. J. Manag. Hist. **17**(2), 165–177 (2011). https://doi.org/10.1108/17511341111112578

Wilsher, S.: Behavior profiling: implications for recruitment and team building. Strateg. Dir. **31**(9), 1–5 (2015). https://doi.org/10.1108/SD-02-2015-0023

Yang, S.B., Choi, S.O.: Employee empowerment and team performance: autonomy, responsibility, information, and creativity. Team Perform. Manag. **15**(5–6), 289–301 (2009). https://doi.org/10.1108/13527590910983549

Yin, Y., Wang, Y., Lu, Y.: Why firms adopt empowerment practices and how such practices affect firm performance? A transaction cost-exchange perspective. Hum. Resour. Manag. Rev. **29**(1), 111–124 (2018). https://doi.org/10.1016/j.hrmr.2018.01.002

Zarifian, P.: Organização e sistema de gestão: à procura de uma nova coerência. Gestão & Produção **4**(1), 76–87 (1997). https://doi.org/10.1590/S0104-530X1997000100004

Zoller, Y.J., Muldoon, J.: Illuminating the principles of social exchange theory with Hawthorne studies. J. Manag. Hist. **25**(1), 47–66 (2019). https://doi.org/10.1108/JMH-05-2018-0026

# Human Resources 4.0: Use of Sociometric Badges to Measure Communication Patterns

Regina Moirano[1(✉)], Marisa A. Sanchez[1], Libor Štěpánek[2], and Gastón Vilches[3]

[1] Departamento de Ciencias de la Administración, Universidad Nacional del sur, Bahía Blanca, Argentina
mas@uns.edu.ar

[2] Masaryk University, Brno, Czech Republic

[3] Departamento de Ingeniería Eléctrica y de Computadoras, Universidad Nacional del Sur, Campus Palihue, Bahía Blanca, Argentina

**Abstract.** Teams are fundamental for innovation and efficient communication may be fundamental for team performance. The aim of this paper is to present the use of sociometric badges in real workplace teams to display how team structure communication patterns can be measured and analyzed. The selected method consists of four case studies of different organizations, where the unit of analysis is represented by a meeting team. Use of this emergent technology allowed the evaluation of vocalization distribution, turn taking frequency and overlapping speech metrics. Main results of the analysis showed that team vocalization tenure corresponds to performance results but vocalization distribution does not, which was unexpected. We found where the average speech segment was high, the meeting overcome the 100% of team vocalization tenure, and the total overlap time was also high. We identified different kinds of overlapping speech. Higher overlapping seems to occur more frequently when total overlap time is high but dominance is low than when total overlap and dominance are both high. Badges metrics also allowed a preliminary identification of three meeting stages. Finally, we identified possible future research lines.

**Keywords:** Sociometric badges · Team interaction · Emergent technologies · Human resources · Innovation

## 1 Introduction

Changes over the last three decades have been highly accelerated owing to the advent of technology and innovations in communications [1] so the role of technology in contemporary organizations plays an increasingly important place [2]. Organizations are called to adapt in order to stand effective, productive and competitive in a volatile, uncertain, complex and ambiguous environment. In this process, a vital feature of modern organizations is working under changing circumstances; they increasingly adopt work teams to coordinate activities [3]. In particular, the development and introduction of innovations in an Industry 4.0 manufacturing environment poses some challenges to human resource management. Industry 4.0 require interdisciplinary project teams [4].

© Springer Nature Switzerland AG 2021
D. A. Rossit et al. (Eds.): ICPR-Americas 2020, CCIS 1407, pp. 265–279, 2021.
https://doi.org/10.1007/978-3-030-76307-7_20

As teams became a common working unit for managing change [5], investigating teams in order to understand team communication patterns is essential. Similarly, comprehension of the collaboration dynamic is becoming more and more important to improve team performance. The role of teams is fundamental for innovation. One of the competences that may enhance innovation processes is communication, understood as creating fluid communication pathways [6]. However, while digital communication is important in the modern workplace, face-to-face interaction still represents a large and important share of organizational communication, information exchange, socialization and informal coordination [7].

Currently, a variety of devices and technologies are used for business administration as a way to guarantee that processes are producing the expected results [8]. Specifically, electronic devices can measure communication patterns among teams [9–14]. For example, it has been argued that a balanced vocalization in between team members can be considered as a good proxy in relation to team performance [12, 13]. Despite these advances mean enormous potential for the strategic role of human resources function [1], there is still a lack of research on the use of this emergent technology in real teamwork settings. Although direct and quantitative approaches are highly desirable [2], teams interaction patterns are still poorly understood [10]. Methods for accurate group dynamics measurement are critical issues to be solved [15].

This research aims to study how team structure communication patterns can be measured and analyzed, to extend current literature over the use of sociometric badges in real settings, and to present challenges and opportunities of this promising emergent technology. Research describe how sociometric badges were used in different kind of teams, such as hackathons competitors [12], recruited specifically for study proposes [15], a crew during mission simulation [16], from first-year undergraduate engineering design course [10], from graduate students [17] or master's students [9]. We did not find studies that describe the use of sociometric badges in a real teamwork setting.

## 2 Background

In order to frame the underpinnings of this case study, we briefly mention theoretical and empirical findings around communication in work teams that seeks for innovation and the use of sociometric badges as a tool to measure that complex phenomenon.

### 2.1 The Use of Sociometric Badges

Traditionally, sociologists have employed human observers, questionnaires or surveys to understand group dynamics for successful intervention [3, 15]. In order to minimize subjectivity [3], the unconscious bias inherent in recording group interaction data [12], remain unobtrusive [7], and responding to the need for automatic tools to measure individual and group behavior in the social sciences [18], researchers have leaned on technology for solutions. There have been used complex measurements like videos, audios, sensors or other methods to understand and enhance group dynamics, with a high cost in data processing and delay in feedback. For addressing the complexity of team and communication dynamics wearable devices are becoming a common research tool for studying social behavior [12].

Sociometric badges are wearable sensors that may measure body movements, interactions, and speech [11]. Sociometric badges can collect unbiased and richer data than traditional methods [15]. They also reduce the large amount of post-processing data and gives researchers a chance to efficiently examine human interaction in granularity that was previously impossible [10], they allow exploring the structure of verbal interactions and provide further insight into complex team interaction process [9]. They can automatically measure individual and collective dynamics in a quantitative way and guarantee privacy due to the fact it is impossible to determine the content of the conversation or identify the speaker from the sociometric data [3].

## 2.2 Team Performance Communication Metrics

Within the use of this emergent technology, we identify three main verbal communication metrics related to team performance in literature, namely vocalization distribution, turn taking frequency, and overlapping speech.

First, the critical factor in a group's dynamics is the extent to which members participate in developing shared ideas [19]. Vocalization distribution is the proportion of the speaking time [9]. There is enough empirical evidence to support that a balanced vocalization distribution is positively correlated with team performance [9, 10, 12, 13]. There is higher team performance when less centralized communication or more sharing of ideas among multiple team members rather than a single or a few members' communication is displayed. Dominant behavior is a key determinant in the formation of a group's social structure, and consequently group communication and dynamics. A dominant participant may have a negative effect on group's dynamics by discouraging participation of other members or imposing their thoughts on the whole group [15]. Although conversely, in practical terms, Paulus and his colleagues [19] observe that typically one or two persons dominate team conversation.

Second, turn taking frequency was defined by Kim and her colleagues [15] as each instance a participant takes over in a conversation either from another participant or from silence, so every turn taking divides a speech segment. A speech segment is any continuous stream of speech from an individual, regardless of interruption or overlap from other participants. A segment will end either by an interruption caused by another participant that resulted in the speaker to stop speaking (which means a turn taking) or by a significant length of silence.

It is not clear how turn taking frequency correlates with results. A recent study by Zhou and her colleagues [17] indicates the contradictions. On one hand, disagreement may appear as a high level of turn taking in meetings, with participants interrupting each other or frequently speaking at the same time [12]. Turn taking was found to be negatively associated with perceived collaboration quality [17]. On the other hand, studies confirm that turn taking correlates positively with the results [4, 20, 21]. The possibility of frequent exchange is closely linked to coordination and collaboration with positive impact on team performance. Higher turn taking rates contribute to better group cohesiveness and satisfaction, generate a higher number of quality ideas and have higher performance [10]. Turn taking was significantly and positively associated with creative fluency, which provides strong support for theoretical suggestions that elaboration is a process that directly enhances group creativity [17].

Third, overlapping speech is the time that two or more members are speaking at the same time [15], is the duration of speech that speaker A and B uttered simultaneously [17]. Overlapping speech may evidence lack of hearing or disagreement, when participants cut each other off or speaking at the same time, so correlates negatively with performance [12, 17]. However, we are cautious on this metric. To Lederman [12] it was difficult to identify if overlaps mean the configuration of subgroups and regarding brief overlaps, Endedijk and her colleagues [9] clarify in their study they could not define whether this consisted of a lot of brief interruptions (e.g., a confirmatory 'yes') or fewer long interruptions (which might be experienced as more disrupting than brief confirmations) and ask for more advanced algorithms to add measures that provide more insight into overlaps effects, i.e. in relation to the length and number of interruptions.

Kim and her colleagues [14] take this emergent technology a step further and develop a Meeting Mediator (MM), a real-time portable system that shows sociometric badge outputs. Their study concludes that the use of MM has a significant effect on overlapping speaking time and interactivity level without distracting the subjects. The sociometric badges also able to detect dominant players in the group and measure their influence on other participants. Most interestingly, they highlight in groups with one or more dominant people, MM effectively reduce the dynamical difference between co-located and distributed collaboration as well as the behavioral difference between dominant and non-dominant people. The authors suggest the system encourages change in group dynamics that may lead to higher performance and satisfaction.

## 3 Method

This is an exploratory research, and its strategy is the elaboration of four different case studies [22]. The unit of analysis is represented by a meeting where the team joint to solve a real problem in the organization setting. First, we describe the instruments used to measure communication metrics and team performance. Second, we present the pilot test and the description implemented to collect the data. Third, we detail how we calculate metrics with the data analysis software. Finally, we describe the team's organizations as teams' context and the characteristics of the task and meeting under study.

### 3.1 Instruments

In order to measure team structure communication patterns and correlate results with team performance, we use two instruments, sociometric badges during a meeting and a questionnaire, after the meeting.

The MIT Media Lab facilitates information and layouts to develop an affordable wearable technology; the Rhythm badges [12]. The badges (Fig. 1) enable to extract vocalization signals highly undersampled, which means they do not record any speech content but are capable of recording whether the user is speaking or not. They can not identify the person who is speaking. The badges are connected to a base station, and send

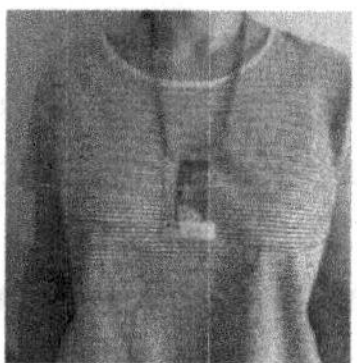

**Fig. 1.** Sociometric badges.

collected information via Bluetooth. This allows measuring the communication metrics described in the analysis section.

A questionnaire with a 6-level Likert scale based on Lewis [21] performance metric is used to identify each team member's perception about what happened throughout the meeting from a referent-shift approach [23]. The questionnaire includes the following points: (a) The team's deliverables were of excellent quality, (b) The team managed time effectively, (c) The team met important deadlines on time (d) The team did a good job at meeting [the expected or client's] needs. The questionnaire is designed to be administered individually, considering that the aggregate perspective of the participants belonging to the group is usually an accepted measure in the literature to measure the collective-team level constructs [24].

### 3.2 Pilot Test

The project, protocol and instruments were studied and approved by the Bioethics Committee of the Municipal Hospital of Bahía Blanca (Argentina). The protocol requires each participant to sign a formal participation consent.

A pilot test was conducted in order to evaluate badges utilization, software recording parameters, and data processing. The questionnaire was validated. The pilot test was conducted in a team of 5 engineers' members at the Universidad Nacional del Sur, which aimed to verify a series of technical issues on a project. The meeting lasted 40 min and allowed to measured, for example, the time it took to sign the participation consent. This pilot case allowed us to adapt the data collection instrument. We have also checked devises functioning, such as if they identify false positives from the software. Improved versions were used in the four cases under study, following the next procedure.

### 3.3 Procedure Description

First, the meeting was planned. Having the organization and leader authorization, we anticipated potential user's details of the technology that was going to be used and sent the consent. The meeting was schedule by the team, as their activity demanded, and researchers were informed. Second, following Lederman [12] guides, we installed the base station in the meeting room, before the meeting. Third and while each member arrived to the meeting room, we asked for the consent signature, we shifted on and helped wearing the devices. This process took 2–4 min, depending on how many members the meeting had.

A researcher assisted during the meetings, who has no eye contact with any participant, focused on checking the data was being collected properly by the software and registered observable interactional dynamics, such as subgroups. Forth, a form was completed with data about the exact time the badges were turn on, the formal time of the beginning and end of the meeting and every event associated with studied metrics (for example, if a member went out of the room, used the phone or a subgroup was set up). Fifth and having finished the meeting, each member handed over the badges. While each member delivered the equipment, the effect of the presence of the researcher was consulted. In all cases they expressed they had had neither distractions nor discomfort. Finally, and after the meeting, the questionnaire was sent by email to all assistants in GoogleForms format.

### 3.4 Data Analysis

In order to process the badges data, we developed a data analysis algorithm for this special purpose, which can recognize which participants were speaking every moment. First, for each participant we extracted a subset of data where we were sure that only that participant was speaking. Second, using statistical tools we studied that subset to define rules to detect in the entire data set when each participant was likely to speak. Finally, analyzing the correlation between the audio signals of each participant, we eliminated false detections due to crosstalk. The environment noise and researcher's notes were considered, such as a bus passing by the window (case 2) or a mobile ringing (case 4). As Lederman's explained [12] we met metric imperfections, due is mathematically impossible to be precise with this technology. However, the devises allowed to access to team dynamics representative metrics.

In order to analyze questionnaire results, we processed responses in an Excel spreadsheet. Considering its 6 level scale, we transform each case average result into percentage.

### 3.5 Metric Definitions

Based on background empiric literature [9, 10, 12, 14, 17] we define metrics and indicators (shown in Table 1 with an asterisk*). Having collected and processed the data, we develop a set of indicators that serve us to complement the analysis (shown in Table 1 with a double asterisk**). Table 1 also shows indicators' calculation and unit of measurement.

### 3.6 Description of Case Studies

We studied four team works, belonging to different organizations. Table 2 present each team organization characteristics.

Organization 1 is an association of professionals that provide management and human resources consulting services in the region, especially to small and medium size firms. Their teams set up to respond to specific client's demands, so flexibility is a valuable competence.

Organization 2 is a public national university, funded in 1956 and represent an educational hub. It must comply with laws, resolutions and administrative decisions issued

**Table 1.** Metric definitions and calculations details

| Metric ID | Definitions | Indicator name & calculation | Unit of measurement |
|---|---|---|---|
| Vocalization distribution | Fraction of time per person that distributes total speaking time during which a member speaks, regardless of interruptions or overlap speech from others | **Vocalization Distribution*:** Total member speaking time divided the meeting time | **Minutes (per member)** |
| Team vocalization distribution | Level of homogeneity or heterogeneity in which the team distributes its vocalization | **Standard Deviation **** considering each member vocalization distribution | **Proxy:** the lower-closest to 0, the better team balance distribution |
| | | **Dominance**:** the maximum participation level divided by the average participation of the other members | **Proxy:** the higher, the more difference is between the dominant speaker and the rest of the members |
| | | **Team Vocalization Tenure**:** time speech occupation over total meeting time | **% of the meeting time:** can range from 0 (no one spoke throughout the meeting) to 100% multiplied the number of members (all spoke throughout the length of the meeting) |
| | | **Clean Vocalization**:** team vocalization tenure subtracting total overlap time percentage | **% of the meeting time** during which a member speaks, without counting interruptions or overlap speech from others |
| Turn taking frequency | A turn is the number of times a member speaks after another member. A turn divides a speech segment | **Turn Taking Frequency*** is the number for each member during the entire meeting | **Count (per member)** |
| | | **Team Turn Taking**:** Total turn taking divided the meeting time | **Proxy:** the higher, the faster the speed of the exchange frequency in shifts |
| | More turn taking frequency means shorter speech segments per member | **Average Speech Segment**:** Average of speaking time per member divided turn taking count per member | **Seconds:** the higher, the longer average duration in the meeting speech segment |
| Overlapping speech | Fraction of time that two or more members are speaking simultaneously | **Overlap Time*:** Fraction of overlap speech time divided the total meeting duration | **% (per member)** |
| | | **Overlap Count**:** number of participation occasions in overlapping speech | **Count (per member)** |
| | | **Total Overlap Time**:** the sum of the member overlap time associated to the meeting time | **% of the meeting time:** the higher, the greater presence of overlap throughout the meeting |
| | | **Average Overlap duration**:** the average of each member difference between each overlap time and overlap count | **Seconds:** the higher, the longer average duration in meeting overlap |

by different government agencies. It has an independent government and budget. The university is organized in departments (17) and research centers (9). There are 23,619 undergraduate students, 1,761 postgraduate students, 3,099 academic staff, 60 undergraduate careers, 64 postgraduate careers, and 581 administrative and technical staff [25].

Organization 3 has emerged as a venture entrepreneurship in 2004. In its beginnings, the firm's activity was centralized in a single point of elaboration, sales and distribution. A short time later and, as a result of the great acceptance of the product proposal, the

**Table 2.** Organizations characteristics.

| No | Sector | Years in business | Staff number |
|---|---|---|---|
| 1 | **Consulting Services** | 10 years | Less than 10 |
| 2 | **Education** | 55 years | More than 500 |
| 3 | **Food** | 15 years | More than 100 |
| 4 | **Gas Distribution** | 74 years | More than 1500 |

structure has expanded: it currently has a production factory and seven distribution and sales locations. It has well established itself as a regional leading gastronomic firm; it has diversified, and currently produces more than 40 products. The organization offers a delivery service to its clients.

Organization 4 is the largest natural gas distributor in Argentina in terms of volume, covering 45% of the country in two contiguous regions. With a complex system of transportation pipelines and distribution networks that exceeds 50,000 linear kilometers in length; they supply more than 2,000,000 users in seven provinces of the country.

## 4 Results

Cross-boundary meetings [26] are regular in organizations as a form to address and solve problems or discuss and define future directions. Each organization planned a meeting in order to solve a current and real problem. Table 3 briefly summarizes the task for which the members were called to the meeting under study and team member characteristics.

Having collected and processed data, we first analyze each case separately. We highlight that the research observations show all four meetings were flowed smoothly and harmonically and no apparent conflict was shown during them. For space reasons, we only include the full description of case 4. We also present main descriptive results of the other three cases. Then, we present a synthesis and compare results in Table 4 and discuss results as preliminary interpretations.

### 4.1 Cases

The eight members meeting was vocally distributed unevenly. Even though member 18 dominated the participation (2,44 dominance) the standard deviation stand at 8,12. As Fig. 2 shows, there were 3 (if not 4) members that stood out. One of them, member number 1, was the facilitator and convener of the meeting, a person that opened the meeting with the agenda items, synthesized agreements achieved during the meeting and gave a formal finishing. Also, case 2 did not show a significant dominance member. In case 1 and 3, one member dominated the vocalization participation, the leader and convener of the meeting.

Figure 3 shows the dominance were both regarding vocalization and turn taking count –as in case 1. This feature was not present in case 2, where some members took

**Table 3.** Team meetings

| No | Context | Objective | Duration | Number of team members | Educational level structure Without //With grade formation |
|---|---|---|---|---|---|
| 1 | A client asks for a recruitment and selection service for a difficult profile to fill, the regular working plan needed to be adjusted | Re-define steps, responsible, schedules… | 50 min | 3 | 1//2 |
| 2 | Academic management top team seeks for an ISO certification, monitoring indicators is mandatory to achieve this organizational goal | Analyze annual KPIs and define next period objectives | 1:45 h | 7 | 3//4 |
| 3 | A new online delivery service is offered in the city. The questions were: What should we do? Should we incorporate these channel sales? What impact would this have over our own delivery personnel? | Analyze pros and conts. Take decisions. Define plan to respond | 1:15 h | 6 | 5//1 |
| 4 | Digital transformation is an imperative in the organization. Some sectors are using paper-based and digital systems to process customers' claims | Analyze change difficulties, synchronize sectors and organize faster migration | 40 min | 8 | 6//2 |

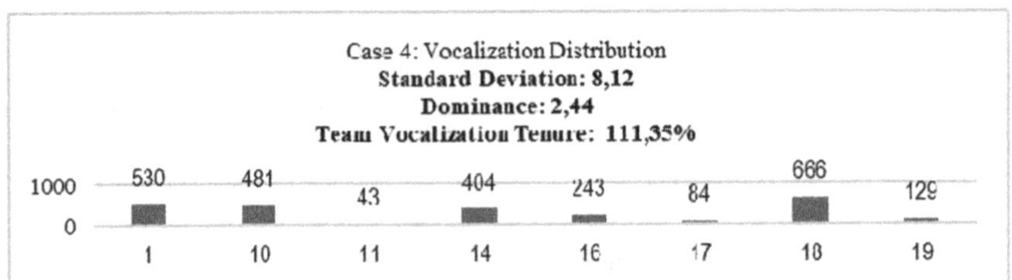


**Fig. 2.** Vocalization distribution indicators of case 4.

shorter turn. The members' sum 111,35% of vocalization over the extent of the meeting, which correspond to the level of the total overlap time (36,43%). Differently, in case 1, the team vocalized among 80,7% of the meeting length, which leaves almost 20% of the meeting time to members silence.

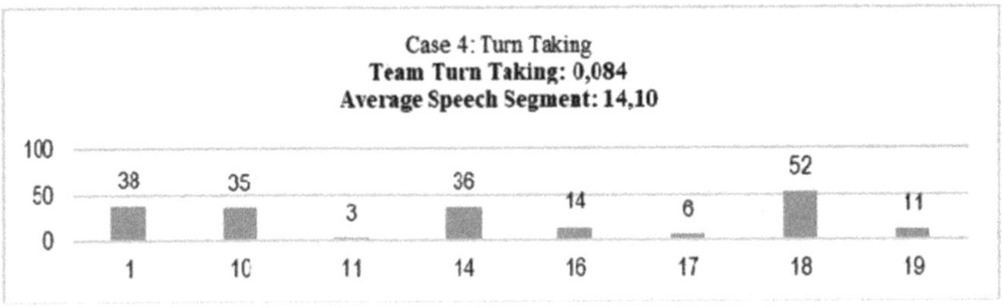


**Fig. 3.** Turn taking frequency indicators of case 4.

Team turn taking speed arose up to 0,085. The slowest was case 2 (0,056) The average speech segment in case 4 was 14,10 s.

The average overlap duration of case 4 was 1,36 s. Besides vocalization, member 18 also contributed more than the other members to overlap count and time (see Fig. 4). The same occurred in case 1, where the member that contributed the more to overlaps was not the leader neither the facilitator. Differently, in cases 2 and 3, vocalization and overlap were balanced. In this case, the dominant member was the only that had an outstanding difference between overlap time and overlap count, meaning overlaps were longer than others'. Case 2 present this same feature among their 3 dominant members, their overlaps were also longer. Differently, in case 1 and 3 all members contributed similarly in times and duration. During case 2 meeting, the researcher observation registered one case of overlap that correspond to a member (n°5) that leaved the room in order to respond a phone call. Considering the data of the entire meeting, it seems this circumstance did not affect results significantly.

Thanks to observer records about case 4, we identified that in general, overlaps appeared as confirmatory interventions or short interruptions (yes, of course, in agreement, a name or reference) or as informal interactions (jokes, laughter). During the meeting, as in case 2 and 3, there were occasions when in dynamic interaction two or more members "compete" to take the turn and one managed to overcome the overlap

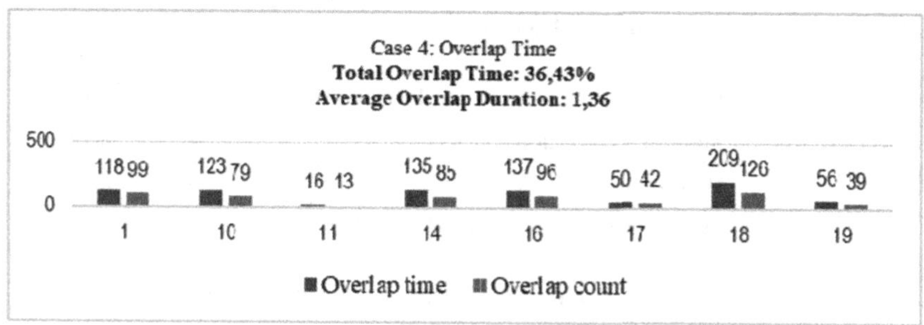


**Fig. 4.** Overlapping speech indicators of case 4.

at the cost of the other member's silence. Subsequently, the turn changed to whoever wanted to contribute. This feature mainly happened in case 3 where the flow of information run between the members quickly, team turn taking was the highest –faster turn- of the four cases (0,099).

The appearance of subgroups in short periods of time, which did not imply disruptive overlaps, was also recorded. In this case, members immediately returned, naturally and without being asked, to configure themselves as the only group. This phenomenon of subgroup assembly and natural return to the unique group was observed in this meeting of eight members, mainly. In case 3, when subgroups were sustained over time and its extent required greater concentration due to the vocalization volume increasing, the leader expressly requested to return to the single group configuration. In these situations, was registered a first functional overlapping speech moment (subgroup) and a second dysfunctional overlapping speech moment (lack of line, organization of speech and general understanding). During case 1 no subgroups were set –which in a meeting of three members would not be possible-. Still, this fact evidence case 1 overlap indicators do not correspond to subgroups.

### 4.2 Team Comparison and Preliminary Interpretations

Table 4 summarizes the four cases. The table includes performance indicator (members' perception about what happened throughout the meeting). According to this perception, all meetings were valued above 75%, the difference between the top (case 1) and the lowest (case 4) is 18,06%.

Based on the theoretical background, we understand that the lower the vocalization distribution standard deviation, the greater the homogeneity; which would indicate greater team efficiency. This indicator simplifies the comparison of the indicators related to the distribution in the vocalization [27]. Although it has been the most proven, we found it did not correlate positively in our four case studies.

The dominance indicator also contributes to this aspect of the meeting analysis, where it reflects the maximum participation level divided by the average participation of the other participants. We found no direct relationship between dominance indicator and meeting performance results neither. A series of factors could be attributed to our findings such as the size of the meeting team, if the meeting had or not a facilitator and the presence of a reference member with the technical knowledge needed to solve

**Table 4.** Team metric results by case

| Metric ID | Indicator | Case 1 | Case 2 | Case 3 | Case 4 |
|---|---|---|---|---|---|
| Team vocalization distribution | **Standard Deviation** | 15.82 | 7.42 | 7.34 | 8.12 |
| | **Dominance** | 2.95 | 1.63 | 2.38 | 2.44 |
| | **Team Vocalization Tenure** | 80.68% | 108.59% | 92.87% | 111.35% |
| | **Clean Vocalization** | 74.63% | 72.84% | 74.02% | 74.92% |
| Turn taking frequency | **Team Turn Taking** | 0.076 | 0.056 | 0.099 | 0.084 |
| | **Average Speech Segment** (seconds) | 9.68 | 18.73 | 8.99 | 14.10 |
| Overlapping speech | **Total Overlap Time** | 6.05% | 35.75% | 18.85% | 36.43% |
| | **Average Overlap Duration** (seconds) | 1.10 | 2.00 | 1.29 | 1.36 |
| Performance result (scale 1–6) | | 5.63 | 5.13 | 4.88 | 4.54 |
| | | 93.75% | 85.25% | 81.25% | 75.69% |

the meeting agenda. In addition, differently from what is indicated by the literature, we found up to 3 dominant members of the team conversation.

We found team vocalization tenure –the extent members use the meeting time to vocalize- were coherent with performance results. As long as the team vocally fits the meeting time (100% of the meeting time or less), the results hold high. To the extent the vocalization percentage exceed 100%, the perception of performance in results decreases. For example in case 4, members vocalized 111,35% and ranked last. Following this reasoning, it is interesting to connect team vocalization tenure with total overlap time percentage. We calculated clean vocalization indicator (vocalization without counting interruptions or overlap speech) This indicator itself did not display any correlation with performance results but, in our cases, where clean vocalization stood around 70–75%, performance results corresponded above 75%.

Similarly, we could not found a positive or negative direct relation in between speed frequency among members' turns nor the average speech segment duration with results. However, and considering the need to identify more clearly the explanation of overlapping speech, we could note that where the average speech segment was high (case 2 and 4) the meeting overcome the 100% of team vocalization tenure, and the total overlap time was also high. Our interpretation was that a faster turn taking could facilitate a better means of overlapping speech in team interaction.

We identified different kinds of overlapping speech, namely informal, disruptive, confirmatory, competitive for a turn and related to subgroups. Subgroups appeared when the meeting has 4 or more members. We found subgroups might have functional or dysfunctional moments; they may disappeared naturally or by demand.

In the four cases, overlapping speech metric did not correlate with performance results directly. However, we can approximate as an interpretation that when total overlap time was high but dominance was low (case 2) results were higher than when total overlap and dominance were both high (case 4). This may indicate that when vocalization was balanced, all members contributed to a synchronized and fluid conversation, meaning the interruptions and overlaps reflected this rhythm of interaction.

The average overlap duration holds below 1,36 s in three cases, and in case 2 it was 2 s. Considering the performance results and the observer notes regarding the fluidity of the meetings, this length was not disrupting in team interaction, with the exception of dysfunctional subgroups moments.

Considering team communication dynamics and thanks to the researcher observation, we offer a preliminary characterization of meeting stages. We identified a first stage, the starting one, where members arrived and located. This stage was characterized with icebreaker interaction, members exchanged informally. The interaction was not structured and dominant members appeared (for instance, the extrovert or the leader) and subgroups were dysfunctional. Disruptive and informal overlaps were observed and turn taking was faster. In a second stage, the developing stage, the group focused their activity on the convening meeting topics and defined a rhythm. Subgroups may appear and the overlapping speech was mainly disruptive, confirmatory or respond to the intention of winning a turn (competitive overlap). The speech segments were longer. Then, as the finishing stage of the meeting, the group returned to unstructured dynamics; commentaries that were not related to the convening meeting topic appeared. During this last stage, overlaps corresponded to informal interaction and turn taking got speed.

## 5 Conclusions and Future Research

Our study extends the understanding team structure communication patterns and provides an important contribution to research in the field of sociometric badges, being the first one that analyzed teams in real work settings. Focus on the use of the badges allowed to measure metrics that have been used previously (vocalization distribution, turn taking frequency and overlapping speech) and new indicators, built with the purpose of exploring the technological possibilities of the devices (dominance, team vocalization tenure and average speech segment, for example).

Based on previous literature, while one might argue that vocalization distribution is a good proxy for team performance, when we examined our four cases, we found that the metric should be analyzed as part of a set of indicators in context. For example, we identify subgroups and overlapping speech key mechanisms of team interactional dynamics.

We can identify at least four lines for future research using sociometric badges. First, the study of the evolution of the same team along different meetings. Second, it would be interesting to focus on technological possibilities of devises to identify and characterize more clearly different meeting stages. Third, the collection of data in a wider set of teams and organizations may allow better comparisons within cases and indicators. Fourth and finally, regarding available technology, there is great chance to incorporate artificial intelligence to give more precision to metrics.

**Acknowledgements.** This work was supported by Secretaría de Ciencia y Tecnología (Universidad Nacional del Sur) under Grant number 24/C055. The Department of Electronic and Computers Engineering (Universidad Nacional del Sur) collaborated in the development of the sociometric devices.

## References

1. Iyer, S.: Understanding the eHRM promise and adoption imperatives. NHRD Netw. J. (2019). https://doi.org/10.1177/2631454119873204
2. Thibeault, J., Wadsworth, K.: Recommend This!: Delivering Digital Experiences that People Want to Share. Wiley, New York (2014)
3. Aloini, D., Covucci, C., Stefanini, A.: Collaboration dynamics in healthcare knowledge intensive processes: a state of the art on sociometric badges. In: Rossignoli, C., Virili, F., Za, S. (eds.) Digital Technology and Organizational Change. LNISO, vol. 23, pp. 213–225. Springer, Cham (2018). https://doi.org/10.1007/978-3-319-62051-0_18
4. Bauer, W., Schuler, S., Hornung, T., Decker, J.: Development of a procedure model for human-centered industry 4.0 projects. Procedia Manufact. **39**, 877–885 (2019)
5. Drach-Zahavy, A., Somech, A.: Understanding team innovation: the role of tream processes and structures. Group Theor. Res. Pract. 111–123 (2001)
6. Shaw, P., Varghese, R.M.: Industry 4.0 and future of HR. J. Manage. **5**(6), 96–103 (2018)
7. Lepri, B., et al.: The SocioMetric badges corpus: a multilevel behavioral dataset for social behavior in complex organizations. In: International Conference on Privacy, Security, Risk and Trust and International Conference on Social Computing (2012). https://doi.org/10.1109/socialcom-passat.2012.71
8. Silva, C., Fortes, D., Nascimento, R.: ICT governance, risks and compliance - a systematic quasi-review. In: Proceedings of the 19th International Conference on Enterprise Information Systems (ICEIS 2017), **3**, 417–424 (2017)
9. Endedijk, M., Hoogeboom, M., Groenier, M., de Laat, S. van Sas, J.: Using sensor technology to capture the structure and content of team interactions in medical emergency teams during stressful moments. Frontline Learn. Res. **6**(3), 123–147 (2018). https://doi.org/10.14786/flr.v6i3.353
10. Chen, H.-E., Miller, S.R.: Can wearable sensors be used to capture engineering design team interactions?: An investigation into the reliability of sociometric badges. In: 29th International Conference on Design Theory and Methodology, vol. 7 (2017). https://doi.org/10.1115/detc2017-68183
11. Yu, D., et al.: Intelligent emergency department: validation of sociometers to study workload. J. Med. Syst. **40**(3) (2016). https://doi.org/10.1007/s10916-015-0405-1
12. Lederman, O.: Hacking innovation-group dynamics in innovation teams. Doctoral dissertation, Massachusetts Institute of Technology (2015)
13. Woolley, A.W., Chabris, C.F., Pentland, A., Hashmi, N., Malone, T.W.: Evidence for a collective intelligence factor in the performance of human groups. Science **330**(6004), 686–688 (2010)
14. Kim, T., Chang, A., Holland, L., Pentland, A. (Sandy): Meeting mediator: enhancing group collaboration using sociometric feedback. In: Proceedings of the 2008 ACM Conference on Computer Supported Cooperative Work, pp. 457–466 (2008)
15. Kim, T., McFee, E., Olguin, D.O., Waber, B., Pentland, A.: Sociometric badges: using sensor technology to capture new forms of collaboration. J. Organ. Behav. **33**(3), 412–427 (2012)

16. Johannes, B., Sitev, A.S., Vinokhodova, A.G., Salnitski, V.P., Savchenko, E.G., Artyukhova, A.E.: Wireless monitoring of changes in crew relations during long-duration mission simulation. PLoS ONE **10**(8), (2015)
17. Zhou, N., Kisselburgh, L., Chandrasegaran, S, Badam, S., Niklas, K, Ramani, K: Using social interaction trace data and context to predict collaboration quality and creative fluency in collaborative design learning environments. Int. J. Hum. Comput. Stud. (2019). https://doi.org/10.1016/j.ijhcs.2019.102378
18. Olguíın, D., Pentland, A.: Sociometric badges: state of the art and future applications. In: IEEE 11th International Symposium on Wearable Computers. Boston, MA (2007)
19. Lebuda, I., Glăveanu, V.P. (eds.): The Palgrave Handbook of Social Creativity Research. PSCC. Springer, Cham (2019). https://doi.org/10.1007/978-3-319-95498-1
20. Paulus, P.B., Dzindolet, M., Kohn, N.W.: Collaborative creativity—group creativity and team innovation. In: Munford, M. (ed.) Handbook of Organizational Creativity, pp. 327–347. Elsevier, Londres (2012)
21. Lewis, K.: Knowledge and performance in knowledge-worker teams: a longitudinal study of transactive memory systems. Manag. Sci. 1519–1533 (2004)
22. Yin, R.: Case Study Research, Design and Methods, COSMOS Corporation, 5th Edn. SAGE (2003)
23. Waller M.J., Okhuysen, G.A., Saghafian, M.: Conceptualizing Emergent States. The Academy of Management Annals (2016)
24. Gibson, C.B., Randel, A.E., Earley, P.C.: Undersantanding group efficacy: an empirical test of multiple assessment methods. Group Organ. Methods **25**, 67–125 (2000). https://doi.org/10.1177/1059601100251005
25. Universidad Nacional del Sur. Direcci6n de Informaci6n Institucional. Anuarios (2018). http://www.uns.edu.ar/contenidos/411/652#anuarios. Accessed 15 May 2019
26. Edmondson, A.C., Harvey, J.-F.: Cross-boundary teaming for innovation: integrating research on teams and knowledge in organizations. Hum. Resour. Manag. Rev. **28**, 347–360 (2018)
27. Harrison, D., Klein, K.J.: What's the difference? Diversity constructs as separation, variety, or disparity in organizations. Acad. Manage. Rev. 1199–1228 (2007)

# Digital Transformation Framework for Adequacy of Maintenance Systems for Industry 4.0

André Luiz Alcântara Castilho Venâncio[1(✉)], Eduardo de Freitas Rocha Loures[1], Fernando Deschamps[1], Ricardo Alexandre Diogo[1], Alysson Felipe Lumikoski[1], and Neri dos Santos[2]

[1] Pontifícia Universidade Católica do Paraná, Rua Imaculada Conceição, Curitiba 1155, Brazil
andre.venancio@pucpr.edu.br

[2] Universidade Federal de Santa Catarina, Campus Universitário Reitor João David Ferreira Lima, S/N, Florianópolis, Brazil

**Abstract.** Manufacturing execution concepts at the highest level of automation and interoperability are proposed by Industry 4.0. This is thanks to a group of information and communication technologies capable of creating integrated and connected systems, known as – cyber-physical systems. This network of integrated systems has been promoting gains in productivity, altering the profile of the workforce and the very conception of the work that will have more cognitive demands, increasing competitive potential. Therefore, organizations are more than ever willing to adapt to this scenario. Aiming to facilitate this digital transformation, avoiding an inefficient allocation of financial and human resources, the present work proposes a framework that provides the necessary guidelines for such a transformation to become viable. This will be done by a series of analyses using multicriteria decision-making methods (MCDM), evaluating the maturity of legacy systems, with a focus in the field of industrial maintenance. Those analysis are also based on the RAMI4.0 architecture, Total Productive Maintenance (TPM), Framework for Enterprise Interoperability (FEI), and will provide companies with guidelines for adequacy plans. To test and validate this framework, a multinational industrial entity belonging to the automotive sector was selected as a case study. The application results confirm that such tool can help in digital transformation projects and indicates adjustments for this to be done in a better way.

**Keywords:** Industry 4.0 · Industrial maintenance · Multicriteria Decision-Making Methods (MCDM) · Interoperability · Digital transformation

## 1 Introduction

In the midst of a highly informational scenario, interoperability is an element to be measured by organizations. Such term represents the capacity of a system to communicate between two or more others, in order to use the shared data and access external functionalities [1]. Among the technologies that seek to consume interoperability in manufacturing, the Internet of Things (IoT), Big Data, Artificial Intelligence (AI), 3D-Printing,

© Springer Nature Switzerland AG 2021
D. A. Rossit et al. (Eds.): ICPR-Americas 2020, CCIS 1407, pp. 280–292, 2021.
https://doi.org/10.1007/978-3-030-76307-7_21

Augmented Reality, Machine to Machine (M2M), Analytics and Cloud Computing stand out. Classified as information and communication technologies (ICTs), they are the basis for Industry 4.0 (I4.0), enabling the emergence of cyber-physical systems. According to [2], some of the benefits that such systems' network have provided to organizations are: increased productivity, alteration of the workforce profile and increased competitive potential. However, for the implementation of those technologies to be assertive, it is necessary that conceptual, technological and organizational requirements are satisfied. As the world experience a time of transition to I4.0, this adaptation recurrently involves legacy systems that [3] characterize as those with high usage times, which are vital to the organization's business, and does not fit into future IT strategies.

Notwithstanding, modernization is not easily prioritized by organizations, similarly for the maintenance sector, which is seen more as an inevitable necessity than as a goal to pursue. However, the role of industrial maintenance has become a strategic element to achieve business objectives [4]. To [5] the maintenance goals involve: Safety, expressed through a higher reliability coefficient of equipment prone to critical failures; Availability when considering the time when the equipment is producing at full capacity; Budget, involving the reduction of maintenance costs. Those goals are easily related to the benefits provided by I4.0.

Given the difficulty evidenced by digital transformation initiatives for legacy systems and the expressed close relation between maintenance and modernization goals, the research developed here sought to answer the following research question: "How to define a technology prioritization plan, in the field of industrial maintenance, in order to adapt legacy systems for Industry 4.0 requirements?". Therefore, focusing on the industrial maintenance area and based on an assessment of qualifying attributes of a given organization, this work seeks to establish a digital transformation framework with a set of models based on Multicriteria Decision Making Methods. Our hypothesis suggests that such a tool will make it possible to elaborate more assertive guidelines, capable of aligning legacy maintenance systems with the vision of highly interoperable manufacture, necessary to fully access the benefits brought by Industry 4.0.

## 2 Theoretical Dimensions and Scientific Scenario

The disruptive technologies envisioned for I4.0 promotes to escalate industrial productivity, putting current economic models in check, foster the growth of industrial organizations, changing the profile of the workforce and ultimately increase the competitiveness of companies [2]. Thus, the proximity with the term "interoperability" is evident because of the prominence of such technologies, which will increase the collaboration between systems, machines and people; that way, enabling greater speed, flexibility and efficiency in production processes, resulting in higher quality at reduced costs. The researches presented here will show how those gains can be achieved in the maintenance sector, before exploring others horizons.

Nonetheless, implementing a system with the maturity level necessary to operate in the Industry 4.0 bias will require a digital transformation project. With this in mind, and understanding that every project must operate with a budget, or maximum expenditure's cost, the present work explores a premise that "not everything which is wanted to be implemented (e.g., maintenance functions of an I4.0 system), can actually be implemented".

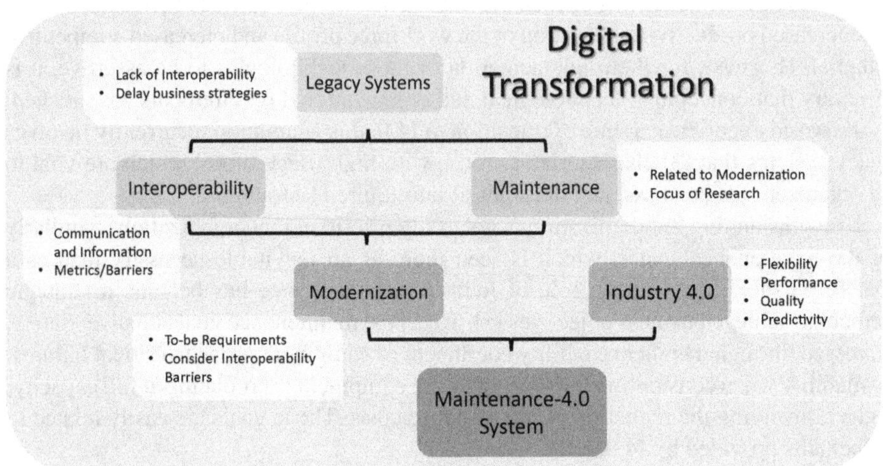


**Fig. 1.** Research elements.

Considering the Fig. 1 elements overview, the researches in this work explores how to adequate maintenance legacy systems requirements for the I4.0 scenario. By hypothesizes, the technologies suggested for implementation will increase the maturity of systems considering interoperability barriers, i.e. without harming subjacent systems in the process of implementing those technologies.

## 2.1 Legacy Systems

Even after three decades of research in modernizing legacy systems, it is notable that many remain in operation. This is due to the fact that these systems are generally very comprehensive [6, 7]. They interoperate with other processes or subsystems, only remain in operation due to their technical complexity of replacement and/or adaptation and criticality in the organization's operations, in such a way that remains in constant activity. Every system is likely to become legacy at some point and its data is characterized as valuable, since its history can be used to understand its behavior in search of optimization [3]. However, to remain competitive, companies must continually change their processes, sometimes radically, and legacy systems can delay modernization processes and directly influence the company's business strategy [8].

## 2.2 Maintenance-4.0 Technologies

The legacy systems scenario researched for this particular work was constrained to industrial maintenance, a sector easily impacted by I4.0. With an exploratory character, a literature review was focused on the Industry 4.0 base technologies that are relevant to the maintenance sector. Here, three research rounds were carried out.

**First Round.** A first research round was sought to understand a general context of I4.0 technologies, and for that, reports with frameworks already formalized in the literature were used. The objective was to gain an overview of I4.0 technologies, in the view of different technology consultants, which were [2, 9–18].

**Second Round.** In the second round, the results from the overviewed technologies were validated in academic articles, focusing on solutions for the maintenance sector. This research round was conducted as follow: (i) was searched the relation between "technology" AND "maintenance" (e.g., Cloud AND Maintenance; or, Augmented Reality AND Maintenance); (ii) only open access articles were searched; (iii) time period from 2014 to 2019 was considered mature since the term "Industry 4.0" appeared by 2011. The research platforms used were: ScienceDirect and Archive Ouverte HAL. In the end, 59 articles were considered.

**Third Round.** Finally, in the third round, I4.0 technologies applied in industrial maintenance were filtered and allocated into categories, regarding more tangible examples of them being applied in manufacture. The whole literature database ended with 87 articles and reports. From it, nine Maintenance-4.0 technology groups were identified: Big Data, Analytics, Artificial Intelligence and Cloud Computing, formalized as cyber-physical subgroup; Advanced Machines, Advanced Materials, Flexible Connection Devices and Digital-to-Real Representation (i.e. encapsulating Digital Twin applied in maintenance activities), formalized as application subgroup; and Sensors (i.e. encapsulating IoT and Smart Sensors, formalized as the bridge to digitalize physical operations).

## 2.3 Models and Requirements

To allocate maintenance into the conformities of I4.0 in an organized way, two different architectures were approached. The first, Framework for Enterprise Interoperability (FEI) from [19], was considered by the premise that interoperability might be a relevant metrics to understand what can or cannot be implemented to a system. This possibility is considered because FEI relates conceptual, technological and organizational barriers linked between the enterprise layers, that could be generated by two systems trying to communicate. Coupled with that, the prerogative that interoperability barriers could difficult the insertion of technology seems feasible, once the legacy system and others adjacent systems/processes may share communication-dependence.

The second is a reference architecture model for Industry 4.0 (RAMI4.0) [18], converging multi-stakeholder views on how I4.0 can be accomplished based on existing communication standards and functional descriptions [20]. Analogously to the FEI, which has a broad organizational view but is not linked to Industry 4.0, the RAMI4.0 presents a similar enterprise's layers perspective. Considering that this research investigates interoperability barriers that might appear by implement I4.0 technologies in maintenance legacy systems, those frameworks were merged (see Fig. 2).

This composed framework was used to promote interoperability elements to an I4.0 referential architecture. The following section explains how industrial maintenance was focused on such a vision.

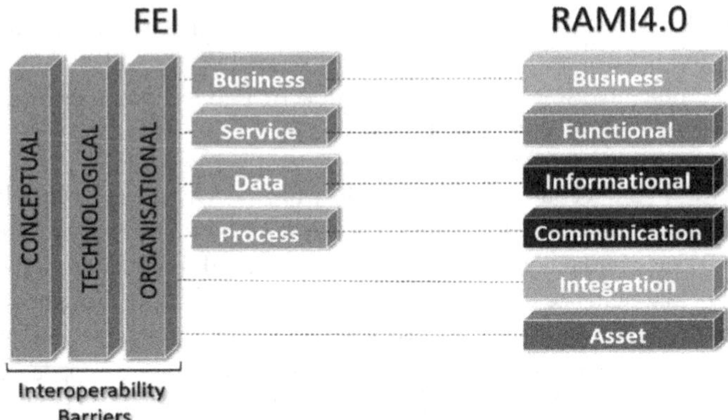


**Fig. 2.** (FEI barriers) × (RAMI4.0 layers) composed frameworks.

**As-Is Functions.** The work [21] describes a study of industrial maintenance through the lens of RAMI4.0 architecture, indicating attributes and functional requirements, which gives the present work directions in how to analyses maintenance systems in an a priori state. Further, the presented case study details this analysis.

**To-Be Model.** In order to achieve I4.0 requirements for the maintenance sector, the Total Productive Maintenance (TPM) six big losses [22], were also considered (see Fig. 3).

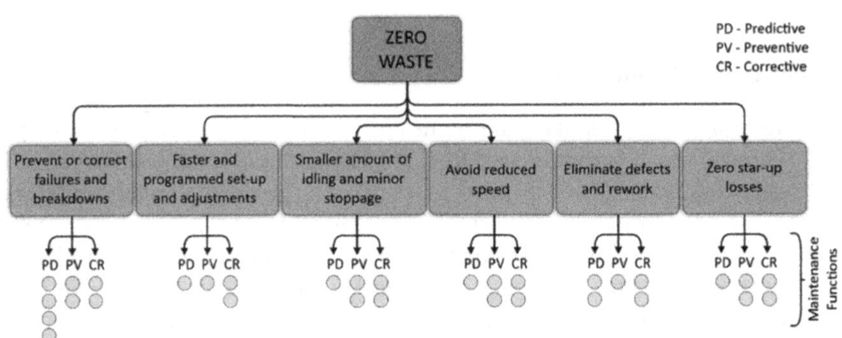


**Fig. 3.** Referential Maintenance-4.0 reference model based on TPM.

For those losses, this model formalizes "courses of action", meaning that for each loss there is a course of action based on I4.0 vision. Moreover, those courses of action are categorized in three main maintenance approaches: predictive, preventive and corrective [23]. Therefore, the spheres tagged as "maintenance functions" represent Maintenance-4.0 enablers for predictive, preventive and corrective approaches. Our premise is that, the full potential of a maintenance system can be achieved with the technology groups proposed in Sect. 2.2.

In resume, aiming to guide maintenance processes to "zero waste" using disruptive technologies, this proposed model serves as a To-be guide, for an As-is analysis [21], due to interoperability barriers. Alternatively, what is needed to implement (i.e. disruptive technologies) according to what is possible to be implemented (i.e. interoperability barriers).

### 2.4 MCDM

Multicriteria decision making/analysis (MCDM/A) methods, emerged in the search for solutions to complex problems that are difficult to measure. In this work they were used as tools for more assertive decisions in systems optimization, also following a couple of referential researches that apply decision making assessments in the dimensions of interoperability, maintenance and Industry 4.0 are [21, 24–28].

There are four important elements that involve the MCDM methods: Set of "alternatives", from which the decision is chosen; set of "criteria", or factors related to making a good decision; the "preferences" of the decision-maker, being clear, the problem becomes more understandable; and the "result" of each choice, measured in terms of criteria according to the decision maker's preferences.

Two different MCDM are used for the three steps framework, detailed in the next section. For the Step 01 and 02, the Analytic Hierarchy Process (AHP) [29] is used in order to derive priorities based on sets of peer comparisons, thus it is structured on the intrinsic ability to ponder their perceptions or ideas hierarchically [30]. In Step 03, the Preference Ranking Organization Method for Enrichment of Evaluations (PROMETHEE II) is used, an interactive method designed to deal with quantitative, qualitative criteria and discrete alternatives. This method can classify alternatives that are difficult to compare due to a commitment to standards of evaluation as non-comparable alternatives [31]. According to [32] this method has been applied in varied fields such as industrial locations, labor planning, investments, medicine, chemistry, tourism and ethics.

## 3 Framework

The framework proposed in this article it is structured in three steps. In Step 01 the AHP method is used to assess the organization's maturity, relating the I4.0 requisites in a maintenance bias. In the Step 02 is also build for the application of AHP method, which will provide the allocation of weights for functions of a Maintenance-4.0 architecture, in order to obtain a selection of the most needed ones. Finally, at Step 03, the PROMETHEE II method will be applied to prioritize the technologies that will best adapt to the functions selected in the previous step (see Fig. 4).

It is expected that after applying the framework, a legacy maintenance system will have its main requirements highlighted, indicating what needs to be improved according to I4.0 technologies. The decision methods analysis considers not only what needs to be implemented to improve the system, but also what is feasible through interoperability barriers.

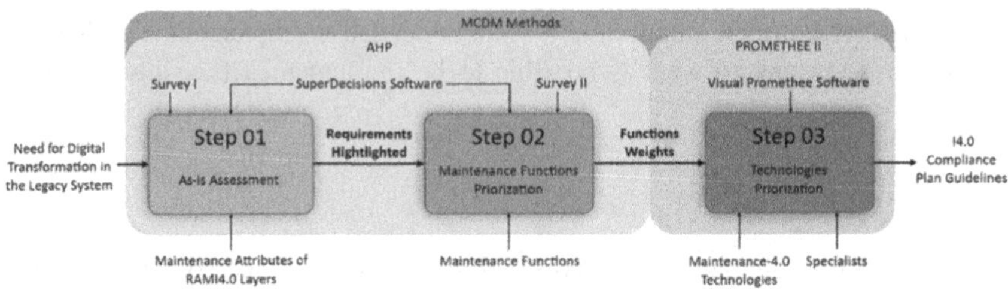


**Fig. 4.** Framework overview.

### 3.1 Maturity Assessment (Step 01)

Once the organization's need to optimize its systems to I4.0 standards is confirmed, in Step 01 an assessment of its maturity in relation to the desired requirements is carried out. For this, engineers and maintainers must be made available, since they will be in the role of decision-makers, answering the proposed survey model. Here the AHP method is executed in the "SuperDecisions" software. Its model reflects a maturity assessment that, through the construction of classification structures from the six layers of RAMI4.0, describes the decomposition of a machine in its structured properties, enabling its virtual mapping (see Fig. 5).

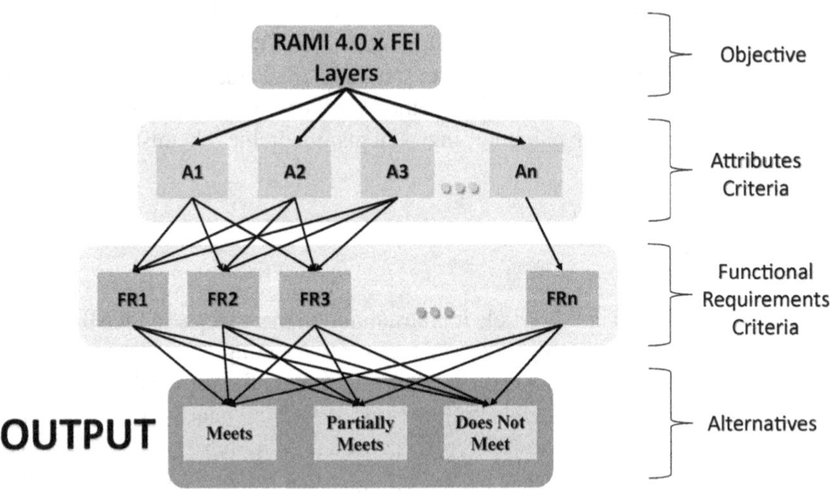


**Fig. 5.** Step 01 – maturity assessment (AHP method model 1).

The name of the analyzed layer will be located at the top level of the decision model, representing the model objective. The intermediate level will consist of attributes and functional requirements belonging to the domain of industrial maintenance, distributed among the six layers to be analyzed [21]. The relation, attributes-requirements, qualifies the analyzed system. In the end, the lower level presents the alternatives: meets, partially meets, and does not meet; related to each functional requirement of the intermediate level.

After performing the peer review survey with the decision-makers, with no inconsistency value identified by the software, it will perform the AHP synthesis in order to rank the three alternatives, thus providing the result of the maturity assessment for each layer of RAMI4.0. When all six layers are evaluated it will be possible to obtain the degree of maturity, related to the requirements of Industry 4.0.

### 3.2 Maintenance-4.0 Functions Prioritization (Step 02)

Having delimited the areas with the greatest lack of industrial maturity in Step 01, the objective of Step 02 is to prioritize the maintenance functions that will be used in the optimization process to implement I4.0 capabilities to the legacy systems analyzed. The AHP will be used again in order to assign weights to the functions accordingly to the preferences from the decision-makers, not synthesizing the alternatives, as done previously (i.e., this step uses the method to gather the functions weights solely, and not to support a final decision) (Fig. 6).

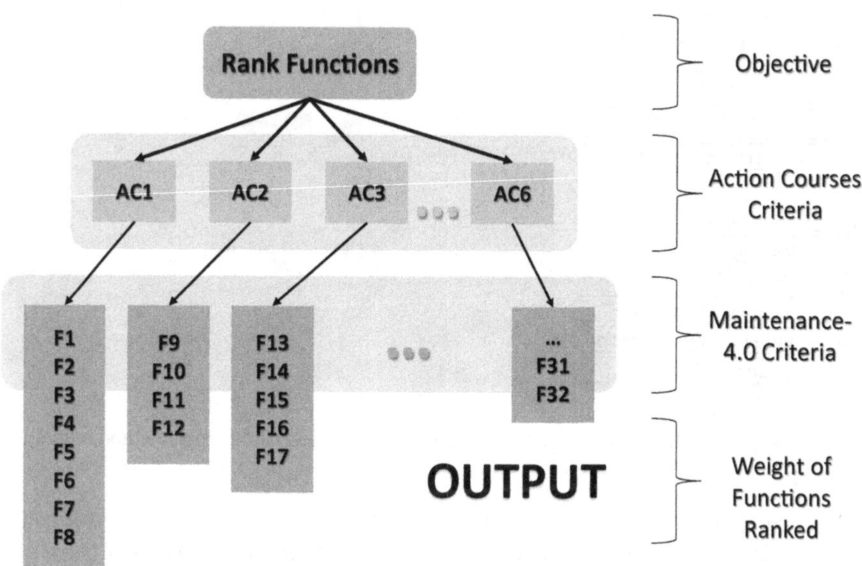


**Fig. 6.** Step 02 – Maintenance-4.0 function's rank (AHP method model 2).

At this stage, the survey reflects the Maintenance-4.0 model (see Sect. 2.3, To-be model), which presents a series of maintenance systems functions and their application in the light of I4.0. Based on the TPM's six main losses, the engineers and maintainers must consider their decisions regarding predictive, preventive and reactive approaches, that will guide maintenance processes to zero waste. At the end of the analysis, each function of the proposed Maintenance-4.0 model is ranked by weights in the AHP method in the "SuperDecisions" software, according to the decision-makers preferences.

### 3.3 Maintenance-4.0 Technologies Prioritization (Step 03)

Based on the maintenance functions weighted in the previous stage, Step 03 objective is the prioritization of I4.0 technologies that best suit those functions. Here, the framework does not require a survey for engineers and maintainers, leaving the role of decision-makers to a specialist. In the case study presented in Sect. 4, one of the authors with access to the organization fill that specialist role. This suggests that is necessary an expert view of possible solutions to be used. The literature review on I4.0 technologies under the maintenance domain was used at this stage.

A decision model using the PROMETHEE II method in "Visual Promethee" software, was built. The weights of each Maintenance-4.0 function, from the previous step, will be inserted and related to the nine technology groups from the literature review. At the end of this analysis the software will provide several types of reports based on the hierarchy carried out by the method, and the specialist is responsible to decide the technologies necessary to cover the maintenance functions (Fig. 7).

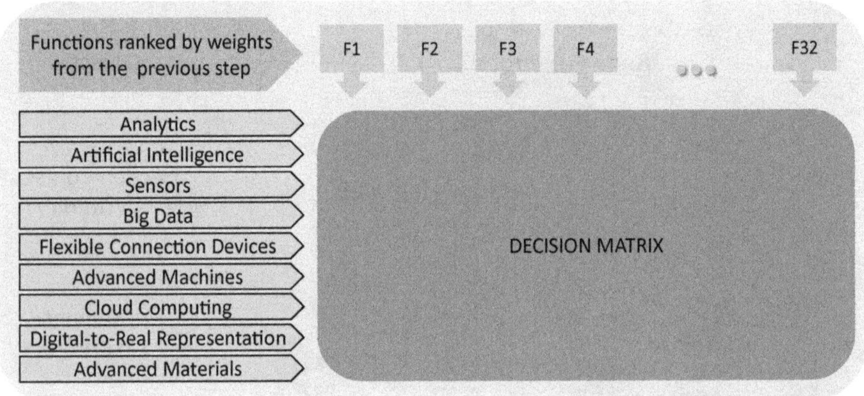


**Fig. 7.** Step 03 – Maintenance-4.0 technologies prioritization (Promethee method model).

After completing all the stages of the framework, there will be enough information for the specialist to develop assertive guidelines for an I4.0 compliance plan. Such plan suggests that: The Maintenance-4.0 technologies selected in the Step 03 enable the functions prioritized in the Step 02, which will act on the diagnosed areas arising from the Step 01 maturity assessment.

## 4 Results

To test the framework, a case study was applied to a multinational vehicle manufacturer. With a presence in more than 120 countries, the manufacturing complex in the southern region of Brazil employs approximately 8 thousand employees and has a production capacity of 320 thousand vehicles per year. We sought an area that offered a wider range of equipment, which is why the recently expanded engine factory has become the best option, mixing a wide range of modern and antique machinery. In Step 01, the industrial maturity assessment made in the "SuperDecisions" software, resulted in the analysis from Fig. 8, according to the maintainers and maintenance managers' preferences.

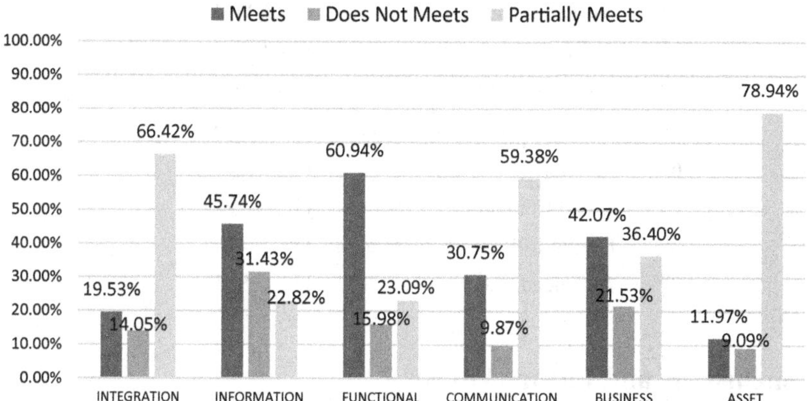


**Fig. 8.** RAMI 4.0 layers individual interoperability assessment from AHP model 1.

Based on RAMI4.0/FEI, and again, following the preferences of managers and maintainers in the role of decision-makers, the Information, Functional and Business layers presented the degree of adequacy "meets" while the Integration, Communication and Asset layers are classified as "partially meets". Also, with a deeper analysis, even though Business and Information layers meets the level of maturity, in both cases the alternatives "does not meets" and "partially meets" together exceeds 50%. This means that the AHP method is pointing out the preference (i.e., acknowledgment) of decision-makers, that the factory is at a level that "meets" the requirements, but with more uncertainty than the "Functional" layer, for example. Such fact highlights the importance of choosing decision-makers that knows their processes and systems.

In the second stage, the maintenance functions of the Maintenance-4.0 model were ranked by relevance level, also following the preference of the same decision-makers from the previous stage. Because this works focused on the application of the decision-making methods for the adequacy of systems, this rank will not be displayed.

For Step 03, one of the authors played the role of decision-makers as a specialist. Using the "Visual Promethee" software, the weights of the functions obtained in Step 02 were inserted, and chosen by their level of need (i.e., syntactic graduation from 1 to 9), measured with the nine groups of Maintenance-4.0 technologies, as alternatives. Table 1 presents the ranking of the nine most relevant maintenance technology groups to meet the functions required to achieve a Maintenance-4.0 system, where phi, represents the preference index used by the method.

It is noticed the predominance of technologies that operate at the Cyber-Physical level: Analytics, A.I. and Big Data, together with Sensors; being responsible to collect, interpret and store a large amount of data, enabling actions that are lacking in the factory, established in Step 02.

**Table 1.** Most relevant technology groups for this analysis.

| Rank | Technology | Method relevance |
|---|---|---|
| 1st | Analytics | Phi 0,4245 |
| 2nd | Artificial intelligence | Phi 0,3455 |
| 3rd | Sensors | Phi 0,2342 |
| 4th | Big Data | Phi 0,1930 |
| 5th | Flexible Connection Devices | Phi 0,1713 |
| 6th | Advanced Machines | Phi 0,1566 |
| 7th | Cloud Computing | Phi -0,1923 |
| 8th | Digital-to-Real Representation | Phi -0,4994 |
| 9th | Advanced Materials | Phi -0,8334 |

## 5 Conclusions and Future Works

The research developed here sought to answer the following research question: "How to define a technology prioritization plan, in the field of industrial maintenance, in order to adapt legacy systems for Industry 4.0 requirements?". For this, a framework was built, based on the RAMI4.0 and FEI architectures, assessing the maturity of legacy systems in the field of industrial maintenance under the requirements of I4.0 and considering interoperability barriers, with the hypothesis of providing assertive guidelines for adequacy plans for manufacturing industries. This need is part of the demand for adaptation to I4.0, where the reconditioning of legacy systems becomes the objective of organizations that seek to assign new functionalities to their equipment through digital transformation/modernization processes. Multicriteria decision-making methods were the elements that encapsulated this work, giving a tooling bias to the framework.

Among the main difficulties encountered are the long questionnaires that need to be filled out in the steps. The professional involved spend a lot of time filling them out, away from his activities. Although the framework has been positively validated, its complexity hinders its use. Many judgments are necessary and it is essential that all the components of the process are clarified in order to avoid mistaken considerations. With this in mind, parallel studies seek to alleviate the complexity of such digital transformation projects and open the scope for guidelines in addition to maintenance. A more mature initiative to design this framework will also be tested, using a different MCDM in the Step 02, solving only the most decisive Maintenance-4.0 functions, with the prerogative that "not everything that is necessary to implement, can actually be implemented – so the focus will be on the decisive aspects". This initiative could reduce the framework complexity focusing on decisive functions only. Further, this could be applied more than once, encountering new decisive functions each time the previous ones were resolved, corroborating with a gradual digital transformation.

Therefore, analogous with the empower of people with decision support tools, we emphasize as a final consideration that, digitalization of information, processes, functions that make up the operations of a business and transforming the business's strategies are necessary, but not enough. Most importantly, digitalization is essentially about technology, but digital transformation is not. The digital transformation is about people. It is how to improve the quality of people's lives at work and how to improve the performance of organizations.

## References


1. Chen, D., Daclin, N.: Framework for Enterprise Interoperability, pp. 77–88 (2006). https://doi.org/10.1002/9780470612200.ch6.
2. The Boston Consulting Group: Industry 4.0 - Future of Productivity and Growth in Manufacturing (2015)
3. Batlajery, B. V, Khadka, R., Saeidi, A.M., Jansen, S., Hage, J.: Industrial Perception of Legacy Software System and their Modernization (2014)
4. Pintelon, L., Parodi-herz, A.: Complex System Maintenance Handbook. Springer, London (2008). https://doi.org/10.1007/978-1-84800-011-7
5. Deac, V., Cârstea, G., Bâgu, C., Pârvu, F.: The modern approach to industrial maintenance management. Informatica Economicâ **14**, 133–144 (2010)
6. Brooke, C., Ramage, M.: Organisational scenarios and legacy systems. Int. J. Inf. Manage. **21**, 365–384 (2001)
7. Ramage, M.: Global perspectives on legacy systems. In: Henderson, P. (ed.) Systems Engineering for Business Process Change: New Directions. Springer, London (2000). https://doi.org/10.1007/978-1-4471-0135-2_19
8. Liu, K., Alderson, A., Sharp, B., Shah, H., Dix, A.: Using Semiotic Techniques to Derive Requirements from Legacy Systems (1998)
9. Capgemini Consulting: Industry 4.0 - The Capgemini Consulting View: Sharpening the Picture beyond the Hype (2014)
10. The Warwick Manufacturing Group: An Industry 4 readiness assessment tool (2017)
11. Deloitte: Industry 4.0 - Challenges and solutions for the digital transformation and use of exponential technologies (2015)
12. PWC: Industry 4.0 : Building the digital enterprise (2016)
13. PWC: The Smart Manufacturing Industry: The Industrial Internet creates new opportunities for Swedish manufacturing companies (2015)
14. Cisco: The Digital Manufacturer Resolving the Service Dilemma (2015)
15. McKinsey & Company: Industry 4.0 at McKinsey's model factories (2016)
16. Acatech: Industrie 4.0 Maturity Index (2017)
17. Roland Berger: Roland Berger 2014 (2014)
18. Plattform Industrie 4.0: Reference Architectural Model Industrie 4.0 (RAMI 4.0). An Introduction (2015)
19. Chen, D., Dassisti, M., Elvesæter, B.: Enterprise Interoperability Framework and knowledge corpus. In: Interoperability Research for Networked Enterprises Applications and Software, pp. 1–44 (2007).
20. Pedone, G., Mezgár, I.: Model similarity evidence and interoperability affinity in cloud-ready Industry 4.0 technologies **100**, 278–286 (2018). https://doi.org/10.1016/j.compind.2018.05.003.
21. Justus, A.D.S., Ramos, L.F.P., Loures, E. de F.R.: A capability assessment model of Industry 4.0 technologies for viability analysis of PoC (Proof of Concept) in an automotive company **7**, 936–945 (2018). https://doi.org/10.3233/978-1-61499-898-3-936.
22. Ahuja, I.P.S., Khamba, J.S.: Total productive maintenance: literature review and directions **25**, 709–756 (2008). https://doi.org/10.1108/02656710810890890

23. Dhillon, B.S.: Engineering maintenance: a modern approach (2002)
24. Ramos, L., Loures, E., Deschamps, F., Venâncio, A.: Systems evaluation methodology to attend the digital projects requirements for Industry 4.0, 1–27 (2019). https://doi.org/10.1080/0951192X.2019.1699666
25. Ruschel, E., Santos, E.A.P., Loures, E. de F.R.: Industrial maintenance decision-making : a systematic literature review, p. 16 (2017)
26. Lazai Junior, M., Loures, E. de F.R., Santos, E.A.P., Szejka, A.L.: Avaliação da gestão da segurança funcional de máquinas na indústria automotiva sob a ótica da interoperabilidade **6**, 3009–3023 (2020). https://doi.org/10.34117/bjdv6n1-218
27. Filho, J.C.B., Piechnicki, F., Loures, E.D.F.R., Santos, E.A.P.: Process-Aware FMEA framework for failure analysis in maintenance **28**, 822–848 (2017). https://doi.org/10.1108/JMTM-11-2016-0150
28. Venâncio, A.L.A.C., Brezinski, G.L., Gorski, E.G., Loures, E. de F.R., Deschamps, F.: System interoperability assessment in the context of Industry 4.0-oriented maintenance activities, 0–5 (2018)
29. Saaty, R.W.: The analytic hierarchy process-what and how it is used. Math. Model. **9**, 161–176 (1987)
30. Forman, E., Peniwati, K.: Aggregating individual judgments and priorities with the Analytic Hierarchy Process **108**, 165–169 (1998). https://doi.org/10.1016/S0377-2217(97)00244-0
31. Athawale, V.M., Chakraborty, S.: Facility location selection using PROMETHEE II method **11**, 16–30 (2010). https://doi.org/10.1504/IJISE.2012.046652
32. Mareschal, B., Brans, J.-P., Figueira, J.R., Greco, S., Ehrogott, M.: Multiattribute Utility Theory (2005)

# Developing a Resource-Based Manufacturing Process Capability Ontology

Arkopaul Sarkar, Dušan Šormaz(✉), David Koonce, and Sharmake Farah

ISE Department, Ohio University, Athens, OH 45701, USA
{sormaz,koonce,sf118714}@ohio.edu

**Abstract.** This paper addresses the challenge of developing an ontology for manufacturing process capability. This is an important task needed to achieve the semantical integration of various manufacturing planning and execution software and hardware (equipment) systems. The research presented has been performed within a framework of the Industrial Ontology Foundry (IOF) in which the authors are active participants. Manufacturing process capability is seen as a composition or combination of capabilities and functions of resources participating in the process. Theoretical foundation for the capability composition is presented. Individual participant resource functions are established using axiomatic design principles. The function and capability of the whole manufacturing processes is derived from individual resources using axiomatic design composition of functions. The overall approach is applied on example prismatic parts, their features, and simple machining processes. For each machining process, the machine tool brings one set of functions, while the cutting tool (mill or drill) brings its own functions. When combined for the manufacturing process those functions become capabilities in the whole process. The approach is illustrated on several feature geometries (holes, slots and pockets).

## 1 Introduction

In recent years, researchers envisioned the adoption of cloud computing and service-oriented architectures (SOA) to achieve a high degree of decentralization and flexibility in distributed manufacturing - commonly termed as Cloud Manufacturing (CM). CM is poised to level the manufacturing industry (Igoe and Mota 2011) by bringing manufacturing resources in reach of inventors, amateur designers, small businesses, and start-ups. CM services (e.g. IaaS, PaaS, and SaaS) aim to virtualize a wide variety of resources: both tangible (hardware, e.g. machine, equipment, server, tools, and material and software e.g. CAD, CAM, and CAPP) and intangibles (e.g. "know-hows", data, standards, employees) (Xu 2012). Smart Manufacturing in the US and Industry 4.0 in Europe also advocate for the deployment of cyber-physical systems, cloud-based 'servitization' of both tangible and intangible resources, and decentralized production to make production systems autonomous, self-maintainable, adaptive, and flexible (Chituc and Restivo 2009).

© Springer Nature Switzerland AG 2021
D. A. Rossit et al. (Eds.): ICPR-Americas 2020, CCIS 1407, pp. 293–306, 2021.
https://doi.org/10.1007/978-3-030-76307-7_22

A major challenge in enabling the interoperability of technically disparate manufacturing resources is the inability to codify and integrate manufacturing knowledge. Manufacturing capability is one type of manufacturing knowledge that is not only complex and elusive but to maintain competitiveness, different industries have traditionally resorted to developing and protecting their unique capability. Manufacturing capabilities have traditionally been formalized by manufacturers in three forms:

1. machinists' handbooks (Oberg et al. 2000) and shop-floor manuals - capture the manufacturing knowledge accumulated by the generational experience and experiments.
2. machine and tool vendors - advertise the capabilities of their products (manufacturing resources).
3. technical and mechanical theory - provide scientific (both rule-based and numerical) models for the capabilities of mechanical devices and related processes. Additionally, manufacturing organizations may also advertise their competency in terms of their production performance and quality of the output in a format that is comprehensible and enticing to the targeted customers.

This paper includes - discussion on research new representations of process capabilities, a brief discussion on representing manufacturing systems with ontologies, the presentation of a limited ontology-based process capability model and a case study of how this ontology can represent a representative selection of manufacturing processes.

## 2 Literature Review

### 2.1 Process Capability in Machining

Process capability is defined as historical and scientific knowledge about manufacturing processes (Chang et al. 2006). The authors define three levels of process capabilities: a universal level, a shop level, and a machine level, where the universal level is governed by materials and physics to provide general knowledge about processes (like twist drilling), but other levels focus on true capability within a certain factory or specific machine (drill press or milling machine) which is bound by both physics of the generalized capability and the constraints the machine and tooling providing the capability. For example, the maximum spindle speed of the drill with a certain diameter bit. Important process capability parameters include shape and size that a process can produce, and dimensions and tolerances that can be produced by various processes. Many manufacturing handbooks contain general capabilities of various manufacturing processes (see for example (Boothroyd et al. 2010)).

Formal analysis of process capability in research on process planning has been usually done by applying some form of decision rules and/or tables. For example, ref (Wang and Li 1991) describes detailed reasoning for hole-making operations. Those procedures have been formalized in the research report (Khoshnevis et al. 1993) and implemented in the research process planning system IMPlanner. Details and examples for hole making and milling are provided in Sormaz et al. (2005) and Sormaz et al. (2018). While this approach gives detailed reasoning concerning process capabilities for achieving required dimensions and tolerances, the research on more rigorous formalisms for process capability from first principles has not been covered. For example, the compatibility of various machines, tools, and manufacturing processes is seen as just a binary relation as shown in Fig. 1. For example, milling and drilling tools can be mounted on milling machine, while turning centers cannot accept milling tools.

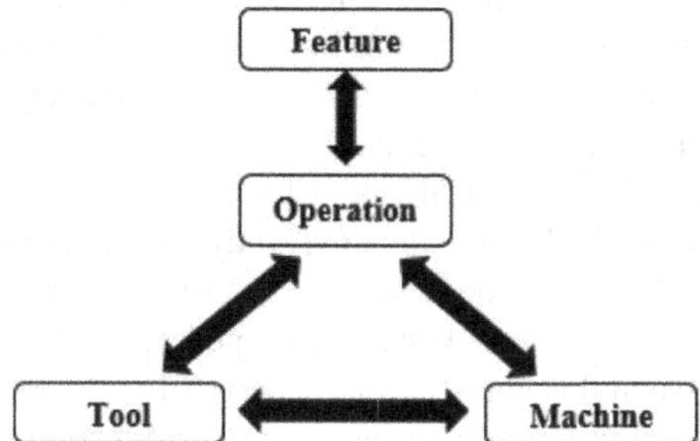


**Fig. 1.** Relations between features, operations, tools, and machines

## 2.2 Ontologies

Ontologies are structured definitions of real-world entities, their hierarchies, and relationships. An ontology can lay a foundation for an application neutral definition to guide the creation, transformation, and integration of data or knowledge within and across organizations and domains. Creating an ontology demands a thorough examination of the systems being modeled and requires adherence to a rigorous methodology that preserves the philosophical constructs of both the system and the modeling methodology. This development process provides insight into the domain, and the resultant ontology can support data processing, storage, translation, and integration. And with the rich and complete representation of a system that an ontology can support, it makes a strong foundation for relational or object-based formats.

While the developer of an ontology should strive to be general and neutral, there will be an implied perspective that the creator (single or group) imparts on their design. Termed as concept orientation, this bias is formed from the creator's perspective and understanding of the system and the ontology methodology being used (Arp et al. 2015). Couple this bias with breadth and depth bounds imposed by the intended purpose of the ontology and opportunity exists for a design that results in usage limitations. When developing an ontology for manufacturing capability, formalizing existing industry biases will necessarily limit generalization and applicability. Any effort to build such an ontology should respect the three levels of capability of encoding already present – machinist's handbooks, organizational capabilities, and unique machine/tool combinations.

Early attempts to build ontologies for manufacturing domain focused on describing products, manufacturing process, resources, and controls which describe the relations among resources and processes. One early work on the core terms of manufacturing was conducted by Borgo and Leitão, who proposed a set of such core concepts related to manufacturing based on foundation ontology DOLCE (Borgo and Leitão 2007). The behavior and function of the manufacturing artifacts were developed in their later works (Borgo and Vieu 2009; Borgo et al. 2006; Borgo et al. 2010). Although these studies established the functionality of products and resources based on behaviorism and provided some of the conceptual guidelines for understanding capability, they are not exclusively about the manufacturing capability. Another ontology called MASON described by Lemaignan et al. also provided a taxonomy of the core elements of manufacturing, such as process, resources including machine, tool, and raw material but generally lacks entities describing their capability (Lemaignan et al. 2006).

Ameri et al. developed an ontological model describing manufacturing processes and resources as services, which included a detailed study on the capability of machine and tool (Ameri et al. 2012). They also conducted a thesaurus-guided discovery of capability related terms (Sabbagh et al. 2018) as well as developed a process and resource selection module leveraging the capability knowledge (Ameri 2013). Lu et al. (2014) developed a taxonomic representation of manufacturing process and resource capabilities directly translating concepts from STEP-NC (ISO) standards using the OntoSTEP methodology proposed by Barbau and Krima (Krima et al. 2009). Luo et al. developed a DL based language called MCDL, which utilized the fuzzy-DL concepts in modeling capability ranges for manufacturing resources (Luo et al. 2013). In a recent publication, Järvenpää et al. described a new ontological model for manufacturing capability, named MaRCo (Järvenpää et al. 2018), which supports capability matchmaking for resource discovery.

The recent development in the ontology application is the Industrial Ontology Foundry (IOF). This effort is based on developing the industrial ontology framework that enables seamless data integration based in semantic marking of data from various applications (such as PLM, ERP, MES, and others). The framework is based on the adoption of the Basic Formal Ontology (BFO)[1] and layered development of industrial ontologies. The initial effort was devoted to building core common terms for all industrial applications (Smith et al. 2019) and the current work is in developing modules for process planning, supply chain, maintenance, and few other areas. The authors of this paper are active participants in the IOF activities and this work is being done in the IOF framework.

## 3 Manufacturing Resource and Process Capabilities Model

### 3.1 Capability of a Resource

Colloquially, the term 'capability' is used in describing the ability of some object to achieve a goal. In manufacturing, objects may be items like consumable or capital resources, e.g. material, machine, tool, fixture, die, mold, vehicles, or non-consumable assets, e.g. land, building, and workers. It is also apparent that the capabilities of a resource or asset can only be described in a particular context, which is composed of two different conditions: there must be a method or process in which the object is to be used and there must be a purpose which is expected to be fulfilled. This observation is demonstrated from the following descriptions of capability:

[1] https://github.com/BFO-ontology/BFO.

1. An LED light bulb can provide more brightness but consume less energy than a CFL.
2. The two pentalobe screws located on the bottom of an iPhone 6 can be unscrewed by a 5-point pentalobe screwdriver.
3. A robotic hand can pick and remove defective parts from an assembly line.

The notion of the capability bears a phenomenological aspect as some capabilities of an artifact manifest in the context of some benefit which some agent intends to derive from the realization of that capability.

Premise 1: For every *c* as Capability, when realized in a Process *p*, some purpose (of an agent) is fulfilled or some intention is realized (of an agent).

Several analyses resort to behaviorism and model this aspect of capability as a behavior of an artifact, ranging from radical stances such as explaining the outcome or trait of an artifact based on its responses caused by some external stimuli interacting with the internal structural state of the artifact to a more teleological interpretation by avoiding to use the internal causal factors of the artifact in the explanation for the behavior of the artifact in a certain process but posit it as an outcome of the process.

Despite the rigor, radical behaviorism faces hurdles when used in practical situations, some of which are discussed in the publication (Sarkar and Šormaz 2019). In this paper, 'capability' is classified as a type of 'disposition' (Spear et al. 2016). Being a subtype of disposition, any instance of capability is by default a realizable entity, which is realized when putting in a specific situation ("special physical circumstances," – BFO). This situation-specific realization, which in a way captures the modal aspect of behavioral tendencies of artifacts, covers premise 1.

We further define capability of the manufacturing resource in terms of the function of the resource. The term 'function', as defined in BFO, contains the basis of etiological selection or artificial creation (Spear et al. 2016), that what the artifact is capable to do is the original purpose of the artifact's existence.

This normative purpose (and the original intention) of an artifact is encapsulated in the functional property of that artifact. Therefore, the premise 1 can be rewritten by using the term function as 'For every *c* as Capability, when realized in a Process *p*, some function *f*, that inheres in the bearer of *c*, is also realized'.

The restatement also assumes the premise that for a capability to be realized, the bearing artifact must participate in a process. This is because capability, being a realizable entity (BFO), can only be realized in a process, of which the bearer object is a participant. Furthermore, the function being a normative purpose of the artifact (Spear et al. 2016), it must be the case that the process, which realizes that function, occurs in a prescribed manner.

In an aim to derive an OWL-friendly definition (Axiom. 1) for capability, we introduce a new relationship appraises, which is further axiomatized below.

Axiom 1: $Capability(c) \equiv Disposition(c) \wedge \exists f\, appraises(c,f) \wedge (Function(f) \vee Role(f))$

Axiom 2: $appraises(c,f) \rightarrow \exists p, \forall o\ inheresIn(c, o) \wedge inheresIn(f, o) \wedge participatesIn(o, p) \wedge realizes(p, c) \rightarrow realizes(p, f)$

The definition given in Axiom 1 restricts the range of relationship appraises to be either function or role. The reason for including the role in the definition is two fold. Although this paper focuses only on the capability of manufacturing resources, such as machines and tools, which are classified under 'material artifact' in BFO, human beings, such as artisans, operators, and workers, and collectives, such as manufacturers, land, and organization, also possess the capability. However, upon being realized in some manufacturing activity, the capabilities do not necessarily realize any function of these agents as it is debatable in the first place that agents can bear such function (abilities are not functions as most of them are acquired in a lifetime of an agent and thus not etiologically selected. They are also not engineered as the physical structure bearing the capability is not modified while acquiring that capability (Merrel et al. 2019)). to another question: which way is the realization of a capability different from the corresponding realization of a function? It is necessary to disseminate the implication 'realizes(p, c) → realizes(p, f)' further. Functions are always end-directed, that one must find the world in a specific state if they are realized in a process. On the other hand, when some capability is realized, the bearing artifact or some other object, which also participate in process p, is expected to be in a certain state(s). What we broadly mean by the statement 'capability appraises function' is that capability demarcates or predicts the extent by which some function is realized. To be specific, the process, which realizes both the capability and the function, brings forth some state of the world which is commonly contained by the state descriptions associated with both.

However, the capability of a resource is always graded, that it is expressed as a range of expected outcomes. It is thus necessary for the predicted range of the capability to be such that it overlaps the set of outcomes specified by the function either completely or partially. For example, a lightbulb being capable of producing warmer color temperature (e.g. there are instances of it producing a color temperature of 2700–3000 K in some occurrences of glowing) implies that the light bulb is suitable for making a room cozy (e.g. it has a function of keeping the color temperature of the room at around 2800 K). Based on the above evidence one may follow Axiom 1 to say that the capability of producing warmer color temperature appraises the function of making the room cozy.

In a an earlier work (Sarkar and Šormaz 2019), we presented a formal axiom to capture this granular notion of capability being realized in a process based on fluent, considering that the state description for the function and the capability, which appraises (labeled as demarcates) the function, is a quality measurement and a set of tolerance limit for that measurement. In practical application, this exposition based on the process-specific behavior of resources is not necessary to talk about the function or capability of resources. Furthermore, the first-order axioms cannot express the function of a system based on its process-specific behavior completely as it is not expressive enough to capture the modality imbued in such models. On the other hand, the teleological approach adopted in Axiom 1 is more straightforward: "A resource has a function by virtue of its capability."

### 3.2 Capability of a System

In the modern manufacturing industry, most operations are conducted on complex systems integrating several types of machines, tools, and equipment. Functional decomposition is a popular reductionist strategy, adopted for modeling complex systems by engineers. This method aims to decompose a higher-order function into sub-functions in such a way that if every such sub-functions is realized then the higher-level function is also realized. One of the utilities of functional decomposition is system design, which is to find a suitable mereological hierarchy of subsystems, assembly, and components, that can be assembled into a complete system, capable of achieving such function. This method is also applied in modeling the system architecture as a logical sequence of the functions, which is commonly recorded as a functional flow block diagram in system analysis (US Department of Defense Systems Management College 2001).

The aggregation among sub-functions mirrors the interdependency of functions as described by Kitamura and Mizoguchi: "A function is achieved by performing(achieving) a series of sub-functions which is called a method of function achievement" (Suh 1998). In their model, the judgment for the realization an upper-level function in a process is derived from comparing the behavior displayed by the system, due to the underlying structure of the system or its component, to the behavior, expected by the function. The crux of this idea is derived from the FBS ontology of (Gero 1990). However, modeling functional decomposition based on process-specific behavior of systems and its components is not trivial as such causal-interaction is often emergent and greatly influenced by the experience of the observer.

Axiomatic design (AD) theory, developed by Nam P. Suh, is an integrated model that can be applied in both system design and analysis. AD uses the functional decomposition method in mapping a set of objectives to solution space. By applying the transformation iteratively, more details of the solution unfold, ultimately generating a mereological hierarchy of the complete system (Suh 1998). AD fits perfectly with the teleological basis of the function we adopted in our definition of capability as it can help in axiomatizing the linear mappings (design matrix) among a set of objectives (functional requirements) and a set of solutions (design parameters of a system). To illustrate the method, the case in example 3 is modeled with AD by constructing the transformation matrix from the capability matchmaking.

The term function used in the definition of capability is based on teleological sense and deviate from the behavioral perspective, adopted in DOLCE (Borgo et al. 2006) and YAMATO (Kitamura and Mizoguchi 2004). However, the teleological interpretation of function is useful in describing the concept of functional decomposition as every sub-function plays some role in achieving the end-goal prophesied by the parent function. We define the relation decomposition based on the result of the process in which these functions are realized without assuming any parthood among its bearers. Although AD links components of a system to the sub-functions, not all types of parthood are described along the functional line.

Axiom 3: $\forall f, f'\ decompose(f', f) \rightarrow \exists f_1 f_2 \ldots f_n \bigwedge_{i=1}^{n} decompose(f_i, f) \wedge f' \in list(f_1 f_2, \ldots f_n) \wedge \exists p p_1 p_2 \ldots p_n realizes(p, f) \wedge \bigwedge_{i=1}^{n} realizes(p_i, f_i) \rightarrow \exists o o_x p_x achieves(p, o) \wedge achieves(p_x, o_x) \wedge o_x = o \wedge generalOccurrentSumOf(p_x, list(p_1, p_2, \ldots p_n))$

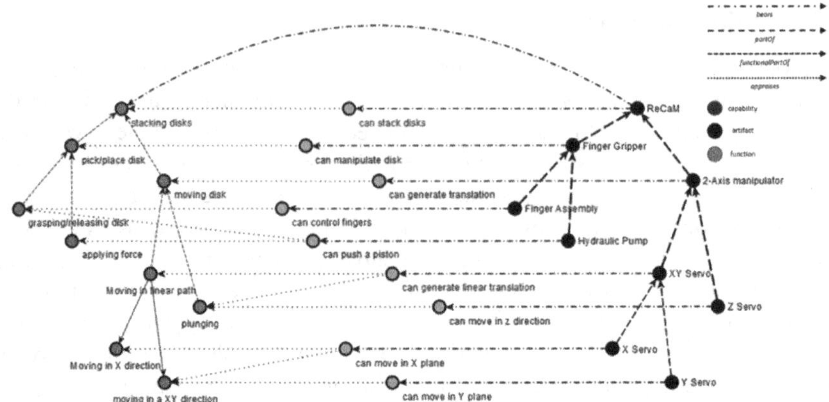


**Fig. 2.** Axiomatic Design principle applied to derive the structural decomposition of a robotic hand which can fulfil the functional requirement of stacking disk on a pallet by collecting them from a conveyor belt.

Axiom 3 states that for every function, that decomposes another function, is a member of a list of functions such that if all functions in the set are realized then the parent function is also realized. This implies that the objective (state description(s)) achieved by the process that realizes the parent function is the same as the objective achieved by the general composition of the processes that realize the sub-functions. It is to be noted that 'list' is a function (not to confuse with the term 'function') which returns a list of the elements given as input. The relationship '*generalSumOf*' holds between the general mereological sum of some occurrents and a list of those occurrents.

In Fig. 2, the primary function of "stacking disk" is realized when two sub-functions "pick/place disk" and "moving disk" is realized. These sub-functions are realized only when processes, such as picking, placing, and moving, occur in a suitable order or in parallel. In other words, the combined effect of these processes results in a disk being stacked.

In the AD method, a design matrix maps the vector of functional requirements to a vector of design parameters, which are components of the purported system that bears the parent function. Each element of the matrix denotes the extent in which a function is 'satisfied' (as a requirement) by a component, that if greater than 0 then the component may bear that function. For example, the vector of functions In Fig. 2, i.e. {'manipulating disk', 'moving disk'}, can be mapped to the vector of components, I.e. {'finger gripper', 'CNC'} using the design matrix A, as,

$$\left\{\begin{matrix} '\mathit{manipulating}\ \text{disk}' \\ '\text{moving disk}' \end{matrix}\right\} = A\left\{\begin{matrix} '\text{finger gripper}' \\ '\text{CNC}' \end{matrix}\right\},\ \mathit{where}\ A = \begin{bmatrix} X & 0 \\ 0 & X \end{bmatrix},$$

X being a non-zero number greater than 0.

However, not every design matrix yields in a valid design as AD dictates that the independence of sub-functions is ensured by using a design matrix which is either diagonal or triangular (Suh 1998). Before axiomatizing the independence axiom, a special parthood relationship among components of a system is given in Axiom 4 based on the functions they bear. This relationship, *functionalPartOf*, is a sub-property of proper parthood but only holds between the parent and child components only when at least one function inhering in the child component decompose some function inhering in the parent component.

Axiom 4: $FunctionalPartOf\left(a', a\right) \rightarrow ContinuantPartOf\left(a', a\right) \wedge \exists ff'\ inheresIn(f, a) \wedge inheresIn\left(f', a'\right) \wedge decompose(f', f)$

The independence axiom of AD requires that the list of sub-functions, which decomposes a parent function, should contain the minimum number of independent functions necessary to satisfy the parent requirement: "each one of the functional requirements can be satisfied without affecting the other functional requirement" (Suh 1998). Therefore, it is not guaranteed that every system containing some components as its functional part is a valid assembly. AD prescribes that the design matrix must be either diagonal or triangular to satisfy the independence axiom. Before we can add that constraint in an ontology, we need to formulate a scheme to derive the matrix on the logical consequence.

As we exposited in the previous section that a function of an artifact exists in virtue of the artifact's capability. For manufacturing resources, they are primarily designed and built to have certain capabilities. Therefore, a resource is suitable for achieving a functional requirement only when it has some capability that appraises such function. The design matrix then can be formulated as

$$A = \{Aij = 0..1 \wedge \mathbb{I}|1{:}\ \exists c\ inheresIn(a_i, c) \wedge appraises(c, f_i), 0{:}\ otherwise\}$$

Based on this definition, an aggregation of components is valid only when every component has some capability which either appraises a function independently without any other capability also appraising the processor along with another capability that again appraises another sub-function independently. We can classify the capabilities in this line as an independent and enabled capability. An enabled capability is a capability that is enabled by either some independent capabilities or enabled capabilities. An enabled capability that enables another enabled capability cannot be enabled by that capability it enables. Therefore, the relationship *enable*, given in Axiom 5, is transitive, non-reflexive, and anti-symmetric.

Axiom 5: $enable(ci, cj) \rightarrow \exists fi\ appraises(cj, fi) \wedge appraises(ci, fi) \rightarrow \exists fofj\ decompose(fi, fo) \wedge decompose(fj, fo) \wedge appraises(ci, fj) \wedge \neg appraises(cj, fj)$

## 4 Case Study

The illustration of a capability ontology will be given on building ontologies for the case of machining prismatic features. Most discrete manufacturing industries have parts of generally prismatic shape (usually produced on milling and drilling machine tools) with so-called prismatic features as components of the part design (shape). Most common features are either direct or derived from holes, slots, and pockets (Shah and Rogers 1988;

Shah and Mantyla 1995) which, when combined, generate a variety of complex shapes. The features and their attributes form the functional requirement for the machine-tool selection. As explained in the earlier sections, machining systems capable of satisfying these functional requirements should have some capability that can appraise the corresponding functions (no distinctions between function and its requirement is made in this study). In Table 1, some of the typical form features and required capabilities of the machine are given.

Obtaining a certain feature is all dependent on a combination of several factors such as a method of tool entry, the primary cutting motion of spindle, and the direction of cutting. For example, a small hole requires plunging, the rotational motion of the spindle, and axial cutting. Following the AD method, a suitable machine should have some components which bear these capabilities individually. However, the overall requirement also needs to be decomposed to deduce the correct assembly and combination of these components. In most of the CNC machining, a suitable tool is also required. In this study, we only focus on drill bits, which are used for making a hole (Fig. 3).

**Table 1.** Machining features

| | Feature | Method of Tool Entry | Primary Cutting Motion (feed) of Spindle | The Direction of Cutting |
|---|---|---|---|---|
| | Closed pocket | Ramping | XY path | Radial Cutting |
| | Large hole (circular pocket) | Helical interpolation | Circular XY | Radial Cutting |
| | small hole | Plunging | Rotation | Axial Cutting |
| | Slot | Ramping | X or Y | Radial Cutting |

The type of drill bit you need typically depends on the type of hole being created. For example, a counterbore hole - Round bottom requires a step drill and a round bottom profile as given in Table 2 and Fig. 4.

The decomposition of function 'drilling small hole' is given in the diagram shown in Fig. 5. The parthood among the corresponding CNC drill machine and a drill bit is also shown. The capabilities presented in Fig. 5 present the ability of a machine/tool combination to produce the defined hole feature.

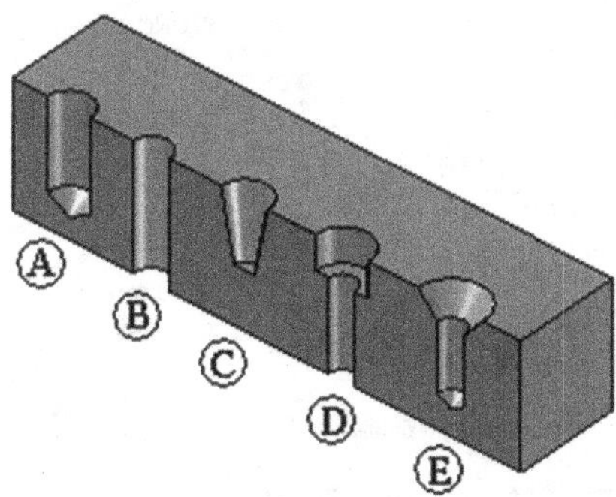


**Fig. 3.** Common hole types (https://www.soliddna.com/SEHelp/ST6/EN/feature_modeling/feat11c.htm)

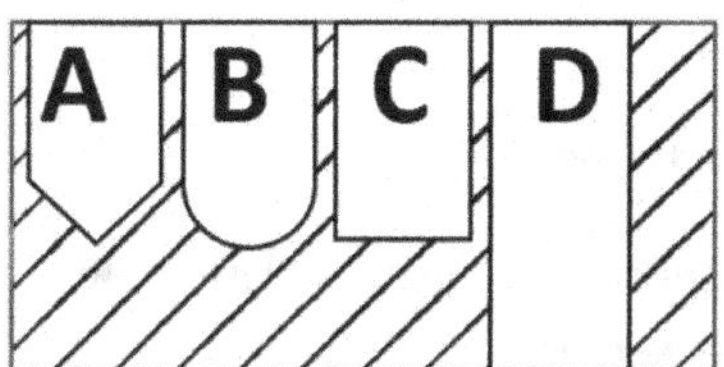


**Fig. 4.** Common types of hole bottoms

**Table 2.** Matching between holes and drilling tools

| Side profile | Type of drill bit | Bottom Profile of hole |
|---|---|---|
| Simple hole - Round bottom | Conventional/Spur/Spade | B |
| Simple holes - Flat bottom | Conventional/Spur/Spade | C |
| Simple holes - V bottom | Conventional/Spur/Spade | A |
| Counterbore holes - Flat bottom | Counterbore | C |
| Counterbore holes - Round Bottom | Step drill | B |
| Counterbore holes - V bottom | Counterbore | A |
| Countersink - Flat bottom | Countersink | C |
| Countersink - Round bottom | Countersink | B |
| Countersink - V bottom | Countersink | A |
| Threaded holes - Flat bottom | Tap drill | C |
| Threaded holes - Round bottom | Tap drill | B |
| Threaded holes - V bottom | Tap drill | A |

The OWL file contains a list of rules which capture the knowledge of what type of capability appraises what type of function (see Fig. 6). Other rules implement the axioms given in the previous section in description logic-friendly way.

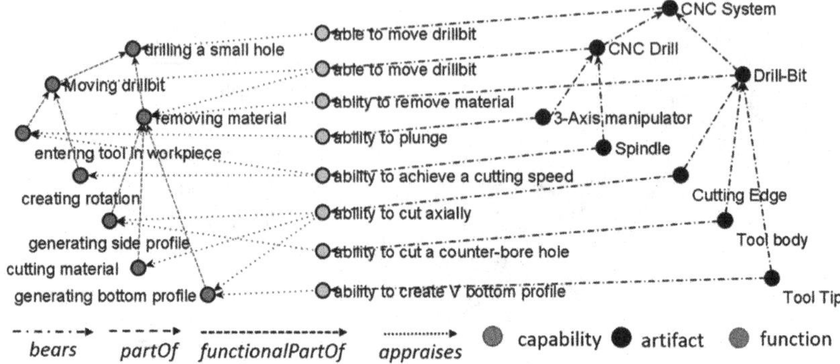


**Fig. 5.** Relations between resources, functions, and capabilities

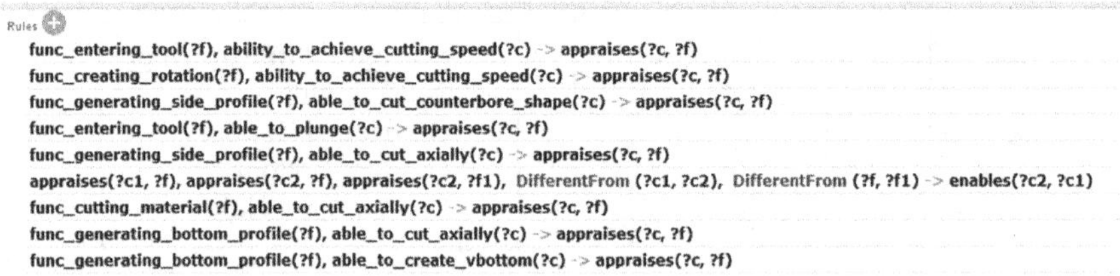

```
Rules
func_entering_tool(?f), ability_to_achieve_cutting_speed(?c) -> appraises(?c, ?f)
func_creating_rotation(?f), ability_to_achieve_cutting_speed(?c) -> appraises(?c, ?f)
func_generating_side_profile(?f), able_to_cut_counterbore_shape(?c) -> appraises(?c, ?f)
func_entering_tool(?f), able_to_plunge(?c) -> appraises(?c, ?f)
func_generating_side_profile(?f), able_to_cut_axially(?c) -> appraises(?c, ?f)
appraises(?c1, ?f), appraises(?c2, ?f), appraises(?c2, ?f1), DifferentFrom (?c1, ?c2), DifferentFrom (?f, ?f1) -> enables(?c2, ?c1)
func_cutting_material(?f), able_to_cut_axially(?c) -> appraises(?c, ?f)
func_generating_bottom_profile(?f), able_to_cut_axially(?c) -> appraises(?c, ?f)
func_generating_bottom_profile(?f), able_to_create_vbottom(?c) -> appraises(?c, ?f)
```

**Fig. 6.** Relation definitions as part of ontology in OWL file

The hierarchy is automatically verified by running the reasoner (Pellet) in Protégé. The result of *enable* property assertions are shown in the Fig. 7.

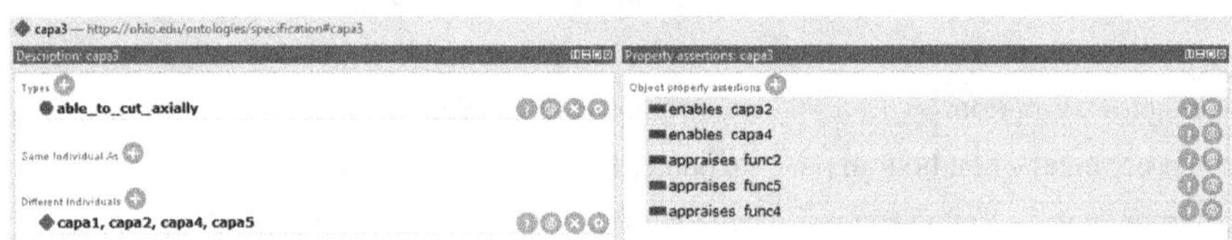


**Fig. 7.** Results of the reasoner processing in OWL

## 5 Conclusions

In this study, we described the manufacturing resource capability as the evaluator of its function. Based on this concept, we derived a logical method to model a machining system by combining suitably capable components, which satisfies the overall function by satisfying the sub-functions. The major contribution of this study is to derive relationships based on the independence axiom of AD. A case study is also conducted to show that suitable CNC machining systems for generating prismatic form features of a part design can be built based on primitive capabilities of machine components and tools. There can be a variety of tools and machines which have similar capabilities. The appraises relationship can be deduced based on numerical ranges. In that sense, many alternative machine and tool combinations may be derived. For finding the best system design, the second axiom of AD, called the information axiom, should be applied. Also, the functional decomposition shown in the case study does not consider the flow of material, energy, and information. These flows are useful for designing the connections among the components. The information axiom and the flow among functions are not covered in this study and will be addressed in a future effort.

## References

Ameri, F.: An Intelligent Process Planning System Based on Formal Manufacturing Capability Models Capability Modeling for Digital Factories (CaMDiF) View project Information Content Measurement View project (2013). https://doi.org/10.1115/DETC2013-13286

Ameri, F., Urbanovsky, C., McArthur, C.: A systematic approach to developing ontologies for manufacturing service modeling. In: CEUR Workshop Proceedings, vol. 886, pp. 1–14 (2012)

Arp, R., Smith, B., Spear, A.D.: Building Ontologies with Basic Formal Ontology. The MIT Press (2015). https://doi.org/10.7551/mitpress/9780262527811.001.0001

Boothroyd, G., Dewhurst, P., Knight, W.: Product Design for Manufacture and Assembly. CRC Press, Boca Raton (2010)

Borgo, S., Carrara, M., Garbacz, P.: Formalizations of functions within the DOLCE ontology. In: Proceedings of TMCE 2010 Symposium, May 2014, pp. 113–126 (2010)

Borgo, S., Carrara, M., Vermaas, P. E., Garbacz, P.: Behavior of a technical artifact: an ontological perspective in engineering. In: Formal Ontology in Information Systems. IOS Press, January 2006

Borgo, S., Leitão, P.: Foundations for a core ontology of manufacturing. In: Sharman, R., Kishore, R., Ramesh, R. (eds.) Ontologies, vol. 14, pp. 751–775. Springer, Boston (2007). https://doi.org/10.1007/978-0-387-37022-4_27

Borgo, S., Vieu, L.: Artefacts in formal ontology. In: Philosophy of Technology and Engineering Sciences, pp. 273–307. Elsevier (2009). https://doi.org/10.1016/B978-0-444-51667-1.50015-X

Chang, T.-C., Wysk, R.A., Wang, H.-P.: Computer-Aided Manufacturing, 3rd edn. Pearson, Upper Saddle River (2006)

Chituc, C.M., Restivo, F.J.: Challenges and trends in distributed manufacturing systems: are wise engineering systems the ultimate answer. In: Second International Symposium on Engineering Systems. MIT (2009)

Gero, J.: Design prototypes: a knowledge representation schema for design. AI Magazine (1990). https://www.aaai.org/ojs/index.php/aimagazine/article/viewArticle/854

Igoe, T., Mota, C.: A strategist's guide to digital fabrication. Strategy+Business **64**, 2010–2011 (2011)

Järvenpää, E., Siltala, N., Hylli, O., Lanz, M.: The development of an ontology for describing the capabilities of manufacturing resources. J. Intell. Manuf. **30**(2), 959–978 (2018). https://doi.org/10.1007/s10845-018-1427-6

Khoshnevis, B., Tan, W., Sormaz, D.: A process selection rule base for hole-making, Gaithersburg, MD (1993)

Kitamura, Y., Mizoguchi, R.: Ontology-based systematization of functional knowledge. J. Eng. Des. **15**(4), 327–351 (2004). https://doi.org/10.1080/09544820410001697163

Krima, S., Barbau, B., Fiorentini, X., Sudarsan, R., Sriram, R.: OntoSTEP: OWL-DL Ontology for STEP. National Institute of Standard and Technology (2009). https://www.nist.gov/customcf/get_pdf.cfm?pub_id=901544

Lemaignan, S., Siadat, A., Dantan, J.-Y., Semenenko, A.: MASON: a proposal for an ontology of manufacturing domain. In: IEEE Workshop on Distributed Intelligent Systems: Collective Intelligence and Its Applications (DIS 2006), pp. 195–200. IEEE (2006). https://doi.org/10.1109/DIS.2006.48

Lu, Y., Shao, Q., Singh, C., Xu, N., Ye, X.: Ontology for manufacturing resources in a cloud environment. Int. J. Manuf. Res. **9**(4), 448 (2014). https://doi.org/10.1504/IJMR.2014.066666

Luo, Y., Zhang, L., Tao, F., Ren, L., Liu, Y., Zhang, Z.: A modeling and description method of multidimensional information for manufacturing capability in cloud manufacturing system. Int. J. Adv. Manuf. Technol. **69**(5–8), 961–975 (2013). https://doi.org/10.1007/s00170-013-5076-9

Oberg, E., Jones, F.D., Horton, H.L., Ryffel, H.H.: Machinery's Handbook (2000)

Sabbagh, R., Ameri, F., Yoder, R.: Thesaurus-guided text analytics technique for capability-based classification of manufacturing suppliers. J. Comput. Inf. Sci. Eng. **18** (2018). https://doi.org/10.1115/1.4039553

Sarkar, A., Šormaz, D.: Ontology model for process level capabilities of manufacturing resources. Procedia Manuf. **39**, 1889–1898 (2019). https://doi.org/10.1016/j.promfg.2020.01.244

Shah, J.J., Rogers, M.T.: Functional requirements and conceptual design of the Feature-Based Modelling System. Comput. Aided Eng. J. **5**(1), 9 (1988). https://doi.org/10.1049/cae.1988.0004

Shah, J., Mantyla, M.: Parametric and Feature-Based CAD/CAM: Concepts, Techniques, and Applications. Wiley, New York (1995)

Smith, B., et al.: A first-order logic formalization of the industrial ontologies foundry signature using basic formal ontology. In Proceedings of the Joint Ontology Workshops (2019)

Sormaz, D., Gouveia, R., Sarkar, A.: Rule based process selection of milling processes based on GD&T requirements. J. Prod. Eng. **21**(2), 19–26 (2018). https://doi.org/10.24867/JPE-2018-02-019

Sormaz, D.N., Khurana, P., Wadatkar, A.: Rule-based process selection of hole making operations for integrated process planning. In: 25th Computers and Information in Engineering Conference, Parts A and B, vol. 3, pp. 983–988. ASME (2005). https://doi.org/10.1115/DETC2005-85082

Spear, A.D., Ceusters, W., Smith, B.: Functions in basic formal ontology. Appl. Ontol. **11**, 103–128 (2016). https://doi.org/10.3233/AO-160164

Suh, N.P.: Axiomatic design theory for systems. Res. Eng. Des. Theory Appl. Concurr. Eng. **10**(4), 189–209 (1998). https://doi.org/10.1007/s001639870001

Wang, H.P., Li, J.K.: Computer-Aided Process Planning, 1st edn. Elsevier Science, Amsterdam (1991)

Xu, X.: From cloud computing to cloud manufacturing. Rob. Comput. Integr. Manuf. **28**(1), 75–86 (2012). https://doi.org/10.1016/j.rcim.2011.07.002

# Development of a Methodology to Analyze Implementation Patterns of Industry 4.0 Technologies

Oscar Quiroga[1] (✉), Samuel Osina[2], and Mariana Díaz[3]

[1] Fac. de Ing. Química (FIQ-UNL), Universidad Nacional del Litoral, 3000 Santa Fe, Argentina
oquiroga@fiq.unl.edu.ar

[2] Arts et Métiers (ParisTech), 75013 Paris, France

[3] Fac. Cs. Económicas (FCE-UNL), Universidad Nacional del Litoral, 3000 Santa Fe, Argentina

**Abstract.** The concept called Industry 4.0 has a very complex technological architecture based on Cyber-Physical Systems (CPS), which represents one of the main concerns in the start-up stage at an industrial level. The effective implementation of Industry 4.0 technologies continues to be a research topic. Some works propose maturity models for the implementation of these technologies, while others focus on the study of their impact on industrial performance. However, there is a lack of studies that provide evidence on the way in which these technologies are implemented in manufacturing companies in general. This fact leads to the following question: What are the current implementation patterns (degree of implementation – maturity stage) of Industry 4.0 technologies in companies? To answer this question, this work proposes the development of a tool for maturity assessment which allows understanding what is required to carry out an effective analysis of implementation of Industry 4.0 technologies in manufacturing companies by studying the degree of implementation – maturity stage in companies from different sectors.

**Keywords:** Industry 4.0 · Maturity assessment tool · Maturity model

## 1 Introduction

The Fourth Industrial Revolution – also named as Industry 4.0 – is characterized by a fusion of advanced technologies that is disrupting almost every industry in every economy [1].

Industry 4.0 is a German strategic initiative whose main objective has been to create smart factories, where manufacturing technologies are updated and transformed through Cyber-Physical Systems (CPS), Internet of things (IoT) and Cloud Computing [2–4].

The fundamental idea of Industry 4.0 is based on integrating manufacturing systems of different smart factories along a value chain, or a value network, in the form of CPS so that data and information can be obtained in real time throughout the value chain, allowing time and precise decision making [5]. Moreover, the CPS are defined

© Springer Nature Switzerland AG 2021
D. A. Rossit et al. (Eds.): ICPR-Americas 2020, CCIS 1407, pp. 307–320, 2021.
https://doi.org/10.1007/978-3-030-76307-7_23

as transformative technologies to manage interconnected systems among their physical assets and their computational capabilities [6]. The competitive nature of today's industry forces more factories to implement high-tech methodologies through recent developments that have resulted in greater availability and affordability of sensors, data acquisition systems, and computer networks [6, 7].

Industry 4.0 comes from the need to improve the means of production in a context of high emergence of new technologies in the field of information technology. It corresponds to a strategic program with the objective of developing production systems, thus improving the efficiency of the industry [2].

An important part of this program fits with the concept of Smart Manufacturing, which corresponds to a great flexibility of the production lines that can adapt to any change in products or conditions, ensuring an increase in quality and productivity [8].

Another part of the program focuses on Smart Working, i.e., the role of human workers that needs to evolve with the help of robots, which could fulfill difficult tasks and also make decisions through communication networks among them [9].

The exchange of information also represents a challenge for this new industry. Whether it is with customers, suppliers or other companies, this concept of Smart Supply Chain allows reducing delivery time and possible falsification of information. Furthermore, communication with other companies contributes to the targeting of their resources on the subject of innovation, working together to limit false leads and accelerate the development of technologies [10].

The last part of the program is given by integrated-embedded technologies which are called Smart Products, whose information is fed back from the final products, with the aim of improving and detecting possible production defects, developing new products and solutions to the customer [11].

Industry 4.0 technologies can be separated into the following layers [12]:

A. Front-end technologies of Industry 4.0 which considers the transformation of manufacturing activities based on emerging technologies (Smart Manufacturing) and the way in which products are offered (Smart Products) [8]. The way in which raw materials and the products are delivered (Smart Supply Chain), as well as the new ways in which workers carry out their activities based on the support of emerging technologies (Smart Working), are also contemplated.
B. Base technologies which comprise technologies that provide connectivity and intelligence for Front-end technologies. This layer makes the Industry 4.0 concept feasible, differentiating it from previous industrial stages.

This work presents the development of a Preliminary Maturity Assessment Tool (PMAT) that is used to identify technologies related to Industry 4.0, and to verify their degree of implementation – maturity stage in companies from different sectors.

This approach is considered as the first step for future research works, and consists of a Questionnaire that includes three types of questions: A) Basic; B) About technologies; and C) Open. The first and second types are closed questions and include a 5-point scale: 1) Non-existent; 2) Low; 3) Medium; 4) High; and 5) Very High. This scale indicates the level and degree of implementation – maturity stage.

The PMAT – Questionnaire is based on two propositions – hypotheses which are presented in Sect. 3 and guide to establish the points of interest and to develop more concrete and clear questions.

Section 2 depicts the research methodology which consists of multiple–case studies, and the propositions – hypotheses (PHs) constitute the units of analysis (UAs). Section 4 shows the analysis of the results, while Sect. 5 presents a discussion by including a comparative analysis of different maturity models of Industry 4.0. Finally, Sect. 6 develops the conclusions.

## 2 Methodology (Multiple – Case Study)

Based on the methodology of multiple – case studies with many units of analysis (UAs) [13], this methodology is proposed as a part of the first phase of a research project that aims to study the implementation of technologies related to Industry 4.0 in the region under analysis. For this purpose, four case studies have been selected representing different industrial sectors (see Fig. 1).

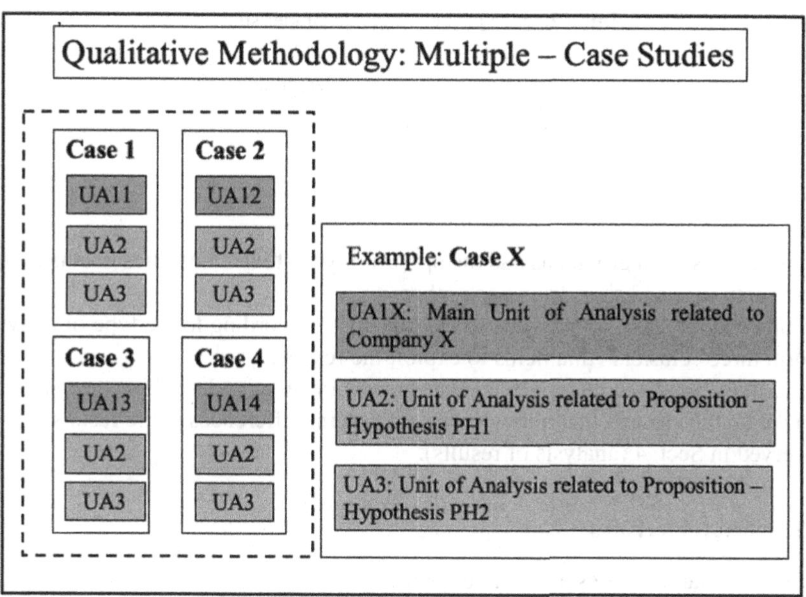


**Fig. 1.** Methodology of multiple – case studies with many units of analysis (UAs).

For each case study three units of analysis are considered, in which two of them are directly connected to the propositions - hypotheses (PHs) that are described in Sect. 3.

The choice of a qualitative methodology allows us to analyze in greater detail certain characteristics related to Industry 4.0 that are being implemented in the region, as well as to give us the first approach for future studies using other types of qualitative or quantitative methodologies.

## 2.1 The Companies

In the current phase of the research project focused on Industry 4.0 in the Santa Fe region, fifteen companies were analyzed to apply the methodological process.

After studying several characteristics, such as size and clusters, four companies were finally selected with which it was possible to complete an exhaustive questionnaire that is described later.

Tables 1 and 2 show the clusters and the size, respectively.

**Table 1.** Industrial clusters of the companies under study.

| Cluster | Percentage |
|---|---|
| Textile | 25 |
| Equipment manufacturing | 25 |
| Software and technology | 50 |

**Table 2.** Size of the companies under study.

| Size | Percentage |
|---|---|
| SME (<100 employees) | 75 |
| Large (>100 employees) | 25 |

Thus, the research group has had the opportunity to study companies that had totally different activities and sizes from each other.

In this way, it was also possible to classify them according to the type of activities and form three "clusters" that helps to exploit the results.

In addition, the research group chose several characteristics describing the companies that make up the clusters that allow to explain several differences in the results that can be observed in Sect. 4 (analysis of results).

## 2.2 The Characteristics

The selected characteristics are detailed below.

**Direct and Indirect Consumption.** The terms direct and indirect consumption of a product indicate, in the case of direct consumption, the direct use of the product by the customer. In other words, the product has no other purposes than those for which it was produced. (Examples: Food, clothes). On the contrary, an indirect consumption of a product means that the product has been produced to be used for another purpose (Examples: Laboratory equipment, packaging).

**Long Life Time.** It is considered that a life span is long when it exceeds one year. Food or drinks, for example, do not fall into this category.

**Maintenance.** A product needs maintenance when an expert must visit the product to check its proper operation. Thus, the product is constantly monitored until the end of its life cycle. (Examples: Mechanical lifts, cars).

**Remote Modification.** A modification is considered remote when a company can make it on a product from its offices, without requiring the intervention of an expert. (Examples: Websites or Software).

The products of each cluster and their characteristics are detailed in Table 3.

**Table 3.** Main characteristics of the clusters.

| Cluster characteristics | Textile | Equipment manufacturing | Software and technology |
|---|---|---|---|
| Direct consumption | X | – | – |
| Indirect consumption | – | X | X |
| Long life time | X | X | X |
| Maintenance | – | X | – |
| Remote Modification | – | – | X |

## 3 The Propositions – Hypotheses and Questionnaire

### 3.1 The Propositions – Hypotheses

The first proposition – hypothesis (PH1) is proposed to establish the link among the presence of technologies related to the concept of Smart Manufacturing, Smart Product, Smart Working and Smart Supply Chain in a company under study, and the degree of implementation – maturity stage of the Industry 4.0 concept in this company. Indeed, all these technologies fit the concept of Industry 4.0; thus, they are essential to reach a high degree of implementation – maturity stage.

The second proposition – hypothesis (PH2) surges as the certain importance of the technologies linked to the term Smart Manufacturing. The other technologies are equally important; however, those related to the term Smart Manufacturing have a special value due to the characterization of the concept of Smart Manufacturing as the heart of Industry 4.0.

Therefore, it can be sustained the fact that a company with many technologies connected to the term Smart Manufacturing has a high degree of implementation – maturity stage of Industry 4.0.

These propositions – hypotheses have guided to establish the points of interest and in this way, to develop more concrete and clear questions, as it is depicted in Fig. 2.

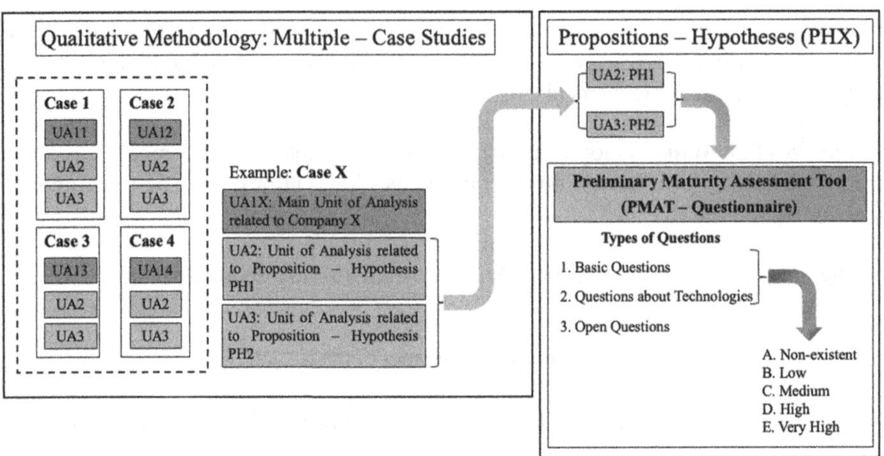


**Fig. 2.** The relationship between the qualitative methodology and the Preliminary Maturity Assessment Tool (PMAT – questionnaire).

## 3.2 Characteristics of the Questionnaire

The objective of the questionnaire is to verify the degree of implementation – maturity stage of Industry 4.0, i.e., for each technology determine whether its implementation is basic or advanced, as well as the use of the new paradigms. Thus, most of the questions refer to estimate the degree of implementation – maturity stage of the technologies and paradigms related to Industry 4.0. Three types of questions are proposed:

- Basic questions.
- Questions about technologies.
- Open questions.

Basic questions provide information about the company, such as its name, focus, and size. These questions help to clearly define what type of company you are dealing with, whether it is large, medium or small; if it produces in series or it develops projects.

The answers make it possible to classify the different companies in a way that is then easier to take advantage of the results.

The questions that refer to technologies help to estimate their degree of implementation – maturity stage in the Industry 4.0 concept. These questions are closed and a 5-point scale has been used to specify the level – stage of knowledge of each employee regarding every technology, since this level – stage is directly related to the actual degree of implementation – maturity stage. The 5 points are as follows:

A) Non-existent.
B) Low.
C) Medium.
D) High.
E) Very High.

Some questions asked whether the company had particular computer systems or it had the previous versions of these systems.

On the other hand, other questions referred to if the employees thought if a technology was feasible (to be able to verify its implementation in the company in another way). Thus, several different questions were available, although they focused only on technologies.

At the end of the questionnaire, the open questions appeared with the objective of explaining the differences in the degree of implementation – maturity stage of different technologies:

- One of these questions concerned measures taken in the past to achieve compliance with the principles of Industry 4.0. The completed responses helped to understand the current situation of the company, and thus, it was possible to verify if these measures were efficient.
- The other one refers to the belief in the efficiency of Industry 4.0 and the need to trace the production system to it. This question was necessary since it indicated the opinion of the employees on this matter, an opinion that is revealed to be important since the employees are who allow the good functioning of the company, and if they do not believe in Industry 4.0, they cannot adapt to use it.

The following section analyzes the responses of the companies and shows if the established clusters are related to each other through common characteristics.

# 4 Analysis of Results

## 4.1 Closed Questions

Table 5 represents the implementation of advanced technologies in Industry 4.0 through the use of colors whose meanings are coded in Table 4.

**Table 4.** Meaning of the color coding.

| Meaning | Colors |
|---|---|
| High implementation | |
| Medium implementation | |
| Low implementation | |
| No implementation | |
| Not known | |

The companies are divided into the clusters chosen above and the results were averaged for each company.

It has been chosen to use colors to describe the degree of implementation – maturity stage of the technologies, so that the differences among the clusters are more visible.

Table 5 shows that the cluster related to the textile sector fits globally with the principles of Industry 4.0. Indeed, with the exception of 3D printing, all advanced technologies are implemented and particularly those that relate to the term Smart Manufacturing, a fact that, according to the propositions – hypotheses previously described, shows a high degree of implementation – maturity stage of Industry 4.0.

**Table 5.** Results of the closed questions (types 1 and 2).

| Cluster | | | Textile | Software and technologies | Equipment manufacturing |
|---|---|---|---|---|---|
| Smart Manufacturing | Vertical integration | Sensors and Actuators | | | |
| | | ERP | | | |
| | | MES | | | |
| | | SCADA | | | |
| | Virtualization | 3D Representation | | | |
| | | Artificial Intelligence | | | |
| | Autonomous robots | Robots along the production chain | | | |
| | | Communication networks among them | | | |
| | Traceability of products | | | | |
| | Flexibility | 3D Printing | | | |
| | | Adaptation of production lines | | | |
| | | Monitoring about energy consumption | | | |
| Smart Working | Remote control of production | | | | |
| | Augmented reality | Augmented reality during manufacturing | | | |
| | | Augmented reality during training of workers | | | |
| Smart Supply Chain | Digital platforms company-clients-suppliers | | | | |
| | Digital platforms among companies | | | | |
| Smart Product | Interconnectivity of products | | | | |
| | Supervision and control of final products | | | | |
| | Data return | | | | |
| Cybersecurity | | | | | |

The cluster related to Software and Technologies companies presents a medium degree of implementation – maturity stage of Industry 4.0, due to the diversity of degree of implementation – maturity stage of each technology. This grade is high for some technologies contained in Smart Manufacturing and low for others. Therefore, this cluster can be assimilated to an intermediate cluster respecting the principles of Industry 4.0.

The cluster that includes Equipment Manufacturing companies shows a low degree of implementation – maturity stage of Industry 4.0. Indeed, very few current technologies are known in these companies, with the exception of the technology that refers to the rapid adaptation of production lines.

Thus, the three clusters describe three implementation states of Industry 4.0 in the region under study: high, medium and low. In this way, it was possible to validate the choice made previously about the distribution of the companies in the three clusters and to analyze the reasons for the differences among them, thanks to the open questions that were proposed in the questionnaire.

From now on, it is called Cluster 1, the cluster associated to the Textile sector; Cluster 2, the one related to Software and Technologies, and Cluster 3, the one linked to Equipment Manufacturing.

### 4.2 Open Questions

The question related to the measures taken in the past regarding the integration of the principles of Industry 4.0 indicates that Clusters 1 and 2 took many measures with the aim of integrating these principles, unlike Cluster 3, which did not take any measure.

The question that refers to the opinion of the company regarding the need for Industry 4.0 shows a lack of knowledge of the companies in the Cluster 3. Indeed, the companies do not seem to know about the existence of this Industry, but they seem inclined to be interested to this concept. On the other hand, Clusters 1 and 2 are both convinced of the importance of matching it.

### 4.3 Analysis of the Results Based on the Characteristics Studied

According to the results obtained, it can be seen that the two characteristics of the final products of a company that induce a high degree of implementation – maturity stage of Industry 4.0 are direct consumption combined with a long-life span. These two characteristics are those of the products produced by the companies of the Cluster 1 and caused industrial needs that resulted in an efficient implementation of the advanced technologies of Industry 4.0.

On the other hand, it was found that indirect consumption induces a good degree of implementation – maturity stage, only if it is combined with a possibility of remote modification (Cluster 2) and not with a need for maintenance (Cluster 3).

These conclusions are clear, but one should be cautious about their implications, due to the limits of the study that are described as follows.

### 4.4 Limits of the Study

The number of companies that answered the questionnaire is the main limit of this study. For the study to be more precise, more responses should have been received, therefore, a better description of the situation could have been guaranteed, adapting or even generating other clusters with other characteristics.

In addition, it has been considered throughout the study that all the technologies of Industry 4.0 were adaptable to each company and all of them were important, whatever the industrial sector. However, there are sectors that do not need certain technologies because they are incompatible with the way companies operate.

For instance, a software company does not need 3D printing to produce its services. Thus, this technology is not going to be raised or considered by the directors of this company, and the absence of this technology does not mean that the company is lacking a way to square with Industry 4.0.

# 5 Discussion

## 5.1 Comparative Analysis

This section proposes a comparative analysis by considering different *maturity models* of Industry 4.0.

Mittal et al. [14] propose a critical comprehensive review of smart manufacturing and Industry 4.0 *maturity models*, and their implications for small and medium-sized enterprises (SMEs), by analyzing their fit and recognizing the specific requirements of SMEs. This work presents features that are characteristic for SMEs and identify research gaps needed to be addressed to successfully support manufacturing SMEs in their progress towards Industry 4.0.

The work of [14] develops an exhaustive analysis by comparing different methodologies, such as *framework*, *maturity model*, *readiness assessment*, and *gap analysis*, which are used to discuss the enterprise journey towards smart manufacturing (SM) and Industry 4.0.

An important observation from the study of [14] is that researchers have different perspectives towards the understanding of *dimensions* [15–19]. For instance, Schuh et al. [15] mentioned resources, information systems, organization structure and organization culture as the *dimensions*. While Schumacher et al. [18] noted strategy, leadership, culture, people, and governance.

Moreover, Mittal et al. [14] indicate that only [20] and [21] mentioned the prerequisites of Industry 4.0 in the form of technologies, toolbox and building blocks. In this group, the work of Frank et al. [12] can be included.

Frank et al. [12] propose a structure of Industry 4.0 technology layers that shows levels of adoption of these technologies and their consequences for the implementation of the Industry 4.0 concept. This structure is synthesized as a *final framework* which represents a *maturity pattern* of the Industry 4.0 implementation. This framework of Industry 4.0 technologies suggests two *layers of adoption*: 1) front-end technologies, and 2) base technologies.

The first layer includes four *dimensions* of Industry 4.0: Smart Manufacturing, Smart Products, Smart Supply Chain and Smart Working. The second layer considers technologies that provide connectivity and intelligence to the front-end technologies. In order to understand the connections among the technologies, as well as to define patterns of adoption of the two layers, a cluster analysis in relation to the companies under study is applied.

According to Schuh et al. [15], the Acatech Industrie 4.0 Maturity Index divides the structure of a company into four structural areas or *dimensions* (Industry 4.0 capabilities): 1. Resources 2. Information systems 3. Organizational structure 4. Culture. Each dimension is defined by two principles that are identified along with the capabilities required for each structural area. The capabilities are geared towards the achievement of the various stages of development and provide manufacturing companies with the foundation to transform into agile organizations. The Acatech Industrie 4.0 Maturity Index has six *maturity stages*: 1. Computerization 2. Connectivity 3. Visibility 4. Transparency 5. Predictive capability 6. Adaptability.

Schumacher et al. [18] propose five *maturity stages*: Likert-scale reaching from 1 (not distinct) to 5 (very distinct). Nine company *dimensions*: 1. Strategy; 2. Leadership; 3. Customers; 4. Products; 5. Operations; 6. Culture; 7. People; 8. Governance; 9. Technology.

Leyh et al. [22] describe the model called SIMMI 4.0, which has five *maturity stages*: 1. Basic digitization level. 2. Cross-departmental digitization. 3. Horizontal and vertical digitization. 4. Full digitization 5. Optimized full digitization. The model SIMMI 4.0 includes four *dimensions*: 1. Vertical integration. 2. Horizontal integration. 3. Digital product development. 4. Cross-sectional technology criteria.

Lichtblau et at. [23] present an Industry 4.0 *maturity model* (IMPULS) which consists of six *stages*: 0. Outsider. 1. Beginner. 2. Intermediate. 3. Experienced. 4. Expert. 5. Top performer. Moreover, [23] depict six *dimensions* which are further (detailed into 18 fields): 1. Strategy and organization. 2. Smart factory. 3. Smart operations. 4. Smart products. 5. Data-driven services. 6. Employees.

Colli et al. [24] propose a digital *maturity model* which takes into consideration a number of existing digital *maturity models* [15, 18, 22, 23].

This maturity model of Colli et al. [24], it is suggested to assess the level of digitization of an organization, and it is composed by six sequential digital *maturity stages*: None, Basic, Transparent, Aware, Autonomous, Integrated.

In order to map the digital capabilities of the organization, these stages are grouped into five areas which are called digital *dimensions*: Governance, Technology, Connectivity, Value creation, Competences.

As an extension of [24], Colli et al. [25] present a contribution that consists of raising the need for contextualization in relation to digital maturity assessments by proposing a novel assessment approach to address this need.

In addition, [25] provide a reference model which is based on the Acatech maturity model [15], chosen for its focus on data and connectivity, which relates *maturity stages* and *dimensions* for mapping the information collected during the assessment process, as well as a more detailed description of the proposed assessment procedure, testing it in three cases and discussing its contextualization capabilities and the key contextual factors that emerged. The proposed approach is intended as a learning tool, based on PBL (Problem Based Learning), to assist professionals in providing company-specific guidelines as a result of a digital *maturity assessment*, and academics in their study of the transition through *digital maturity stages* in many contexts.

Finally, this work has proposed the development of a Preliminary Maturity Assessment Tool (PMAT) which is used to identify technologies related to Industry 4.0, and to verify their *degree of implementation – maturity stages* in companies from different sectors.

The PMAT is based on two propositions – hypotheses which are connected to two *dimensions* of Industry 4.0, related to front-end-technologies: 1. Smart Manufacturing. 2. Smart Products. 3. Smart Supply Chain. 4. Smart Working.

The PMAT consists of a Questionnaire with three types of questions. The first and second types include a 5-point scale: 1. Non-existent. 2. Low. 3. Medium. 4. High. 5. Very High. This scale indicates five *degree of implementation – maturity stages.*

### 5.2 Possible Future Work

A possible future work could be an improvement of the PMAT, i.e. the SMART (Skilled–Maturity Assessment Research Tool) by means of increasing the propositions – hypotheses, thus, to achieve an enrichment of the number of *dimensions* related to Industry 4.0 front-end-technologies, as well as their base technologies.

Moreover, the SMART with an improved Questionnaire with other types of questions could be proposed. Therefore, a discussion with each company could be established with the aim of adapting the questions to the language of the company and the type of products it provides so that employees can answer the questions more precisely and exhaustively. In order to indicate five *degree of implementation – maturity stages*, the 5-point scale (1. Non-existent. 2. Low. 3. Medium. 4. High. 5. Very High) could also be included.

Finally, in a new study it is possible to investigate in depth, what are the current patterns of adoption of technologies related to Industry 4.0 that are implemented in the companies of the region under study.

## 6 Conclusions

This work developed a Preliminary Maturity Assessment Tool (PMAT). The PMAT founds on two propositions – hypotheses and consists of a questionnaire that is based on three types of questions, where the two first types include a 5-point scale.

The PMAT is used to identify technologies related to Industry 4.0, as well as to verify their degree of implementation – maturity stages in companies from different sectors.

What stands out from this study is the willingness of companies to adopt Industry 4.0 as a production paradigm. Even companies that did not know the term or the concept showed willingness to take an interest in the matter and adopt it in the near future.

The proposed questionnaire (PMAT) made it possible to investigate the degree of implementation - maturity stages of the technologies involved in different companies, and in this way, to have a point of reference on the current state of knowledge and the measures adopted by the companies.

Therefore, this first study was very encouraging for future research projects related to the patterns of implementation of Industry 4.0 technologies in companies in the region under study.

## References

1. Chong, G., Zhiying, J., Ding, D.: The Emerging Business Models. World Scientific Publishing Company, Singapore (2020)
2. Kagermann, H., Wahlster, W., Helbig, J.: Recommendations for implementing the strategic initiative industrie 4.0: securing the future of german manufacturing industry. Final report (2013)
3. Schwab, K.: The Fourth Industrial Revolution, 1st edn. World Economic Forum, Geneva (2017)
4. Quiroga, O.D., Zeballos, L.J.: Cyber-physical systems and manufacturing paradigms. In: XVIII Semana da Engenharia de Produção e Mecânica Sul-Americana, pp. 1–8. Asociación de Universidades del Grupo Montevideo (AUGM) (2018)
5. Liu, Y., Xu, X.: Industry 4.0 and cloud manufacturing: a comparative analysis. J. Manuf. Sci. Eng. **139**(3), (2017)
6. Lee, J., Bagheri, B., Kao, H.: A cyber-physical systems architecture for Industry 4.0-based manufacturing systems. Soc. Manuf. Eng. (SME) Manuf. Lett. **3**, 18–23 (2015)
7. Lee, J., Lapira, E., Bagheri, B., Kao, H.: Recent advances and trends in predictive manufacturing systems in big data environment. Manuf. Lett. **38**, 41 (2013)
8. Dalenogare, L., Benitez, G., Ayala, N., Frank, G.: The expected contribution of industry 4.0 technologies for industrial performance. Int. J. Prod. Econ. **204**, 383–394 (2018)
9. Stock, T., Obenaus, M., Kunz, S., Kohl, H.: Industry 4.0 as enabler for a sustainable development: a qualitative assessment of its ecological and social potential. Process Saf. Environ. Protect. **118**, 254–267 (2018)
10. Lin, H.W., Nagalingam, S.V., Kuik, S.S., Murata, T.: Design of a global decision support system for a manufacturing SME: towards participating in collaborative manufacturing. Int. J. Prod. Econ. **136**(1), 1–12 (2012)
11. Tao, F., Cheng, J., Qi, Q., Zhang, M., Zhang, H., Sui, F.: Digital twin-driven product design, manufacturing and service with big data. Int. J. Adv. Manuf. Technol. **94**(9–12), 3563–3576 (2018)
12. Frank, A., Dalenogare, L., Ayala, N.: Industry 4.0 technologies: implementation patterns in manufacturing companies. Int. J. Prod. Econ. **210**, 15–26 (2019)
13. Yin, R.: Case Study Research and Applications: Design and Methods, 6th edn. SAGE Publications, Thousand Oaks (2018)
14. Mittal, S., Khan, M.A., Romero, D., Wuest, T.: A critical review of smart manufacturing & industry 4.0 maturity models: implications for small and medium-sized enterprises (SMEs). J. Manuf. Syst. **49**, 194–214 (2018)
15. Schuh, G., Anderl, R., Gausemeier, J., ten Hompel, M., Wahlster, W.: Industrie 4.0 maturity index. In: Managing the Digital Transformation of Companies (Acatech Study) Herbert Utz Verlag, Munich (2020)
16. Lichtblau, K., Stich, V., Bertenrath, R., Blum, M., Bleider, M., Millack, A., et al.: IMPULSindustrie 4.0-readiness. Impuls-stiftung des VDMA, Aachen-köln (2018)
17. Geissbauer, R., Vedso, J., Schrauf, S.: Industry 4.0: building the digital enterprise: 2016 global industry 4.0 survey: price water house coopers (2016)
18. Schumacher, A., Erol, S., Sihn, W.: A maturity model for assessing industry 4.0 readiness and maturity of manufacturing enterprises. Procedia CIRP **52**, 161–166 (2016)
19. Jung, K., Kulvatunyou, B., Choi, S., Brundage, M.P.: An overview of a smart manufacturing system readiness assessment. In: Nääs, I., et al. (eds.) APMS 2016. IAICT, vol. 488, pp. 705–712. Springer, Cham (2016). https://doi.org/10.1007/978-3-319-51133-7_83
20. Scremin, L., Armellini, F., Brun, A., Solar-Pelletier, L., Beaudry, C.: Towards a framework for assessing the maturity of manufacturing companies in industry 4.0 adoption. In: Analyzing the Impacts of Industry 4.0 in Modern Business Environments, pp. 224–254 (2018)

21. Lee, J., Jun, S., Chang, T.W., Park, J.A.: Smartness assessment framework for smart factories using analytic network process. Sustainability **9**(5), 794–808 (2017)
22. Leyh, C., Schäffer, T., Ble, K., Forstenhäusler, S.: SIMMI 4.0 – a maturity model for classifying the enterprise-wide IT and software landscape focusing on industry 4.0. In: Proceedings of the Federated Conference on Computer Science and Information Systems, Gdansk, Poland, pp. 1297–1302 (2016)
23. Lichtblau, K., et al.: IMPULS - Industrie 4.0- Readiness. Impuls-Stiftung des VDMA, Aachen-Köln (2015)
24. Colli, M., Madsen, O., Berger, U., Moller, C., Vejrum Waehrens, B., Bockholt, M.: Contextualizing the outcome of a maturity assessment for industry 4.0. IFAC Int. Fed. Autom. Control Papers On Line **51-11**(5), 1347–1352 (2018)
25. Colli, M., Berger, U., Bockholt, M., Madsen, O., Moller, C., Vejrum Waehrens, B.: A maturity assessment approach for conceiving context-specific roadmaps in the industry 4.0 era. Annu. Rev. Control **48**, 165–177 (2019)

# Digital Twin and PHM for Optimizing Inventory Levels

Joceir Chaves(✉), Eduardo F. R. Loures, Eduardo A. P. Santos, Julio C. Silva, and Ricardo Kondo

Pontifical Catholic University of Paraná, Curitiba, Brazil
{eduardo.loures,eduardo.portela}@pucpr.br

**Abstract.** Intelligent manufacturing associated with the advent of industry 4.0 marks a new era associated with technological solutions. Physical and virtual systems (CPS) are employed to improve processes and operations in a highly competitive environment. In this context, digital twin (DT) and asset health prognosis and management (PHM) play a fundamental role in manufacturing, allowing, through monitoring and analysis by physical and virtual means, to maximize the use of assets with estimated remaining life (RUL) and minimize unscheduled interruptions. This article proposes an extension to this context addressing the optimization of the inventory of materials and spare parts, representing a significant value that is often immobilized without expected use, causing a financial impact on organizations. It's evaluated in the literature the use of DT and PHM related to the control of spare parts and gaps found, and opportunities to be addressed.

**Keywords:** Digital twin · PHM · RUL · Spare parts

## 1 Introduction

MRO (Maintenance, Repair, and Operations) activities are considered essential for the proper operation of industrial assets, and consequently, superior performance in the manufacture of products. Although widely recognized as a critical activity, its operating cost has a significant financial impact on manufacturing processes, creating a trade-off between the need and the resulting cost. In addition to the values resulting from planning, execution, and repair activities, there is also an increase in spare parts inventory cost, which is not always transparent in organizations.

However, with the advent of new information technologies that enabled the emergence of the internet of things (IoT) and the 4th industrial revolution (industry 4.0), they support the development of solutions to maximize the result of MRO activities at a reasonable cost to manufacture.

One of the potential solutions that have received significant attention from researchers, players, and the industry is the Digital Twin (DT). The concept was initially developed in the aerospace field [1] and with a growing number of publications in complex applications such as maintenance of aircraft and power plants [2]. Although in a smaller number of publications, recent research efforts have been found using the DT

© Springer Nature Switzerland AG 2021
D. A. Rossit et al. (Eds.): ICPR-Americas 2020, CCIS 1407, pp. 321–329, 2021.
https://doi.org/10.1007/978-3-030-76307-7_24

approach for industrial manufacturing and maintenance [1, 3, 4] and aligned with the concept of smart manufacturing.

While DT plays an essential role in developing intelligent manufacturing, PHM (Prognostics and Health Management) represents a breakthrough in condition-based maintenance (CBM) techniques, going beyond monitoring the status of a component or system of equipment. The combination of historical data, online monitoring, and the use of algorithms allow PHM to estimate the remaining service life (RUL) of a component, in addition to predicting future failures [5–8].

There is an increasing number of research on these topics with different objectives, such as increasing the performance and reliability of the assets used in manufacturing and reducing costs related to maintenance activities. Although there is concern about the costs incurred in maintaining the equipment's operability, few efforts are made to extend this benefit to the inventory of spare parts, impacting manufacturing's operational cost, not always in a transparent manner.

This paper proposes a review of studies carried out using the DT and PHM approach to optimize inventory costs and evaluate existing gaps and opportunities that can be explored. The rest of the article is organized in the following order: Sect. 2 presents the research methodology used for this study, in Sect. 3 a review of the concepts of DT and PHM found in the literature, in Sect. 4, the works found with adherence are discussed the researched topic, and in Sect. 5 the study is concluded, and proposals for future work are presented.

## 2 Methodology

Two different searches were made to find the theoretical framework for the proposed theme to conduct the study, using the databases from ScienceDirect and CAPES (Coordination for the Improvement of Higher Education Personnel), the second one added in the second stage. In the first stage, the following combined terms were used: ("digital twin" AND MRO) OR (PHM AND MRO) OR ("smart manufacturing" OR MRO) considering the search for terms throughout the article and with a date greater than 2010. They were 45 articles found in the ScienceDirect database with search terms.

An additional search was made combining all terms and taxonomy between the main search terms due to the number of publications found. The investigation was restricted to the title, abstract, and keywords in the ScienceDirect database and the subject in the CAPES database. The terms used in the search were as follows: ("digital twin" OR "smart manufacturing" OR "industry 4.0") AND (PHM OR prognostics OR "predictive maintenance") AND (MRO OR "spare parts" OR maintenance). Fifty-seven articles were found in the ScienceDirect database and 22 in the CAPES database. In both stages, articles published in English and with a publication date greater than 2010 were considered.

After removing duplicate articles and outside the context of manufacturing and industry, 115 remained, of which the content was assessed to ascertain the adherence to the proposed theme. To consider that the paper had adherence to the theme, the DT and or PHM approach directly or indirectly connected with the spare parts inventory was taken into account. Thus, the citation of the inventory or cost of parts was considered a direct adherence to the theme, and the prediction of the useful life of components with a focus on cost was considered an indirect way since this information can be used as one of the indicators for optimization of inventory levels. As a result, 34 articles were selected for the research, as shown (Fig. 1).

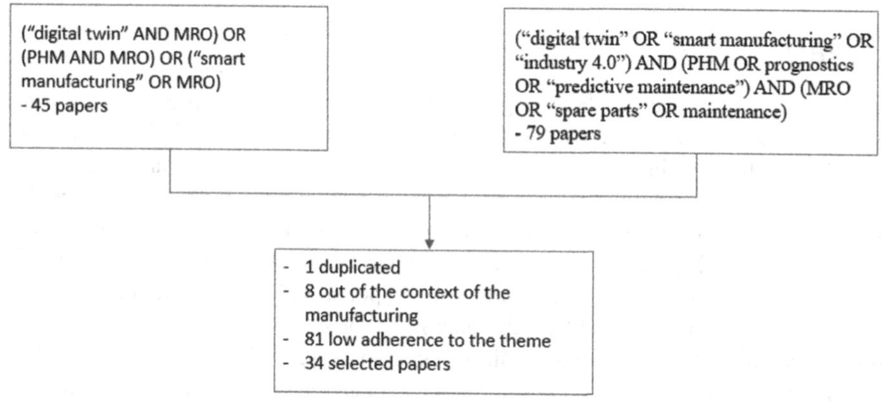


**Fig. 1.** Methodology applied in research

# 3 Related Technologies

## 3.1 Digital Twin

Despite being positively related, DT technology concepts emerged before Industry 4.0 in a presentation at Michigan College in 2002 [9] as a proposal for the PLM (Product Life Cycle) concept and have since received significant attention from researchers. In general, DT is the virtual representation of a physical object or system capable of simulating its physical counterpart's characteristics and behavior, in addition to both being mutually connected [3]. Among the various definitions of DT, [3] states that the most commonly used was proposed in 2012 by Glaessegen and Stargel, which describes that DT consists of three parts: physical products, virtual products, and the connections between them. However, one of the first definitions was given by NASA [1], where DT is an integrated multi-physical, multi-scalable probabilistic simulation of a product or system that uses the best available physical models to reflect the life of its corresponding twin.

Although DT has an expressive approach in complex applications related to aircraft, power generating plants, wind turbines [2], and aerospace engineering [10], attention has been focused on new applications in industrial manufacturing such as products, processes, and operations as well as interoperability between them [11]. The DT approach in high-performance industrial equipment and systems is also broad, both for manufacturing processes [1, 3, 12–14] and for predictive maintenance [15–18], but not limited to these application areas, including the entire product life cycle [4].

DT uses the resources of CPS (cyber-physical systems) to implement solutions within intelligent manufacturing, integrating physical systems as assets and production processes with the elements of the virtual environment [19] for processing and simulation. [20] defines CPS as an environment where natural and human-made systems are tightly integrated with computing, communication, and control systems. CPS is a vital part of industry 4.0, promoting advanced real-time data connectivity and intelligent management with a high analytical capacity [21].

### 3.2 PHM

Preventive maintenance has been a great ally in manufacturing processes, ensuring greater availability of productive assets with a minimum of unscheduled interruptions [22]. Although it has its positive points, it has a relatively high cost, and it does not prevent all failures resulting from maintenance. This is because, with its characteristic of inspections or exchanges of components programmed for time (TBM), it is not always proven sufficient [23]. Inspections cover only a part of the characteristics necessary to ensure the proper functioning of systems and components. The programmed exchange does not always manage to use the full potential of the component, which can often be changed in advance, generating an additional cost to the process [24].

With the advance in inspection technologies, many preventive maintenance activities have been gradually replaced by predictive or condition-based maintenance (CBM), where inspections with the aid of equipment can measure the health of the monitored component or assembly, highlighting potential points of failure in advance [24]. Techniques such as vibration analysis, thermography, and ferrography allow equipment´s condition analysis by planning scheduled interventions, reducing the risk of interruptions during its operation. However, many of these activities are also scheduled at intervals of time, coupled with the fact that a specialist is needed to analyze, which does not make an adequate estimate of the remaining useful life (RUL) [19].

Considered an evolution of CBM, PHM (Prognostics and Health Management) uses tools and algorithms with predictive characteristics to forecast the remaining useful life and identify possible points of failure [25]. The advantage of predictive maintenance based on the equipment's health factors is being able to work with an early prognosis with an indication of what and when to be treated, assisting in strategic maintenance planning decisions [26]. Within the context of intelligent manufacturing and industry 4.0, PHM plays an essential role in predictive maintenance, seeking to maintain the reliability of assets with the correct estimate of RUL throughout its life cycle [27] by assessing the degradation of systems and monitored components.

According to [19], the term prognosis and health management has its origins in the field of medicine but was adopted by the maintenance and management of assets due to its similarity in its semantics, since PHM is based on the monitoring of important signals through sensors at strategic points and your data stored for analysis. Although PHM is considered a predictive maintenance model, it addresses a set of techniques and methods used to generate forecasts and support maintenance management [27], such as monitoring the condition in real-time and future estimates based on historical and updated data, in addition to estimated remaining life [8].

## 4 DT and PHM for MRO and Inventory Management

In the literature, it is possible to find a broad approach to applying DT and PHM in the context of manufacturing and maintenance, aiming to increase the performance of assets and their reliability and provide better predictability of the actions taken by maintenance management. The application of these technologies in the era of industry 4.0 leads to a reduction in organizations' costs. However, not all benefits are still being explored. Besides, to support planning maintenance activities and reduce unscheduled downtime, the RUL estimate obtained through DT and PHM can also benefit the planning and optimization of the spare parts inventory. The high cost of many replacement items, which are often stored without anticipated use, impact organizations financially since the item's value is immobilized and cannot be converted into capital for use, in addition to the costs of inventory management.

Within the literature used in this study, [28] Kumar et al. it is the only one to directly consider the use of predictive maintenance and the estimate of RUL as a factor for reducing the inventory of materials and spare parts. Even when not mentioned directly, RUL can be considered an essential element for the management of inventory levels, [29] proposes a proof of concept of using DT in machines with low availability of data acquisition through the application of parallel monitoring of main components for estimating RUL.

When the application area is the industry, it is possible to observe the use of DT and PHM in complex equipment such as machining centers and milling machines with numerical control (CNC), [15] proposes a predictive hybrid maintenance system with DT based on the model and oriented to data to estimate and monitor the RUL of the main components, [12] also presents a study with similar modeling to predict the condition of machine tools to analyze and predict the degradation of cutting tools. [23] Einabadi et al. propose a dynamic predictive maintenance method for industry 4.0, based on real-time data and artificial neural networks (ANN) for fault prognosis, reducing unnecessary maintenance interventions, exploiting the life of components as much as possible, and reducing or preventing unplanned downtime scheduled. Still, in the context of complex assets, Tao et al. [30] propose an application in DT based on PHM for life cycle control of wind turbines. The applied case study presented by [30] addresses efficiency issues and indirectly the costs involved due to the complexity of the activities due to the characteristics of the equipment. Table 1 presents an overview of the main studies related to the topic.

Much research with an approach to estimating RUL is also found in aircraft maintenance. In addition to the complexity, it is extremely critical due to the safety involved and is even monitored by regulatory authorities [31]. However, in addition to the differentiated expectations of this activity, it is possible to find applications aimed at more significant control of replacement items [31–33] and the useful life of the components in use [32–34].

**Table 1.** Overview of the main studies related to the researched topic

| Authors | Dimensions | Framework | Main approaches |
|---|---|---|---|
| Roy et. al (2016) | Maintenance | | Analysis and monitoring of degradation of electrical and mechanical components in general under MRO and PHM context. It addresses obsolescence, autonomous maintenance, and integrated planning |
| Li et. al (2020) | Aircraft Maintenance | x | Addresses PHM project design based on the definition of stakeholder expectations |
| Lamoureux et. al (2014) | Aircraft Maintenance | | Validation of PHM indicators before implementation |
| Werner et. al (2019), Booyse et. al (2020), Nuñez et. al (2017), Susto et. al (2018), Cattaneo et. al (2019) | Maintenance | | Strategy for the implementation of predictive maintenance in the industry using DT and PHM with a focus on RUL |
| Raman et. al (2018) | Aerospace engineering | | Addresses the importance of DT in MRO in modern manufacturing methods |
| Luo et. al (2020) | Maintenance | x | Application of DT oriented to hybrid data for predictive maintenance and RUL estimation |
| Short et. al (2019), Aivaliotis et. al (2019) | Maintenance | | DT application with modeling for predictive maintenance |
| Sénéchal et. al (2019) | Maintenance | x | Addresses CPS, forecasts for maintenance activities with a focus on sustainability |
| Einabadi et. al (2019) | Maintenance | | Presents the use of sensing and neural networks in the context of industry 4.0 for estimating RUL |
| Traini et. al (2019), Ruiz-Sarmiento et. al (2020) | Maintenance | x | Machine learning applied to predictive maintenance to reduce costs and to focus on RUL |
| Ardila et. al (2020) | Maintenance | | Proposes an information system for PHM based on three pillars: Heterogeneity, integration, and usability of search |
| Kumar et. al (2017) | Maintenance | x | Big data applied to CBM for greater accuracy in estimating RUL and cost reduction of spare parts |
| Tao et. al. (2018) | Power generation | | Application of PHM-oriented DT for highly complex maintenance assets |
| Kraft et. al (2017), Liu et. al (2018) | Aircraft Maintenance | | DT focused on MRO applications |

## 5 Conclusions

With the advent of Industry 4.0 and related technologies that drive the development of new techniques and methods for superior and sustainable manufacturing performance, it is possible to realize the need to change paradigms and behaviors in this new industrial era, where IoT is increasingly present. Many approaches in literature and industry already demonstrate the first moves towards change with promising proposals and results. However, much still needs to be explored in the context of intelligent manufacturing.

DT and PHM play an essential role in asset performance and predictability. This work sought to relate the studies that have connections with the inventory of materials and spare parts for manufacturing assets. Within the results found, an extensive direct relationship with the theme was not identified. However, as the RUL estimate is a concern in most of the researched papers, it is correct to say that indirectly the optimization of inventory levels can be addressed.

As future work, a greater connection between DT and PHM approaches with the inventory of materials and replacement items is suggested, increasing the potential for results in the organization by extending the connection in the asset value chain. The maturation phase of new technologies is still beginning, and there is excellent potential to be explored soon.

## References


1. Zhu, Z., Liu, C., Xu, X.: Visualisation of the digital twin data in manufacturing by using augmented reality. Procedia CIRP **81**, 898–903 (2019). https://doi.org/10.1016/j.procir.2019.03.223
2. Tao, F., Sui, F., Liu, A., Qi, Q., Zhang, M., Song, B., et al.: Digital twin-driven product design framework. Int. J. Prod. Res. **25**, 1–19 (2018). https://doi.org/10.1080/00207543.2018.1443229
3. Qi, Q., Tao, F., Zuo, Y., Zhao, D.: Digital twin service towards smart manufacturing. Procedia CIRP **72**, 237–242 (2018). https://doi.org/10.1016/j.procir.2018.03.103
4. Schleich, B., Dittrich, M.-A., Clausmeyer, T., Damgrave, R., Erkoyuncu, J.A., Haefner, B., et al.: Shifting value stream patterns along the product lifecycle with digital twins. Procedia CIRP **86**, 3–11 (2019). https://doi.org/10.1016/j.procir.2020.01.049
5. Roy, R., Stark, R., Tracht, K., Takata, S., Mori, M.: Continuous maintenance and the future – foundations and technological challenges. CIRP Ann. **65**(2), 667–688 (2016). https://doi.org/10.1016/j.cirp.2016.06.006
6. Li, R., Verhagen, W.J.C., Curran, R.: Stakeholder-oriented systematic design methodology for prognostic and health management system: stakeholder expectation definition. Adv. Eng. Inform. **43**, (2020). https://doi.org/10.1016/j.aei.2020.101041
7. Lamoureux, B., Mechbal, N., Massé, J.-R.: A combined sensitivity analysis and kriging surrogate modeling for early validation of health indicators. Reliab. Eng. Syst. Saf. **130**, 12–26 (2014). https://doi.org/10.1016/j.ress.2014.03.007
8. Werner, A., Zimmermann, N., Lentes, J.: Approach for a holistic predictive maintenance strategy by incorporating a digital twin. Procedia Manuf. **39**, 1743–1751 (2019). https://doi.org/10.1016/j.promfg.2020.01.265
9. Rajesh, P.K., Manikandan, N., Ramshankar, C.S., Vishwanathan, T., Sathishkumar, C.: Digital twin of an automotive brake pad for predictive maintenance. Procedia Comput. Sci. **165**, 18–24 (2019). https://doi.org/10.1016/j.procs.2020.01.061
10. Raman, V., Hassanaly, M.: Emerging trends in numerical simulations of combustion systems. Proc. Combust. Inst. **37**, 2073–2089 (2018). https://doi.org/10.1016/j.proci.2018.07.121
11. Bao, J., Guo, D., Li, J., Zhang, J.: The modelling and operations for the digital twin in the context of manufacturing. Enterp. Inf. Syst. **21;13**(4), 534–556 (2019). https://doi.org/10.1080/17517575.2018.1526324
12. Qiao, Q., Wang, J., Ye, L., Gao, R.X.: Digital twin for machining tool condition prediction. Procedia CIRP **81**, 1388–1393 (2019). https://doi.org/10.1016/j.procir.2019.04.049

13. Melesse, T.Y., Pasquale, V.D., Riemma, S.: Digital twin models in industrial operations: a systematic literature review. Procedia Manuf. **42**, 267–272 (2020). https://doi.org/10.1016/j.promfg.2020.02.084
14. Misrudin, F., Foong, L.C.: Digitalization in semiconductor manufacturing- simulation forecaster approach in managing manufacturing line performance. Procedia Manuf. **38**, 1330–1337 (2019). https://doi.org/10.1016/j.promfg.2020.01.156
15. Luo, W., Hu, T., Ye, Y., Zhang, C., Wei, Y.: A hybrid predictive maintenance approach for CNC machine tool driven by Digital Twin. Robot. Comput. Integr. Manuf. **65**, (2020). https://doi.org/10.1016/j.rcim.2020.101974
16. Booyse, W., Wilke, D.N., Heyns, S.: Deep digital twins for detection, diagnostics and prog-nostics. Mech. Syst. Signal Process. **140**, (2020). https://doi.org/10.1016/j.ymssp.2019.106612
17. Short, M., Twiddle, J.: An industrial digitalization platform for condition monitoring and predictive maintenance of pumping equipment. Sens. Basel **31;19**(17) (2019). https://doi.org/10.3390/s19173781
18. Aivaliotis, P., Georgoulias, K., Arkouli, Z., Makris, S.: Methodology for enabling digital twin using advanced physics-based modelling in predictive maintenance. Procedia CIRP **81**, 417–422 (2019). https://doi.org/10.1016/j.procir.2019.03.072
19. Nuñez, D.L., Borsato, M.: An ontology-based model for prognostics and health management of machines. J. Ind. Inf. Integr. **6**, 33–46 (2017). https://doi.org/10.1016/j.jii.2017.02.006
20. Bagheri, B., Yang, S., Kao, H.-A., Lee, J.: Cyber-physical systems architecture for self-aware machines in industry 4.0 environment. IFAC-PapersOnLine **48**(3), 1622–1627 (2015). https://doi.org/10.1016/j.ifacol.2015.06.318
21. Sénéchal, O., Trentesaux, D.: A framework to help decision makers to be environmentally aware during the maintenance of cyber physical systems. Environ. Impact Assess. Rev. **77**, 11–22 (2019). https://doi.org/10.1016/j.eiar.2019.02.007
22. Weiss, B.A., Sharp, M., Klinger, A.: Developing a hierarchical decomposition methodology to increase manufacturing process and equipment health awareness. J. Manuf. Syst. **48**, 96–107 (2018). https://doi.org/10.1016/j.jmsy.2018.03.002
23. Einabadi, B., Baboli, A., Ebrahimi, M.: Dynamic predictive maintenance in industry 4.0 based on real time information: case study in automotive industries. IFAC-PapersOnLine **52**(13), 1069–1074 (2019). https://doi.org/10.1016/j.ifacol.2019.11.337
24. Traini, E., Bruno, G., D'Antonio, G., Lombardi, F.: Machine learning framework for predictive maintenance in milling. IFAC Papers OnLine **52**(13), 177–182 (2019). https://doi.org/10.1016/j.ifacol.2019.11.172
25. Ardila, A., Martinez, F., Garces, K., Barbieri, G., Sanchez-Londono, D., Caielli, A., et al.: XRepo - towards an information system for prognostics and health management analysis. Procedia Manuf. **42**, 146–153 (2020). https://doi.org/10.1016/j.promfg.2020.02.044
26. Susto, G.A., Schirru, A., Pampuri, S., Beghi, A., De Nicolao, G.: A hidden-Gamma model-based filtering and prediction approach for monotonic health factors in manufacturing. Control Eng. Pract. **74**, 84–94 (2018). https://doi.org/10.1016/j.conengprac.2018.02.011
27. Ruiz-Sarmiento, J.-R., Monroy, J., Moreno, F.-A., Galindo, C., Bonelo, J.-M., Gonzalez-Jimenez, J.: A predictive model for the maintenance of industrial machinery in the context of industry 4.0. Eng. Appl. Artif. Intell. **87**, (2020). https://doi.org/10.1016/j.engappai.2019.103289
28. Kumar, A., Shankar, R., Thakur, L.S.: A big data driven sustainable manufacturing framework for condition-based maintenance prediction. J. Comput. Sci. **27**, 428–439 (2017). https://doi.org/10.1016/j.jocs.2017.06.006
29. Cattaneo, L., Macchi, M.: A digital twin proof of concept to support machine prognostics with low availability of run-to-failure data. IFAC-PapersOnLine **52**(10), 37–42 (2019). https://doi.org/10.1016/j.ifacol.2019.10.016

30. Tao, F., Zhang, M., Liu, Y., Nee, A.Y.C.: Digital twin driven prognostics and health management for complex equipment. CIRP Ann. **67**(1), 169–172 (2018). https://doi.org/10.1016/j.cirp.2018.04.055
31. Liu, Y., Wang, T., Zhang, H., Cheutet, V., Shen, G.: The design and simulation of an autonomous system for aircraft maintenance scheduling. Comput. Ind. Eng. **137**, (2019). https://doi.org/10.1016/j.cie.2019.106041
32. Kraft, J., Kuntzagk, S.: Engine fleet-management: the use of digital twins from a MRO perspective. Volume 1: aircraft engine; fans and blowers; marine; honors and awards. ASME. V001T01A007 (2017). https://doi.org/10.1115/gt2017-63336
33. Liu, Z., Meyendorf, N., Mrad, N.: The role of data fusion in predictive maintenance using digital twin. In: AIP Conference Proceedings, vol. 1949, p. 020023 (2018). https://doi.org/10.1063/1.5031520
34. Zio, E., Fan, M., Zeng, Z., Kang, R.: Application of reliability technologies in civil aviation: lessons learnt and perspectives. Chin. J. Aeronaut. **32**(1), 143–158 (2018). https://doi.org/10.1016/j.cja.2018.05.014

# Strategies for Flexibility in Production Systems in Industry 4.0: A Framework for Characterization

Diana C. Tascón[1(✉)] and Gonzalo Mejía[2]

[1] Faculty of Engineering, Universidad Distrital Francisco José de Caldas, 110231588 Bogotá, Colombia
dctasconh@udistrital.edu.co

[2] Faculty of Engineering, Universidad de la Sabana, Campus Universitario Puente del Común, 140013 Chía, Colombia

**Abstract.** This article presents a proposal for the characterization of strategies for increasing flexibility of production systems in the context of industry 4.0. For this, a bibliometric analysis was carried out (specifically a study of co-occurrences), from which the categories included in the proposal are defined. This research seeks to provide a framework for the evaluation and analysis of flexibility strategies of productive systems within industry 4.0. The results of the bibliometric analysis, five proposed categories were established: flexibility objective, level in the process engineering, level in the manufacturing process, required technologies and links in the supply chain involved. Along with these categories possible levels were also proposed.

**Keywords:** Industry 4.0 · Flexibility · Adaptability · Strategies categorization · Production systems

## 1 Introduction

Current global manufacturing operations are demanding increasingly stringent requirements, such as tighter deadlines, competitive inventory levels, uncertain demand management, process standardization, and product diversity [1]. This translates into constant and new challenges for manufacturing systems, as they are compelled to stay at the forefront of management strategies to efficiently articulate recent technological developments in their work.

Industry 4.0, hereinafter I4.0, represents a shift towards interconnected manufacturing processes where individual entities within the supply chain communicate with each other to achieve both greater adaptability and responsiveness in manufacturing, as well as more efficient manufacturing which involves a reduction in production costs [1].

It is relevant to have references that, within the framework of Industry 4.0, contribute to organizations in the evaluation and selection of strategies to improve their adaptability and flexibility without exceeding increasingly tighter budgets; a fundamental input for the evaluation and selection processes is characterization.

© Springer Nature Switzerland AG 2021
D. A. Rossit et al. (Eds.): ICPR-Americas 2020, CCIS 1407, pp. 330–341, 2021.
https://doi.org/10.1007/978-3-030-76307-7_25

This article, based on a review of the bibliography, seeks to propose a framework for the characterization of the strategies proposed to increase the flexibility of productive systems in I4.0.

To do this, the starting point is the approach to the conceptual references that support the characterization action, then the methodological proposal is described and developed, thus the results are presented (the characterization proposal), finally the derived conclusions are outlined.

## 2 Conceptual Referents

The objective of this section is to provide a framework for the characterization of strategies based on classification criteria. The theory of classification is defined as the principles that govern the organization of objects into groups according to their similarities and differences or their relationship to a set of criteria [2]. During this section, the bases that supported the choice of criteria used to characterize the strategies are given. In general terms, the references considered include aspects related to the components of I4.0, its design principles, strategies to increase flexibility of production systems, and the relationships between I4.0 and decision support systems.

### 2.1 Components and Design Principles of Industry 4.0

In recent years there has been a hype in the study of the possibilities offered by the fourth industrial revolution for the manufacturing sector. There are state-of-the-art studies [3–7], management model proposals [8–13], as well as studies that analyze the flexibility of manufacturing through elements of I4.0 [3, 14–16]. Multiple strategies have emerged to bring manufacturing to a higher level in terms of its adaptability and productivity; the need to select the appropriate strategies for the transition to Industry 4.0 from a multidisciplinary perspective is also mentioned [17]. This sustained and growing interest in this field has been consolidated since its beginning of the last decade, when the term was introduced at the Hannover fair, acquiring greater relevance as of 2013 when an official and detailed report of the concept and its perspectives for German industry [18], was also presented at the Hannover fair.

The main difference between Industry 4.0 and its predecessors is the interaction of autonomous agents in decentralized structures that connect with each other through decision centers [19]. Among the technologies provided by I4.0 that are most relevant to decision-making processes are: cloud computing, Internet of Things, (IoT), Big Data and RFID (Radio-frequency identification) connections [19–21].

The Cyber-Physical Systems (CPS), understood as systems of collaborating computational entities that are in close connection with the surrounding physical environment and its ongoing processes [22], and that have the autonomous capacity to exchange information, trigger actions and control each other independently [18], have paved the way for the transition from centralized control to decentralized control architectures in manufacturing [23].

The Cyber-Physical Production Systems (CPPS) are part of the systems on which the use of industry 4.0 is proposed in manufacturing. Monostori [24] defines CPPS as an autonomous set of elements and cooperative subsystems interconnected in a way that all stages of the production process are covered, from the manufacturing plant to the logistics networks. On the principles of design in I4.0, a reference of the work done by Hermann, M. Pentek, T., & Otto B., [25], which is synthesized in Fig. 1.

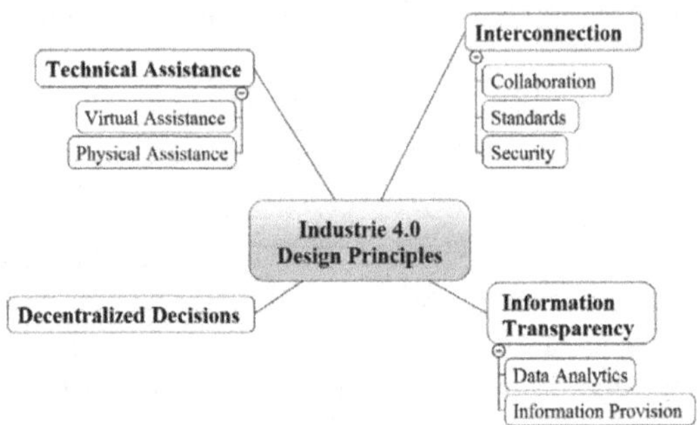


**Fig. 1.** Design principles of industry 4.0. Source: [25]

### 2.2 Flexibility of Manufacturing Systems in Industry 4.0

Flexibility is defined in the Cambridge dictionary as the ability to easily change or be changed depending on the situation [26]; within its synonyms are adaptability and modifiability. The term adaptability is defined as the ability or willingness to change to adapt to different conditions [27].

The need for production systems to adapt to market requirements without incurring excessive costs or spending an excessive amount of resources [15], has given continuous and incremental relevance to those strategies that under concepts and principles such as flexibility, adaptability, reconfigurability, evolution of the control architecture and agile decision making, tend to improve the adaptability in manufacturing systems.

Aspects such as balance between flexibility and efficiency and the importance of having a common understanding of I4.0 (semantic degree) between the parties involved have been studied [28]; different forms of flexibility have been investigated and evaluated [15, 16, 29–31], adaptability has been studied from components or elements of I4.0 i.e., Cloud Computing, IoT, Smart Factory [20, 32, 33].

### 2.3 Relationships Between Decision Support Systems and Industry 4.0

Decision Support Systems (DSS) are computer models and applications to facilitate business decision-making in the management, operations and planning of organizations [30, 34]. One of the benefits of CPPS is the possibility of directly connecting sensors and actuators in the production plant with a high-level decision-making system [19]. This allows the provision of real-time data to the DSS that allow the production system to react quickly to unforeseen events [19].

Current developments and future trends in DSS for manufacturing and supply chain management are based on CPS [35, 36]. The rapid growth of business analytics provides opportunities for operations research to solve both high complexity and large scale optimization problems and/or with uncertainty and stochastic nature, as well as emerging problems in the supply chain and operations management [35].

# 3 Proposal and Methodological Development

From a bibliometric analysis approach, a systematic review of the literature was carried out to analyze documents in which strategies to take advantage of I4.0 technologies in improving adaptability in manufacturing systems are studied or proposed.

Based on this, the related categories and levels were offered. This could serve as a starting point for evaluating strategies and supporting decision-making.

## 3.1 Bibliometric Analysis

The bibliometric analysis was developed following the stages synthesized in Fig. 2, for this it was taken as a basis what was presented in the references section where some common elements that provide inputs for this review were found.

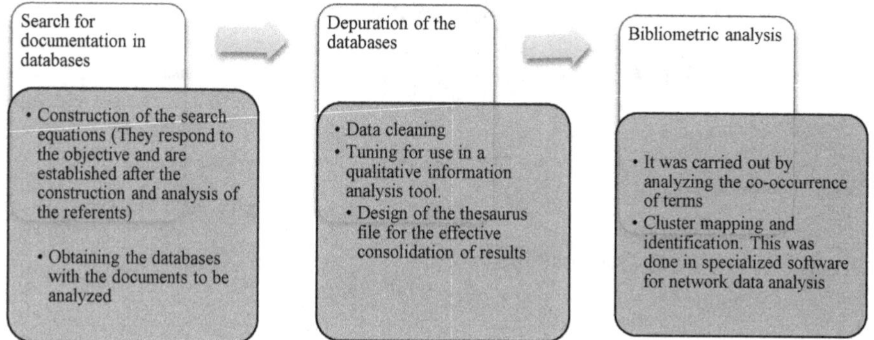


**Fig. 2.** Stages of bibliometric analysis

**Search and Debugging of Documents in Databases.** The search was made in the SCOPUS database.

Table 1 summarizes some of the most representative search equations used, the co-occurrence maps for this selection of search equations will also be shown.

**Bibliometric Analysis (Analysis of Co-occurrences).** A term co-occurrence analysis was run in VOSViwer, through the generation of maps (networks) of data (keywords), with co-occurrence for the analyzed documents. The results obtained are shown in Fig. 3, 4 and 5. At the bottom of each figure, the parameters for generating the map and the number of resulting links are indicated.

In summary, the analysis of co-occurrences gives us evidence of the interrelationships that exist between the search terms and the frequency with which they appear, in each

**Table 1.** Search equations

| No. | Search equation | Observations |
|---|---|---|
| 1. | (Adaptability OR flexibility OR adaptive AND manufacturing OR reconfigurability OR reconfigurable) AND (manufacturing AND systems OR production AND systems) AND (industry 4.0 OR forth AND industrial AND revolution OR industrie 4.0) | Search fields: title, summary, and keywords. Documents found 75 |
| 2. | (Cyber-physical AND systems OR cyber-physical AND production AND systems) AND (manufacturing AND systems OR production AND systems) AND (industry 4.0 OR forth AND industrial AND revolution OR industrie 4.0) AND (decision AND support AND systems) | Search fields: title and keywords. Documents found 21 |
| 3. | (Industry 4.0 OR Industrie 4.0) AND (manufacturing systems OR Production systems) AND (adaptability OR flexibility OR Reconfigurability OR agility) | Search fields: title and keywords. Documents found 19 |

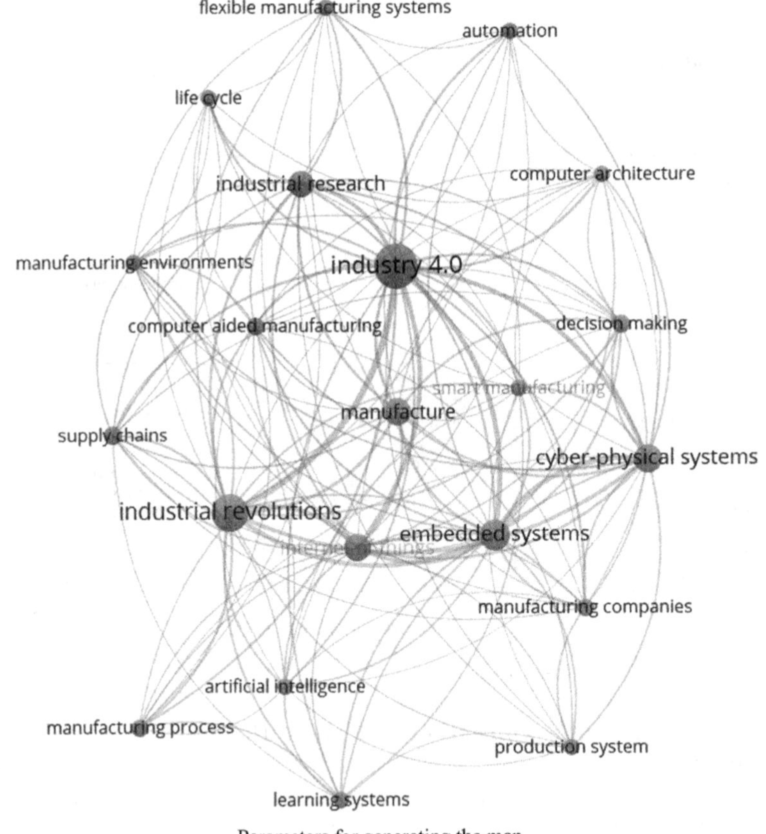


Parameters for generating the map.
Keywords: From author ☐ indexed ☐ all ☒
Minimum of co-occurrences: 5
Resulting links: 150

**Fig. 3.** Co-occurrence map, search equation 1

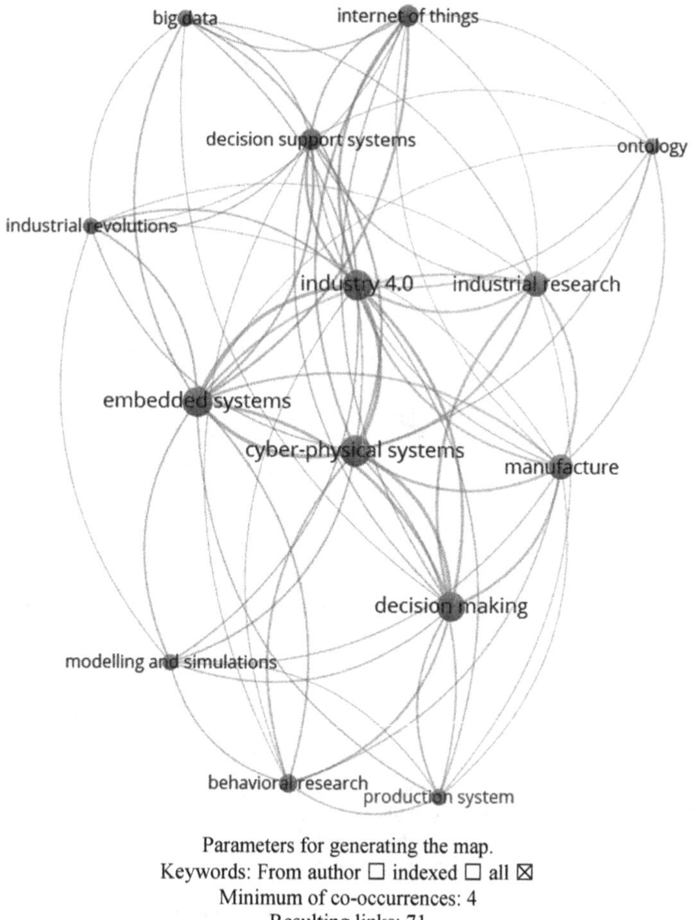


Parameters for generating the map.
Keywords: From author ☐ indexed ☐ all ☒
Minimum of co-occurrences: 4
Resulting links: 71

**Fig. 4.** Co-occurrence map, search equation 2

map the size of the nodes is directly related to the number of documents that contain the terms indicated in the search fields, the arcs are generated from the "minimum of co-occurrences" that was parameterized, the colors illustrate the existence of document clusters.

**Analysis of Relevant Documents for the Establishment of Categories.** After reviewing those articles in which the co-occurrence of terms is more relevant for the fulfillment of our objective, the categories and levels shown in Table 2, were established. This table is a synthesis of the categories together with some of their possible levels, the levels are also derived from the bibliometric review. The reference column indicates the sources that support the proposal and that are also suggested to deepen the levels proposed in each category.

As an example, the strategy followed to define the name of the category "flexibility objective" as well as its levels, was the analysis of documents in which key terms like

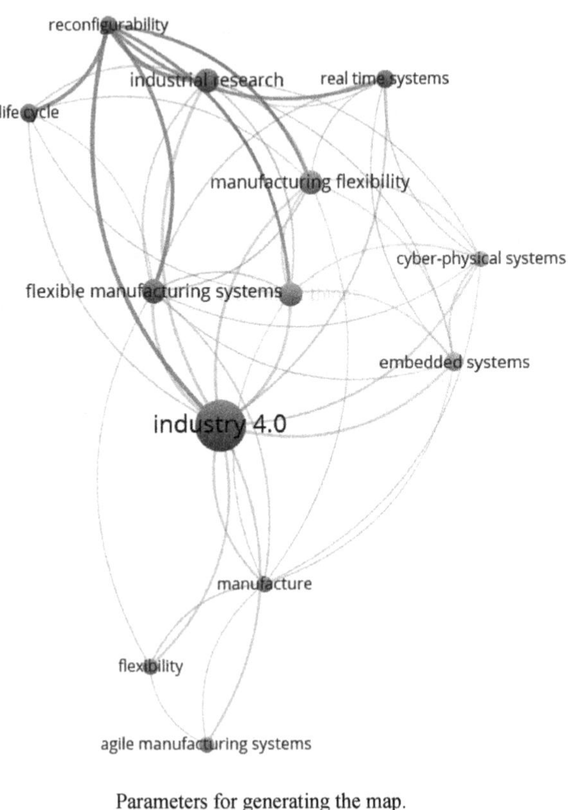


Parameters for generating the map.
Keywords: From author ☐ indexed ☐ all ☒
Minimum of co-occurrences: 3
Resulting links: 103

**Fig. 5.** Co-occurrence map, search equation 3

"flexibility" were presented more frequently, documents which in turn presented high co-occurrence of terms with other aspects included in the categorization (i.e., required technologies or links in the supply chain involved). In particular for this category (flexibility objective) the study carried out by Ivanov *et al.* [16], on new drivers of flexibility for manufacturing, supply chain and service operations, was particularly relevant. It mentions the following aspects of interest to implement flexibilization strategies: 1) risks of interruption, resilience, redundancy and ripple effect in the supply chain, 2) digitization, smart operations and electronic supply chains, 3) sustainability and capacity response, and 4) provider integration and behavioral flexibility [16]; they also state that within the topics covered in the articles they considered in their study, are the following: 1) flexibility of operations with the help of new information technologies, 2) supply chains and manufacturing systems re-configurability, 3) flexibility of operations and business analysis, 4) flexibility of operations and additive manufacturing 5) flexibility of operations and lean management 5) flexibility of operations and strategic sourcing 7) flexibility, resilience and

**Table 2.** Proposed categories and their references

| Category | Possible levels | Ref. |
|---|---|---|
| Flexibility goal<br>Level in process engineering | Increase agility | [15, 18] |
| | Compensate for temporal variabilities | [18] |
| | Deal with disruption risks | [16] |
| | Deal with ripple effect risks | [16, 39] |
| | Improve sustainability | [16, 38] |
| | Product design | [18, 40] |
| Level in the manufacturing process | Production planning | [18, 41, 42] |
| | Service planning | [18] |
| | Manufacturing management | [49–51] |
| Required technologies | Process control | [35, 42, 47] |
| | Production control | [23, 41, 47, 48] |
| | Plant floor | [3, 43–46] |
| | IoT | [18, 21] |
| Supply chain participants involved | IA | [18, 50, 52] |
| | Cloud computing | [8, 21, 46, 52] |
| | RFID | [21, 35] |
| | Blockchain | [35, 50] |
| | Suppliers | [37, 39] |
| | Factory | [39, 53, 54] |
| | Distributors | [39] |
| | Retailers | [39] |

risk management 8) flexibility in service 9) flexibility of the supply chain. Characteristics that served as references for what was proposed, in an equivalent way, the other categories and levels presented in Table 2, in which, as previously indicated, the considered references are indicated.

## 4 Conclusion

Starting from a bibliometric review, it was possible to identify a set of categories related to strategies that, using elements of Industry 4.0, tend to improve the flexibility of production systems, which subsequently favors the analysis, configuration, and choice of these in concordance with the objectives and resources of an organization.

## References

1. Lam, C., van Velthoven, M.H., Meinert, E.: Application of Internet of Things in cell-based therapy delivery: protocol for a systematic review. JMIR Res. Protoc. **9**, (2020). https://doi.org/10.2196/16935
2. Classification theory—Britannica. https://www.britannica.com/science/classification-theory
3. Rossit, D.A., Tohmé, F., Frutos, M.: Production planning and scheduling in cyber-physical production systems: a review. Int. J. Comput. Integr. Manuf. **32**, 385–395 (2019). https://doi.org/10.1080/0951192X.2019.1605199
4. Osterrieder, P., Budde, L., Friedli, T.: The smart factory as a key construct of industry 4.0: a systematic literature review. Int. J. Prod. Econ. **221** (2020). https://doi.org/10.1016/j.ijpe.2019.08.011
5. Ng, T.C., Ghobakhloo, M.: Energy sustainability and industry 4.0. In: IOP Conference Series: Earth and Environmental Science. Institute of Physics Publishing (2020)
6. Xu, B., Shen, J., Liu, S., Su, Q., Junjui, J.: Research and development of electro-hydraulic control valves oriented to industry 4.0: a review. Chin. J. Mech. Eng. **33**, 29 (2020). https://doi.org/10.1186/s10033-020-00446-2
7. Chiappetta Jabbour, C.J., Fiorini, P.D.C., Ndubisi, N.O., Queiroz, M.M., Piato, É.L.: Digitally-enabled sustainable supply chains in the 21st century: a review and a research agenda. Sci. Total Environ. **725** (2020). https://doi.org/10.1016/j.scitotenv.2020.138177
8. Xu, X.: From cloud computing to cloud manufacturing. Robot. Comput. Integr. Manuf. **28**, 75–86 (2012). https://doi.org/10.1016/j.rcim.2011.07.002
9. Jain, A., Jain, P.K., Chan, F.T.S., Singh, S.: A review on manufacturing flexibility. Int. J. Prod. Res. **51**, 5946–5970 (2013). https://doi.org/10.1080/00207543.2013.824627
10. Yadav, G., Luthra, S., Jakhar, S.K., Mangla, S.K., Rai, D.P.: A framework to overcome sustainable supply chain challenges through solution measures of industry 4.0 and circular economy: an automotive case. J. Clean. Prod. **254** (2020). https://doi.org/10.1016/j.jclepro.2020.120112
11. Tao, F., Zhang, L., Liu, Y., Cheng, Y., Wang, L., Xu, X.: Manufacturing service management in cloud manufacturing: overview and future research directions. J. Manuf. Sci. Eng. Trans. ASME. **137** (2015). https://doi.org/10.1115/1.4030510
12. Ivanov, D., Tsipoulanidis, A., Schnberger, J.: Global Supply Chain and Operations Management: A Decision-Oriented Introduction into the Creation of Value. Springer, Berlin (2019)
13. Cachon, G.P., Fisher, M.: Supply chain inventory management and the value of shared information. Manage. Sci. **46**, 1032–1048 (2000). https://doi.org/10.1287/mnsc.46.8.1032.12029
14. Wang, S., Wan, J., Zhang, D., Li, D., Zhang, C.: Towards smart factory for industry 4.0: a self-organized multi-agent system with big data based feedback and coordination. Comput. Netw. **101**, 158–168 (2016). https://doi.org/10.1016/j.comnet.2015.12.017
15. Fragapane, G., Ivanov, D., Peron, M., Sgarbossa, F., Strandhagen, J.O.: Increasing flexibility and productivity in industry 4.0 production networks with autonomous mobile robots and smart intralogistics. Ann. Oper. Res., 1–19 (2020). https://doi.org/10.1007/s10479-020-03526-7
16. Ivanov, D., Das, A., Choi, T.-M.: New flexibility drivers for manufacturing, supply chain and service operations. Int. J. Prod. Res. **56**, 3359–3368 (2018). https://doi.org/10.1080/00207543.2018.1457813
17. Kaya, I., Erdoğan, M., Karaşan, A., Özkan, B.: Creating a road map for industry 4.0 by using an integrated fuzzy multicriteria decision-making methodology. Soft Comput., 1–26 (2020). https://doi.org/10.1007/s00500-020-05041-0

18. Kagermann, H., Wolfgang, W., Helbig, J.: Securing the future of German manufacturing industry. Recommendations for implementing the strategic initiative INDUSTRIE 4.0. Final report of the industrie 4.0 working group. Plattf. Ind. 4.0., 1–78 (2013)
19. Rossit, D., Tohmé, F.: Scheduling research contributions to smart manufacturing. Manuf. Lett. **15**, 111–114 (2018). https://doi.org/10.1016/j.mfglet.2017.12.005
20. Liu, Y., Tong, K.Di, Mao, F., Yang, J.: Research on digital production technology for traditional manufacturing enterprises based on industrial Internet of Things in 5G era. Int. J. Adv. Manuf. Technol. (2019). https://doi.org/10.1007/s00170-019-04284-y
21. Zhong, R.Y., Lan, S., Xu, C., Dai, Q., Huang, G.Q.: Visualization of RFID-enabled shopfloor logistics Big Data in cloud manufacturing. Int. J. Adv. Manuf. Technol. **84**, 5–16 (2016). https://doi.org/10.1007/s00170-015-7702-1
22. Monostori, L., et al.: Cyber-physical systems in manufacturing. CIRP Ann. **65**, 621–641 (2016). https://doi.org/10.1016/j.cirp.2016.06.005
23. Boccella, A.R., Centobelli, P., Cerchione, R., Murino, T., Riedel, R.: Evaluating centralized and heterarchical control of smart manufacturing systems in the era of industry 4.0. **10**, 755 (2020). https://doi.org/10.3390/app10030755
24. Monostori, L., Procedia, C.: Cyber-physical production systems: roots, expectations and R&D challenges. Procedia CIRP **17 SRC**, 9–13 (2014). https://doi.org/10.1016/j.procir.2014.03.115
25. Hermann, M., Pentek, T., Otto, B.: Design principles for industrie 4.0 scenarios. In: Proceedings of the Annual Hawaii International Conference on System Sciences, pp. 3928–3937. IEEE Computer Society (2016)
26. Flexibility – significado, definici6n en el Cambridge English Dictionary. https://dictionary.cambridge.org/es-LA/dictionary/english/flexibility
27. Adaptability – significado, definici6n en el Cambridge English Dictionary. https://dictionary.cambridge.org/es-LA/dictionary/english/adaptability
28. Cheng, C.H.C.-H., Guelfirat, T., Messinger, C., Schmitt, J.O.J.O., Schnelte, M., Weber, P.: Semantic degrees for industrie 4.0 engineering: deciding on the degree of semantic formalization to select appropriate technologies. In: Proceedings of the 2015 10th Joint Meeting of the European Software Engineering Conference and the ACM SIGSOFT Symposium on the Foundations of Software Engineering, ESEC/FSE 2015, pp. 1010–1013. Association for Computing Machinery, Inc, New York (2015)
29. Bortolini, M., Galizia, F.G.F.G., Mora, C.: Reconfigurable manufacturing systems: Literature review and research trend. J. Manuf. Syst. **49**, 93–106 (2018). https://doi.org/10.1016/j.jmsy.2018.09.005
30. Abdi, M.R., Labib, A.W., Delavari Edalat, F., Abdi, A.: Evolution of MS paradigms through industrial revolutions. Integrated Reconfigurable Manufacturing Systems and Smart Value Chain, pp. 17–42. Springer, Cham (2018). https://doi.org/10.1007/978-3-319-76846-5_2
31. Bortolini, M., Faccio, M., Galizia, F.G.F.G., Gamberi, M., Pilati, F.: Design, engineering and testing of an innovative adaptive automation assembly system. Assem. Autom. **40**, 531–540 (2020). https://doi.org/10.1108/AA-06-2019-0103
32. Oztemel, E., Gursev, S.: Literature review of industry 4.0 and related technologies (2020). https://doi.org/10.1007/s10845-018-1433-8
33. Olalere, I.O., Olanrewaju, O.A.: Optimising production through intelligent manufacturing. In: E3S Web of Conferences. EDP Sciences (2020)
34. Babiceanu, R.F., Seker, R.: Big Data and virtualization for manufacturing cyber-physical systems: a survey of the current status and future outlook. Comput. Ind. **81**, 128–137 (2016). https://doi.org/10.1016/j.compind.2016.02.004
35. Panetto, H., Iung, B., Ivanov, D., Weichhart, G., Wang, X.: Challenges for the cyber-physical manufacturing enterprises of the future. Annu. Rev. Control **47**, 200–213 (2019). https://doi.org/10.1016/j.arcontrol.2019.02.002

36. Zhuge, H.: Semantic linking through spaces for cyber-physical-socio intelligence: a methodology (2011). www.elsevier.com/locate/artint
37. Abdi, M.R., Labib, A.W., Delavari Edalat, F., Abdi, A.: Integrated Reconfigurable Manufacturing Systems and Smart Value Chain. Springer, Cham (2018). https://doi.org/10.1007/978-3-319-76846-5
38. Ante, G., Facchini, F., Mossa, G., Digiesi, S.: Developing a key performance indicators tree for lean and smart production systems. IFAC-PapersOnLine **51**, 13–18 (2018). https://doi.org/10.1016/j.ifacol.2018.08.227
39. Ivanov, D., Dolgui, A., Sokolov, B.: The impact of digital technology and industry 4.0 on the ripple effect and supply chain risk analytics. Int. J. Prod. Res. **57**, 829–846 (2019). https://doi.org/10.1080/00207543.2018.1488086
40. Pereira Pessôa, M.V., Jauregui Becker, J.M.: Smart design engineering: a literature review of the impact of the 4th industrial revolution on product design and development. Res. Eng. Des. (2020). https://doi.org/10.1007/s00163-020-00330-z
41. Gomes, M., Silva, F., Ferraz, F., Silva, A., Analide, C., Novais, P.: Developing an ambient intelligent-based decision support system for production and control planning. In: Madureira, A.M., Abraham, A., Gamboa, D., Novais, P. (eds.) ISDA 2016. AISC, vol. 557, pp. 984–994. Springer, Cham (2017). https://doi.org/10.1007/978-3-319-53480-0_97
42. Trstenjak, M., Cosic, P.: Process planning in industry 4.0 environment. Procedia Manuf. **11**, 1744–1750 (2017). https://doi.org/10.1016/j.promfg.2017.07.303
43. Romero-Silva, R., Hernández-López, G.: Shop-floor scheduling as a competitive advantage: a study on the relevance of cyber-physical systems in different manufacturing contexts. Int. J. Prod. Econ. **224** (2020). https://doi.org/10.1016/j.ijpe.2019.107555
44. Turker, A.K., Aktepe, A., Inal, A.F., Ersoz, O.O., Das, G.S., Birgoren, B.: A decision support system for dynamic job-shop scheduling using real-time data with simulation. Mathematics **7**, 278 (2019). https://doi.org/10.3390/math7030278
45. Dolgui, A., Ivanov, D., Sethi, S.P., Sokolov, B.: Scheduling in production, supply chain and industry 4.0 systems by optimal control: fundamentals, state-of-the-art and applications. Int. J. Prod. Res. **57**, 411–432 (2019)
46. Liu, Y., Wang, L., Wang, X.V., Xu, X., Zhang, L.: Scheduling in cloud manufacturing: state-of-the- art and research challenges. Int. J. Prod. Res. **57**, 4854–4879 (2019). https://doi.org/10.1080/00207543.2018.1449978
47. Wenzelburger, P., Allgöwer, F.: A petri net modeling framework for the control of flexible manufacturing systems. IFAC-PapersOnLine **52**, 492–498 (2019). https://doi.org/10.1016/j.ifacol.2019.11.111
48. Grassi, A., Guizzi, G., Santillo, L.C., Vespoli, S.: A semi-heterarchical production control architecture for industry 4.0-based manufacturing systems. Manuf. Lett. **24**, 43–46 (2020). https://doi.org/10.1016/j.mfglet.2020.03.007
49. Kunath, M., Winkler, H.: Integrating the Digital Twin of the manufacturing system into a decision support system for improving the order management process. Procedia CIRP **72**, 225–231 (2018)
50. Olsen, T.L., Tomlin, B.: Industry 4.0: opportunities and challenges for operations management. Manuf. Serv. Oper. Manage. **22**, 113–122 (2020). https://doi.org/10.1287/msom.2019.0796
51. Marques, M., Agostinho, C., Zacharewicz, G., Goncalves, R., Zacharewicz, G., Jardim-Gonçalves, R.: Decentralized decision support for intelligent manufacturing in industry 4.0. JAISE-J. Ambient Intell. Smart Environ. **9**, 299–313 (2017). https://doi.org/10.3233/AIS-170436
52. Wan, J., Yang, J., Wang, Z., Hua, Q.: Artificial intelligence for cloud-assisted smart factory. IEEE Access **6**, 55419–55430 (2018). https://doi.org/10.1109/ACCESS.2018.2871724

53. Sun, J., Yamamoto, H., Matsui, M.: Horizontal integration management: an optimal switching model for parallel production system with multiple periods in smart supply chain environment. Int. J. Prod. Econ. **221** (2020). https://doi.org/10.1016/j.ijpe.2019.08.010
54. Seif, A., Toro, C., Akhtar, H.: Implementing industry 4.0 asset administrative shells in mini factories. Procedia Comput. Sci. **159**, 495–504 (2019)

# Integrated Production and Maintenance Planning: A Systematic Literature Review

Nicollas Luiz Schweitzer de Souza(✉), Lúcio Galvão Mendes, Eugênio Strassburguer Rovaris, Enzo Morosini Frazzon, and Lynceo Falavigna Braghirolli

Production and Systems Engineering Department, Federal University of Santa Catarina, Florianópolis, Brazil

nicollas.schweitzer@posgrad.ufsc.br, lucio.galvao@ifsc.edu.br, eugenio.rovaris@grad.ufsc.br, {enzo.frazzon, lynceo.braghirolli}@ufsc.br

**Abstract.** Production and maintenance planning are interconnected problems, so that their joint solution can enhance system performance. This paper develops a systematic review on the integrated production and maintenance planning on manufacturing systems. The content analysis included sixty-nine papers, embracing aspects of production system modelling, optimization mechanism, practical applications and research directions. The findings indicate the predominance of theoretical studies, the use of genetic algorithm for the optimization engine, in which costs and makespan are the variables of the objective function. Future research directions point to applications for real scenarios, predictive maintenance and models with increasing complexity.

**Keywords:** Production planning and control · Maintenance scheduling and production planning · Maintenance models and services · Modelling and decision making in complex systems · Modelling of manufacturing operations · Production scheduling · Meta-heuristics

## 1 Introduction

In production environments, machines are not always available as they are subject to periods of downtime due to planned and unplanned shutdowns [1, 2]. Considering the dynamic and stochastic nature of failure occurrence, production planners and managers have complex decisions to deal with on a daily basis [3, 4]. Aiming to speed up production, recommended maintenance intervals are often poorly planned [1].

Production and maintenance scheduling have attracted great practical and academic interest, however those problems are commonly treated as independent [5, 6]. Notwithstanding, literature has shown that developing the production planning and control integrated with the maintenance planning can benefit several production performance indicators [7, 8]. Indeed, manufacturing systems generally show a natural interdependence

© Springer Nature Switzerland AG 2021
D. A. Rossit et al. (Eds.): ICPR-Americas 2020, CCIS 1407, pp. 342–356, 2021.
https://doi.org/10.1007/978-3-030-76307-7_26

between production and maintenance activities, which raises the question of how to optimize them jointly to avoid conflicts and improve system performance [9].

A large number of studies about production and maintenance were carried out, however, almost all of them considered these as two independent problems. Only a few studies tried to combine the two issues to solve them simultaneously [7]. In addition, the criticality of resources may change depending on the production planning being conducted. Therefore, there is a need to coordinate production and maintenance planning tasks, which incorporate potential cost minimization [9].

The objective of this article is to perform a systematic review of the literature regarding the current models that integrate production planning and control with maintenance planning. This systematic review can allow the acknowledgment of existing studies in the field and, the identification of research gaps, helping to prevent the production of duplicate works [10]. Besides, it can provide several insights from the state-of-the-art literature, including common practices and research directions.

The systematic review intends to answer three research questions about the subject: (i) how did the studies on integrated production planning model the production system and the optimization process? (ii) which studies modelled real scenarios for integrated production and maintenance planning? (iii) what are the main directions for future research?

The remaining of this paper is structured as follows. Section 2 shows the research method of the systematic review and bibliometric analysis. Section 3 shows a brief bibliometric analysis, followed by the systematic review in Sect. 4, which is divided accordingly to the research questions. Finally, Sect. 5 presents the discussion and proposals for further research.

## 2 Research Method

The conducted systematic literature review followed the Preferred Reporting Items for Systematic review and Meta-Analysis (PRISMA) approach, which is proposed by [11] and is the most commonly used for reporting literature review [12]. It was first applied in health care studies but is also suitable for many areas, including industrial engineering [13, 14].

The paper selection step used the following search logic: (TITLE-ABS-KEY((“Produ* plann*” AND maint*) OR (“Produ* Sched*” AND maint*) OR (“Manufac* plann*” AND maint*) OR (“Maint* plann*” AND (Produ* OR manufacture*)) OR (“Maint* Sched*” AND (produ* OR manufacture*))). The search was performed in two databases: Scopus (www.scopus.com) and Web of Science (www.webofknowledge.com).

The process for selecting papers is shown in Fig. 1. After the identification of the papers, duplicates were removed. The criteria for inclusion and exclusion of papers are presented in Table 1, following the ones used by [15], and were applied by reading titles, abstracts and keywords only. The remaining papers in the last stage were eligible for the content analysis, i.e. the qualitative synthesis.

From the articles included in the content analysis, we performed a brief bibliometric analysis pursuing to better understand the scientific activity through time and among

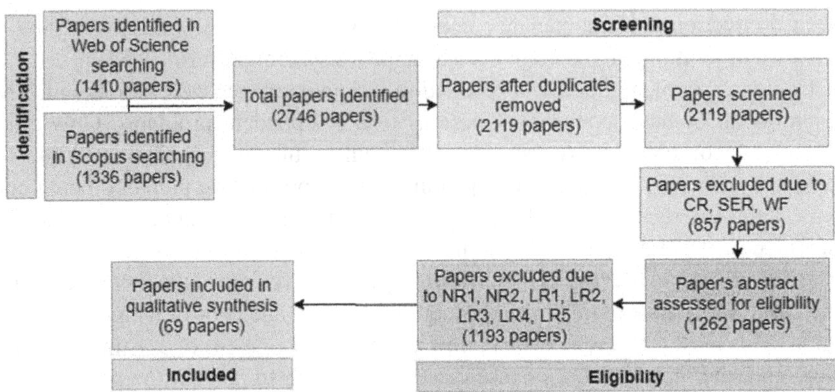


**Fig. 1.** Systematic review process.

**Table 1.** Inclusion and exclusion criteria.

| x | Criteria | Criteria explanation |
|---|---|---|
| Inclusion | Closely related (CR) | Time period: prior to April 2020; Articles that were published until 2014 were included only with more than 5 citations according to Bibliometrix |
| Exclusion | Search engine reason (SER) | The article has only the title, abstract and keywords in English, but not the full text in English |
| | Without full-text (WF) | The full-text of the article is not available |
| | Non-related (NR) | NR1: It is not an academic article; |
| | | NR2: It is not aligned with "integrated production planning model with maintenance" |
| | Loosely related (LR) | The article does not focus on discussing the development of an integrated production planning model with maintenance. More specifically: |
| | | LR1: It is used only as an example; |
| | | LR2: It is used only as part of its future research direction; |
| | | LR3: It is used only as a quoted expression; |
| | | LR4: It is used only in keywords or references; |
| | | LR5: The research that does not address the context |

different journals and research groups. Such analysis was aided by the Bibliometrix package developed in R programming language [16], more specifically using the Biblioshiny interface and the RStudio© integrated development environment.

## 3 Bibliometric Analysis

From the analysis of the bibliometric data, Fig. 2 shows the temporal evolution of the publication of the articles included in the analysis of this research. The first article was published in 2000, and starting in 2010, there is a tendency of publication volume increase. Although the largest number of publications are in the last decade, the number of published articles has fluctuated between 1 and 11 per year. It is also noteworthy that the year 2020 is incomplete because of the date when the search was performed.

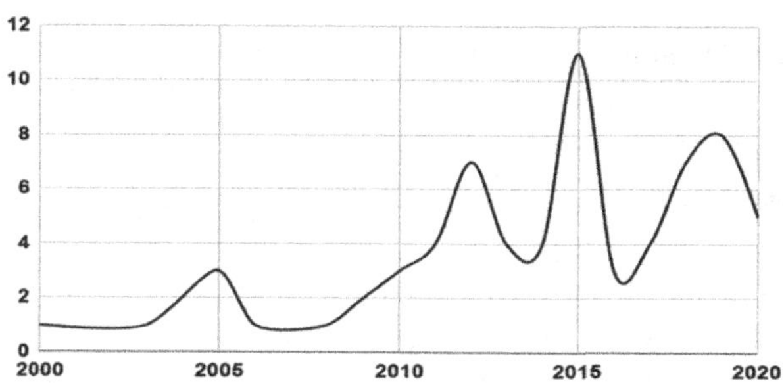


**Fig. 2.** Increase in publication quantity per year.

Most of the selected papers were published in the International Journal of Production Research with 9 occurrences [17–25], in sequence, International Journal of Advanced Manufacturing Technology with 5 papers [26–30]. The paper [31] has the highest number of citations (135) in other studies. The second with the highest number (90) of citations [32] is by the same author. Both are articles about potential studies for the topic presented here. Considering the scientific production by countries, we can highlight that China is ahead with 21 published papers, next we have Canada with 11 and France with 8.

## 4 Systematic Literature Review

This section addresses the research questions: (i) how did the studies on integrated production planning model the production system and the optimization process? (ii) which studies modelled real scenarios for integrated production and maintenance planning? (iii) what are the main directions for future research?

### 4.1 How Did the Studies on Integrated Production Planning Model the Production System and the Optimization Mechanism?

Integrated production and maintenance planning have been treated as an optimization problem, having been addressed by several solution methods. Table 2 shows the quantity and types of methods that were used to solve the models in the literature.

There is a predominance (18,1%) in the use of genetic algorithms (GA), which also appears in approaches combined with other heuristics. These studies that applied the use of a genetic algorithm conceptually validate their application, since they achieved good

**Table 2.** Resolution methods.

| Resolution method | Paper |
|---|---|
| Branched exploration algorithm | [33] |
| Genetic algorithm of approximate non-dominant classification II | [39] |
| Best fit descending algorithm | [40] |
| Optimization based on biogeography | [46] |
| Optimization of hybrid genetic algorithm | [5, 61] |
| Hybrid ant colony optimization | [5] |
| Mixed integer linear programming | [17, 50] |
| Differential evolution method | [47] |
| Annealing simulation method | [47] |
| Randomized research method | [47, 48] |
| Experiences project | [48] |
| Artificial immune algorithm | [70] |
| Simulated annealing algorithm | [38, 65] |
| Harmony search algorithm | [50] |
| Vibration damping optimization algorithm | [50] |
| Pseudo-polynomial programming algorithms | [71] |
| Evolutionary genetic algorithms multiobjective | [74] |
| Non-dominated genetic classification algorithm II | [20, 21, 27, 74] |
| Algorithm based on multi-objective optimization of ant colonies | [72] |
| Pareto ant colony optimization | [27, 72] |
| Stepwise optimization method | [52, 53] |
| Policy improvement algorithm | [19] |
| Dynamic programming | [25] |
| Dantzig-Wolfe decomposition technique | [54] |

*(continued)*

**Table 2.** (*continued*)

| Resolution method | Paper |
|---|---|
| Continuous resource task network | [55, 79] |
| Chaotic particle filling optimization algorithm | [76] |
| Clonal immune selection algorithm | [59] |
| Mixed linear model | [24, 37, 64, 67, 68] |
| Hybrid particle swarm optimization | [27] |
| Pareto force evolutionary algorithm II | [20, 27] |
| Linear quadratic model | [18] |
| Branch and limit algorithm | [21, 31] |
| Optimization-based simulation | [46] |
| Genetic algorithm | [1, 22, 23, 30, 34, 35, 41, 46, 49, 51, 57, 69, 73, 75, 77, 78] |
| Other heuristics | [7, 9, 19, 26, 28, 29, 32, 36, 38, 42–45, 56, 58, 60, 63, 66] |

results, but still need better development, according to the authors. The use of heuristics (other heuristics) created by the authors themselves is also relevant, with 22,9% of the total. The literature is quite comprehensive and there are several methods of resolution, all with different characteristics and applicability. Comparing methods present in the literature can be a good question to find the best result. We can also analyze that in some papers, more than one method was used.

The desired objective in each approach present in the literature also varies. Cost-related indicators were used as key performance indicator (KPI) for optimization in 71% of all articles. The use of cost as a KPI in the optimization method facilitates the aggregation of different variables in the composition of the best solution, such as cost of maintenance, production and delivery delay to the customer. Among them, few articles address the cost of quality, delay and stock in addition to the production and maintenance costs. The second most used KPI was makespan (29%), and system availability is considered in four papers.

In some articles, in order to model the useful life behavior of machines and production systems, or their deterioration process, different forms of representation are used. In Fig. 3 we analyzed the articles that included machine deterioration processes (55% of the total). The predominance was of criteria adopted and elaborated by the authors themselves. Another one with greater use was the use of Weibull Distribution and its combination with Gamma Distribution.

In Fig. 4, the articles were classified according to the type of productive system modeled. The predominance of single machine models is evident. Nowadays, with all

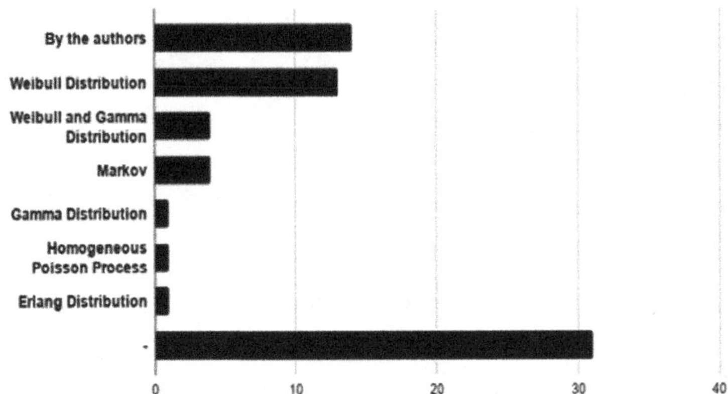


**Fig. 3.** Number of papers in relation to the deterioration method.

the technological advances, the production processes become more complex every day and with significant machine interaction.

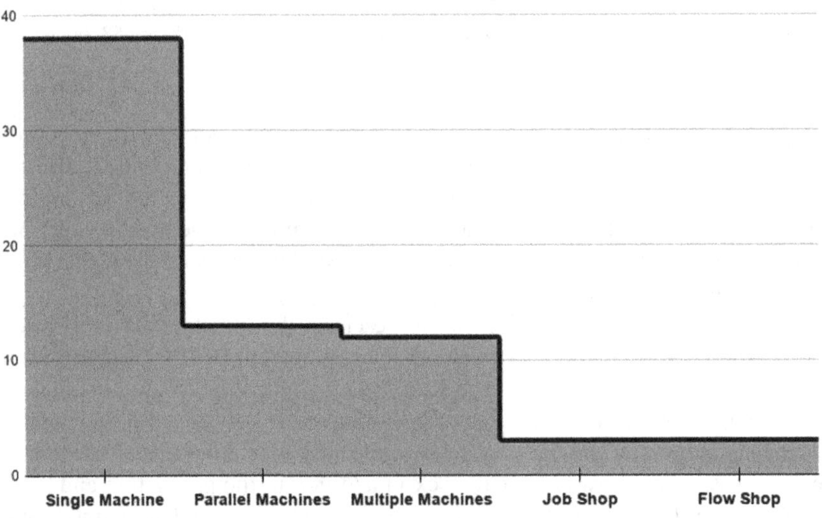


**Fig. 4.** Number of articles in relation to the type of production of the model.

### 4.2 Which Studies Modelled Real Scenarios for Integrated Production and Maintenance Planning?

Only the papers [7, 17, 28, 33–38] developed and applied a model considering a practical case of application, i.e. 13%. The rest of the papers (87%) did not apply in a real production scenario [1, 5, 9, 19–27, 30, 39–79]. The same authors who applied in real scenarios, reinforce the need for this type of application in order to validate the efficiency of the method, so that it can be replicated by other industries and start to be better improved.

Among these studies applied in real production systems, the models [34, 35] used the genetic algorithm as a resolution method, while the rest used other heuristic methods of resolution. Moreover, these authors still point out the need for an approach that uses methods able to adapt to different scenarios. These applied cases brought relevant KPI's improvements. In Table 3, we can see the main gains of the models applied in a real scenario.

**Table 3.** Gains from applying the method.

| Reference | Results |
|---|---|
| [7] | Reduction in general production costs, because rescheduling efforts, rework and scrap, as well as inventory were reduced |
| [17] | Costs reduced by up to 63.5% compared to the company's schedule |
| [28] | Information for the managers to be able to make decisions, the paper did not bring an exact result |
| [33] | Charts to show that, by applying the method, costs can be minimized and production maximized |
| [34] | Limited their study in an injection company related to the maintenance of these molds, obtaining a 44.9% reduction of the production delay |
| [35] | The results obtained validate the efficiency of the new heuristic method |
| [36] | Minimization of operating costs and maximization of the company's profit |
| [37] | It allows you to decide which products need to be produced and the exact amount of production in each period and the best date for maintenance can be set. Production quantities were also influenced by the cost of inventory and the cost of back orders for different products |
| [38] | 80% savings. The numerical example indicates that an integrated model proposal performs better than stand-alone models |

### 4.3 What Are the Main Directions for Future Research?

52 out of the 69 papers present future proposals for research to be developed. Many authors refer to the need for future research to focus on real cases for practical validation [5, 31, 41, 43, 47, 53, 61, 63, 71, 76, 78].

Also, many papers point to the demand for applications in more complex production scenarios, e.g. with different behavior patterns and production systems that use more than one machine [1, 20, 29, 32, 33, 35, 37–41, 50, 63, 66, 68, 74, 75]. Such demand was also observed in our systematic review, as explored in the first research question of this work. In [5, 63] the model developed is limited to a scenario with deterministic behavior; thus, the authors report the need for extension in unplanned and planned environments with stochastic situations.

For [9, 20, 22, 23, 33, 37, 39, 44, 50, 66, 77] a strategic predictive maintenance policy needs to be better used in models so that it can allow the prediction of equipment deterioration in an environment with complexity to optimize machine availability. In [7, 21, 26, 43, 67, 68] the authors state that methods of predicting maintenance failures should be better considered in new models.

For [17, 29, 49, 66], there is a necessity to improve the procedures for choosing and applying heuristics, since different choices may bring different results for a same model. In [20, 27, 45, 64, 73] the authors indicate meta-heuristics investigation for solving their models because it has potential for effective gain. In article [46] the use of optimization-based simulation approach has had good results and has a potential for application in other contexts.

The study by [72] highlights the importance of developing other methods that evaluate the application of genetic algorithm hybridization, since it presents potential to efficiently solve these problems. In [58, 74], the authors encourage the study of more complex methods such as multi-objective optimization of ant colonies to solve new models as well as perform a comparison with Pareto Ant Colony Optimization, in [22, 24], they intend in future research to solve problems with greater complexity using the relaxation method and hybrid considerations. In the same way, the genetic algorithm has potential for new research because it presents good results in initial tests. In summary, we can see in Fig. 5 a summary of the main directions of future research reported here in this section.

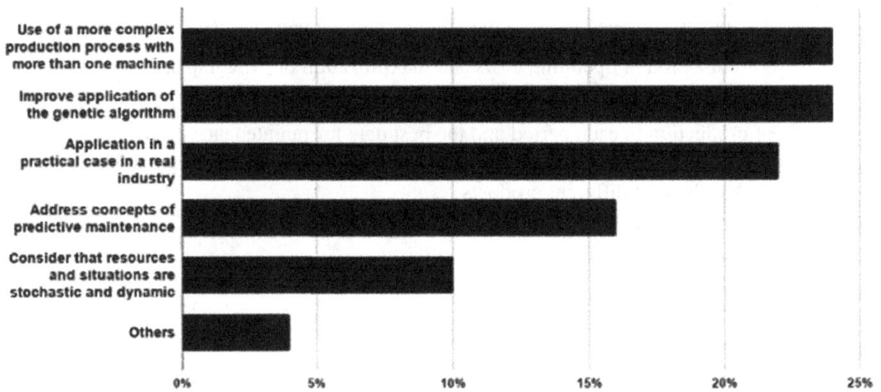


**Fig. 5.** Percentage of the main future researches reported by the authors.

## 5 Discussion

The conducted systematic review addressed the research questions and contributed to a better understanding of the field, considering modelling aspects, applications and research directions.

In summary, by analyzing the directions indicated by the papers in the portfolio and the trend of the papers over time, it is possible to identify some opportunities. Despite the predominance of the use of preventive maintenance for the integration of production and maintenance planning, the development of predictive approaches has increased.

There is extensive use of different heuristics in the optimization problem. The use of hybrid approaches appears as an opportunity to be further developed in this topic. As for the study environment, research that considers the production systems as a whole (most research considers only a single machine), and the application in real environments can still be better developed and only one paper was found that used an optimization-based simulation approach.

Analyzing other aspects and selecting other keywords to apply the literature review methodology of the proposed subject may be a good way, trying to seek other results. The interest in seeking the development of a model that integrates production planning and control with maintenance in a way that can be replicated in various contexts of the productive environment ends up being a challenge for future models.

In a next study by the authors of this paper, the application of these final analyzes in a conceptual and practical context is motivated, aligning these two requirements and being able to bring through the application positive results that may contribute to the subject.

## References


1. Ghaleb, M., Taghipour, S., Sharifi, M., Zolfagharinia, H.: Integrated production and maintenance scheduling for a single degrading machine with deterioration-based failures. Comput. Ind. Eng. **143**, 106432 (2020). https://doi.org/10.1016/j.cie.2020.106432
2. de Souza, N.L.S., Stradioto Neto, L.A., Rossato, I.D.F., da Silva, B.S., Henkes, J.A.: Aplicação de ferramenta da indústria 4.0 em um caso com alguns cenários nacionais e internacionais. Revista Gestão & Sustentabilidade Ambiental **9**(2), 120 (2020). https://doi.org/10.19177/rgsa.v9e22020120-140
3. Frazzon, E.M., Kück, M., Freitag, M.: Data-driven production control for complex and dynamic manufacturing systems. CIRP Ann. **67**(1), 515–518 (2018). https://doi.org/10.1016/j.cirp.2018.04.033
4. de Souza, L.S., Rossato, I.D.F., Henkes, J.A.: Uma análise das estratégias de produção mais limpa e eficiência energética em uma indústria de equipamentos odontológicos. Revista Gestão & Sustentabilidade Ambiental **8**(3), 639 (2019). https://doi.org/10.19177/rgsa.v8e32019639-659
5. Boudjelida, A.: On the robustness of joint production and maintenance scheduling in presence of uncertainties. J. Intell. Manuf. **30**(4), 1515–1530 (2017). https://doi.org/10.1007/s10845-017-1303-9
6. Alimian, M., Ghezavati, V., Tavakkoli-Moghaddam, R.: New integration of preventive maintenance and production planning with cell formation and group scheduling for dynamic cellular manufacturing systems. J. Manuf. Syst. **56**, 341–358 (2020). https://doi.org/10.1016/j.jmsy.2020.06.011
7. Glawar, R., et al.: An approach for the integration of anticipative maintenance strategies within a production planning and control model. Proc. CIRP **67**, 46–51 (2018). http://dx.doi.org/10.1016/j.procir.2017.12.174
8. Aghezzaf, E.-H., Khatab, A., Tam, P.L.: Optimizing production and imperfect preventive maintenance planning's integration in failure-prone manufacturing systems. Reliab. Eng. Syst. Saf. **145**, 190–198 (2016). https://doi.org/10.1016/j.ress.2015.09.017

9. Wang, L., Lu, Z., Ren, Y.: Integrated production planning and condition-based maintenance considering uncertain demand and random failures. Proc. Inst. Mech. Eng. B J. Eng. Manuf. **234**(1–2), 310–323 (2019). https://doi.org/10.1177/0954405419852479
10. Baek, S., Yoon, D.Y., Lim, K.J., Cho, Y.K., Seo, Y.L., Yun, E.J.: The most downloaded and most cited articles in radiology journals: a comparative bibliometric analysis. Eur. Radiol. **28**(11), 4832–4838 (2018). https://doi.org/10.1007/s00330-018-5423-1
11. Moher, D., Liberati, A., Tetzlaff, J., Altman, D.G.: Preferred reporting items for systematic reviews and meta-analyses: the PRISMA statement. Int. J. Surg. **8**(5), 336–341 (2010). https://doi.org/10.1016/j.ijsu.2010.02.007
12. Fink, A.: Conducting Research Literature Reviews: From the Internet to Paper. Sage Publications, Los Angeles (2019)
13. Triska, Y., Frazzon, E.M., Silva, V.M.D.: Proposition of a simulation-based method for port capacity assessment and expansion planning. Simul. Model. Pract. Theory **103**, 102098 (2020). https://doi.org/10.1016/j.simpat.2020.102098
14. Uhlmann, I.R., Frazzon, E.M.: Production rescheduling review: Opportunities for industrial integration and practical applications. J. Manuf. Syst. **49**, 186–193 (2018). https://doi.org/10.1016/j.jmsy.2018.10.004
15. Liao, Y., Deschamps, F., de Freitas Rocha Loures, E., Ramos, L.F.P.: Past, present and future of Industry 4.0 - a systematic literature review and research agenda proposal. Int. J. Prod. Res. **55**(12), 3609–3629 (2017). https://doi.org/10.1080/00207543.2017.1308576
16. Aria, M., Cuccurullo, C.: Bibliometrix: an R-tool for comprehensive science mapping analysis. J. Informetrics **11**(4), 959–975 (2017). https://doi.org/10.1016/j.joi.2017.08.007
17. Chansombat, S., Pongcharoen, P., Hicks, C.: A mixed-integer linear programming model for integrated production and preventive maintenance scheduling in the capital goods industry. Int. J. Prod. Res. **57**(1), 61–82 (2018). https://doi.org/10.1080/00207543.2018.1459923
18. Hajej, Z., Rezg, N., Gharbi, A.: Quality issue in forecasting problem of production and maintenance policy for production unit. Int. J. Prod. Res. **56**(18), 6147–6163 (2018). https://doi.org/10.1080/00207543.2018.1478150
19. Aramon Bajestani, M., Banjevic, D., Beck, J.C.: Integrated maintenance planning and production scheduling with Markovian deteriorating machine conditions. Int. J. Prod. Res. **52**(24), 7377–7400 (2014). https://doi.org/10.1080/00207543.2014.931609
20. Wang, S.: Bi-objective optimisation for integrated scheduling of single machine with setup times and preventive maintenance planning. Int. J. Prod. Res. **51**(12), 3719–3733 (2013). https://doi.org/10.1080/00207543.2013.765070
21. Wang, S., Liu, M.: A branch and bound algorithm for single-machine production scheduling integrated with preventive maintenance planning. Int. J. Prod. Res. **51**(3), 847–868 (2013). https://doi.org/10.1080/00207543.2012.676683
22. Wong, C.S., Chan, F.T.S., Chung, S.H.: A joint production scheduling approach considering multiple resources and preventive maintenance tasks. Int. J. Prod. Res. **51**(3), 883–896 (2013). https://doi.org/10.1080/00207543.2012.677070
23. Wong, C.S., Chan, F.T.S., Chung, S.H.: A genetic algorithm approach for production scheduling with mould maintenance consideration. Int. J. Prod. Res. **50**(20), 5683–5697 (2012). https://doi.org/10.1080/00207543.2011.613868
24. Najid, N.M., Alaoui-Selsouli, M., Mohafid, A.: An integrated production and maintenance planning model with time windows and shortage cost. Int. J. Prod. Res. **49**(8), 2265–2283 (2010). https://doi.org/10.1080/00207541003620386
25. Zied, H., Sofiene, D., Nidhal, R.: Optimal integrated maintenance/production policy for randomly failing systems with variable failure rate. Int. J. Prod. Res. **49**(19), 5695–5712 (2011). https://doi.org/10.1080/00207543.2010.528063

26. Kouedeu, A.F., Kenné, J.-P., Dejax, P., Songmene, V., Polotski, V.: Production and maintenance planning for a failure-prone deteriorating manufacturing system: a hierarchical control approach. Int. J. Adv. Manuf. Technol. **76**(9–12), 1607–1619 (2014). https://doi.org/10.1007/s00170-014-6175-y
27. Berrichi, A., Yalaoui, F.: Efficient bi-objective ant colony approach to minimize total tardiness and system unavailability for a parallel machine scheduling problem. Int. J. Adv. Manuf. Technol. **68**(9–12), 2295–2310 (2013). https://doi.org/10.1007/s00170-013-4841-0
28. Pan, E., Liao, W., Xi, L.: A joint model of production scheduling and predictive maintenance for minimizing job tardiness. Int. J. Adv. Manuf. Technol. **60**(9–12), 1049–1061 (2011). https://doi.org/10.1007/s00170-011-3652-4
29. Pan, E., Liao, W., Xi, L.: Single-machine-based production scheduling model integrated preventive maintenance planning. Int. J. Adv. Manuf. Technol. **50**(1–4), 365–375 (2010). https://doi.org/10.1007/s00170-009-2514-9
30. Yulan, J., Zuhua, J., Wenrui, H.: Multi-objective integrated optimization research on preventive maintenance planning and production scheduling for a single machine. Int. J. Adv. Manuf. Technol. **39**(9–10), 954–964 (2007). https://doi.org/10.1007/s00170-007-1268-5
31. Cassady, C.R., Kutanoglu, E.: Integrating preventive maintenance planning and production scheduling for a single machine. IEEE Trans. Reliab. **54**(2), 304–309 (2005). https://doi.org/10.1109/tr.2005.845967
32. Cassady, C.R., Kutanoglu, E.: Minimizing job tardiness using integrated preventive maintenance planning and production scheduling. IIE Trans. **35**(6), 503–513 (2003). https://doi.org/10.1080/07408170304416
33. Hajej, Z., Rezg, N., Askri, T.: Joint optimization of capacity, production and maintenance planning of leased machines. J. Intell. Manuf. **31**(2), 351–374 (2018). https://doi.org/10.1007/s10845-018-1450-7
34. Batubara, S., Marie, I.A., Cattelya, F.H.: Minimizing mean weighted expected tardiness by integrating preventive maintenance and production scheduling using genetic algorithm. In: International Conference on Industrial Engineering and Operations Management, pp. 3242–3253, Bangkok, Thailand (2019)
35. Fnaiech, N., Fitouri, C., Varnier, C., Fnaiech, F., Zerhouni, N.: A new heuristic method for solving joint job shop scheduling of production and maintenance. IFAC-PapersOnLine **48**(3), 1802–1808 (2015). https://doi.org/10.1016/j.ifacol.2015.06.348
36. Liu, X., Wang, W., Zhang, T., Zhai, Q., Peng, R.: An integrated non-cyclical preventive maintenance and production planning model for a multi-product production system. In: 2015 IEEE International Conference on Industrial Engineering and Engineering Management (IEEM) (2015). https://doi.org/10.1109/ieem.2015.7385696
37. Liu, X., Wang, W., Peng, R.: An integrated production and delay-time based preventive maintenance planning model for a multi-product production system. Eksploatacja i Niezawodnosc – Maint. Reliab. **17**(2), 215–221 (2015). https://doi.org/10.17531/ein.2015.2.7
38. Pandey, D., Kulkarni, M.S., Vrat, P.: A methodology for joint optimization for maintenance planning, process quality and production scheduling. Comput. Indus. Eng. **61**(4), 1098–1106 (2011). http://dx.doi.org/10.1016/j.cie.2011.06.023
39. Chen, X., An, Y., Zhang, Z., Li, Y.: An approximate nondominated sorting genetic algorithm to integrate optimization of production scheduling and accurate maintenance based on reliability intervals. J. Manuf. Syst. **54**, 227–241 (2020). https://doi.org/10.1016/j.jmsy.2019.12.004
40. Xu, S., Dong, W., Jin, M., Wang, L.: Single-machine scheduling with fixed or flexible maintenance. Comput. Ind. Eng. **139**, 106203 (2020). https://doi.org/10.1016/j.cie.2019.106203
41. Shalaby, M.F., Gadallah, M.H., Almokadem, A.: Optimization of production, maintenance and inspection decisions under reliability constraints. J. Eng. Sci. Technol. **14**(6), 3551–3568 (2019)

42. Ao, Y., Zhang, H., Wang, C.: Research of an integrated decision model for production scheduling and maintenance planning with economic objective. Comput. Ind. Eng. **137**, 106092 (2019). https://doi.org/10.1016/j.cie.2019.106092
43. Rivera-Gómez, H., Lara, J., Montaño-Arango, O., Hernández-Gress, E.S., Corona-Armenta, J.R., Santana-Robles, F.: Joint production and repair efficiency planning of a multiple deteriorating system. Flex. Serv. Manuf. J. **31**(2), 446–471 (2018). https://doi.org/10.1007/s10696-018-9313-2
44. Wang, L., Lu, Z., Han, X.: Joint optimal production planning and proactive maintenance policy for a system subject to degradation. J. Qual. Maint. Eng. **25**(2), 236–252 (2019). https://doi.org/10.1108/jqme-11-2016-0068
45. Alimian, M., Saidi-Mehrabad, M., Jabbarzadeh, A.: A robust integrated production and preventive maintenance planning model for multi-state systems with uncertain demand and common cause failures. J. Manuf. Syst. **50**, 263–277 (2019). https://doi.org/10.1016/j.jmsy.2018.12.001
46. Rahmati, S.H.A., Ahmadi, A., Karimi, B.: Developing simulation based optimization mechanism for novel stochastic reliability centered maintenance problem. Scientia Iranica (2017). https://doi.org/10.24200/sci.2017.4461
47. Guiras, Z., Hajej, Z., Rezg, N., Dolgui, A.: Comparative analysis of heuristic algorithms used for solving a Production and Maintenance Planning Problem (PMPP). Appl. Sci. **8**(7), 1088 (2018). https://doi.org/10.3390/app8071088
48. Rivera-Gómez, H., Gharbi, A., Kenné, J.-P., Montaño-Arango, O., Hernández-Gress, E.S.: Subcontracting strategies with production and maintenance policies for a manufacturing system subject to progressive deterioration. Int. J. Prod. Econ. **200**, 103–118 (2018). https://doi.org/10.1016/j.ijpe.2018.03.004
49. Ettaye, G., El Barkany, A., Jabri, A., El Khalfi, A.: Optimizing the integrated production and maintenance planning using genetic algorithm. Int. J. Eng. Bus. Manage. **10**, 184797901877326 (2018). https://doi.org/10.1177/1847979018773260
50. Mehdizadeh, E., Aydin Atashi Abkenar, A.: Preventive maintenance effect on the aggregate production planning model with tow-phase production systems: modeling and solution methods. Eng. Rev. Međunarodni časopis namijenjen publiciranju originalnih istraživanja s aspekta analize konstrukcija, materijala i novih tehnologija u području strojarstva, brodogradnje, temeljnih tehničkih znanosti, elektrotehnike, računarstva i građevinarstva **38**(1), 30–50 (2018)
51. Biondi, M., Sand, G., Harjunkoski, I.: Optimization of multipurpose process plant operations: a multi-time-scale maintenance and production scheduling approach. Comput. Chem. Eng. **99**, 325–339 (2017). https://doi.org/10.1016/j.compchemeng.2017.01.007
52. Jing, Z., Hua, J., Yi, Z.: Multi-objective integrated optimization problem of preventive maintenance planning and flexible job-shop scheduling. In: Proceedings of the 23rd International Conference on Industrial Engineering and Engineering Management 2016, pp. 137–141. Atlantis Press, Paris (2017)
53. Zahedi, Rojali, Yusriski, R.: Stepwise optimization for model of integrated batch production and maintenance scheduling for single item processed on flow shop with two machines in JIT environment. Proc. Comput. Sci. **116**, 408–420 (2017). https://doi.org/10.1016/j.procs.2017.10.081
54. Le Tam, P., Aghezzaf, E.H., Khatab, A., Hieu Le, C.: Integrated production and imperfect preventive maintenance planning: an effective MILP-based relax-and-fix/fix-and-optimize method. In: 6th International Conference on Operations Research and Enterprise Systems (ICORES 2017), pp. 483–490. SCITEPRESS–Science and Technology Publications, Lda (2017)

55. Vieira, M., Liu, S., Pinto-Varela, T., Barbosa-Póvoa, A.P., Papageorgiou, L.G.: Optimisation of maintenance planning into the production of biopharmaceuticals with performance decay using a continuous-time formulation. In: 26th European Symposium on Computer Aided Process Engineering, pp. 1749–1754 (2016). https://doi.org/10.1016/b978-0-444-63428-3.50296-4
56. Kumar, S., Lad, B.K.: Effect of maintenance resource constraints on flow-shop environment in a joint production and maintenance context. In: 2016 IEEE International Conference on Industrial Engineering and Engineering Management (IEEM) (2016). https://doi.org/10.1109/ieem.2016.7797954
57. Liao, W., Zhang, X., Jiang, M.: An optimization model integrated production scheduling and preventive maintenance for group production. In: 2016 IEEE International Conference on Industrial Engineering and Engineering Management (IEEM) (2016). https://doi.org/10.1109/ieem.2016.7798015
58. Lee, S., Prabhu, V.V.: A dynamic algorithm for distributed feedback control for manufacturing production, capacity, and maintenance. IEEE Trans. Autom. Sci. Eng. **12**(2), 628–641 (2015). https://doi.org/10.1109/tase.2014.2339281
59. Chen, X., Xiao, L., Zhang, X.: A production scheduling problem considering random failure and imperfect preventive maintenance. Proc. Inst. Mech. Eng. O J. Risk Reliab. **229**(1), 26–35 (2014). https://doi.org/10.1177/1748006x14545834
60. Assid, M., Gharbi, A., Hajji, A.: Production planning and opportunistic preventive maintenance for unreliable one-machine two-products manufacturing systems. IFAC-PapersOnLine **48**(3), 478–483 (2015). https://doi.org/10.1016/j.ifacol.2015.06.127
61. Fakher, H.B., Nourelfath, M., Gendreau, M.: Hybrid genetic algorithm to solve a joint production maintenance model. IFAC-PapersOnLine **48**(3), 747–754 (2015). https://doi.org/10.1016/j.ifacol.2015.06.172
62. Ho, V.T., Hajej, Z., Le Thi, H.A., Rezg, N.: Solving the production and maintenance optimization problem by a global approach. In: Modelling, Computation and Optimization in Information Systems and Management Sciences, pp. 307–318. Springer, Cham (2015)
63. Chen, X., Xiao, L., Zhang, X., Xiao, W., Li, J.: An integrated model of production scheduling and maintenance planning under imperfect preventive maintenance. Eksploatacja i Niezawodność, 17, 70–79 (2015)
64. Yalaoui, A., Chaabi, K., Yalaoui, F.: Integrated production planning and preventive maintenance in deteriorating production systems. Inf. Sci. **278**, 841–861 (2014). https://doi.org/10.1016/j.ins.2014.03.097
65. Fitouhi, M.-C., Nourelfath, M.: Integrating noncyclical preventive maintenance scheduling and production planning for multi-state systems. Reliab. Eng. Syst. Saf. **121**, 175–186 (2014). https://doi.org/10.1016/j.ress.2013.07.009
66. Pan, E., Liao, W., Xi, L.: A single machine-based scheduling optimisation model integrated with preventive maintenance policy for maximising the availability. Int. J. Ind. Syst. Eng. **10**(4), 451 (2012). https://doi.org/10.1504/ijise.2012.046301
67. Nourelfath, M., Châtelet, E.: Integrating production, inventory and maintenance planning for a parallel system with dependent components. Reliab. Eng. Syst. Saf. **101**, 59–66 (2012). https://doi.org/10.1016/j.ress.2012.02.001
68. Fitouhi, M.-C., Nourelfath, M.: Integrating noncyclical preventive maintenance scheduling and production planning for a single machine. Int. J. Prod. Econ. **136**(2), 344–351 (2012). https://doi.org/10.1016/j.ijpe.2011.12.021
69. Uzun, A., Ozdogan, A.: Maintenance parameters based production policies optimization. J. Qual. Maint. Eng. **18**(3), 295–310 (2012). https://doi.org/10.1108/13552511211265884
70. Reza Golmakani, H., Namazi, A.: Multiple-route job shop scheduling with fixed periodic and age-dependent preventive maintenance to minimize makespan. J. Qual. Maint. Eng. **18**(1), 60–78 (2012). https://doi.org/10.1108/13552511211226193

71. Benmansour, R., Allaoui, H., Artiba, A., Iassinovski, S., Pellerin, R.: Simulation-based approach to joint production and preventive maintenance scheduling on a failure-prone machine. J. Qual. Maint. Eng. **17**(3), 254–267 (2011). https://doi.org/10.1108/13552511111157371
72. Berrichi, A., Yalaoui, F., Amodeo, L., Mezghiche, M.: Bi-Objective Ant Colony Optimization approach to optimize production and maintenance scheduling. Comput. Oper. Res. **37**(9), 1584–1596 (2010). https://doi.org/10.1016/j.cor.2009.11.017
73. Nourelfath, M., Fitouhi, M.-C., Machani, M.: An integrated model for production and preventive maintenance planning in multi-state systems. IEEE Trans. Reliab. **59**(3), 496–506 (2010). https://doi.org/10.1109/tr.2010.2056412
74. Berrichi, A., Amodeo, L., Yalaoui, F., Châtelet, E., Mezghiche, M.: Bi-objective optimization algorithms for joint production and maintenance scheduling: application to the parallel machine problem. J. Intell. Manuf. **20**(4), 389–400 (2008). https://doi.org/10.1007/s10845-008-0113-5
75. Jin, Y.-L., Jiang, Z.-H., Hou, W.-R.: Integrating flexible-interval preventive maintenance planning with production scheduling. Int. J. Comput. Integr. Manuf. **22**(12), 1089–1101 (2009). https://doi.org/10.1080/09511920903207449
76. Leng, K., Ren, P., Gao, L.: A novel approach to integrated preventive maintenance planning and production scheduling for a single machine using the chaotic particle swarm optimization algorithm. In: 2006 6th World Congress on Intelligent Control and Automation (2006). https://doi.org/10.1109/wcica.2006.1713491
77. Kianfar, F.: A numerical method to approximate optimal production and maintenance plan in a flexible manufacturing system. Appl. Math. Comput. **170**(2), 924–940 (2005). https://doi.org/10.1016/j.amc.2004.12.030
78. Sortrakul, N., Nachtmann, H.L., Cassady, C.R.: Genetic algorithms for integrated preventive maintenance planning and production scheduling for a single machine. Comput. Ind. **56**(2), 161–168 (2005). https://doi.org/10.1016/j.compind.2004.06.005
79. Vassiliadis, C.G., Vassiliadou, M.G., Papageorgiou, L.G., Pistikopoulos, E.N.: Simultaneous maintenance considerations and production planning in multi-purpose plants. In: Annual Reliability and Maintainability Symposium, 2000 Proceedings, International Symposium on Product Quality and Integrity (Cat. No.00CH37055) (n.d.). https://doi.org/10.1109/rams.2000.816312

# Future Research Agenda to Understanding the Sustainable Business Model in Industry 4.0

Grazielle Fatima Gomes Teixeira, Osiris Canciglieri Junior(✉), and Anderson Luis Szejka

Polytechnic School of the Pontifical Catholic University of Paraná – PUCPR, R. Imaculada Conceição, 1155 - Prado Velho, Curitiba CEP, 80215-901, PR, Brazil
osiris.canciglieri@pucpr.br

**Abstract.** All over the world, Industry 4.0 (I4.0) and Sustainable Development (SD) have progressively gained the interest of scholars, politicians, and other parts of society. Besides beings two of the most debated topics of the last decades; they also have overlaps between their independent research fields. Some examples are reductions of environmental impacts and improvements in production technologies. This integration of technologies and sustainable advances within an industrial context can enable a set of important competitiveness forces, which results can reflect in business improvement . However, the link between I4.0, SD, and business still needs a broader understanding. Basing on this perspective, this paper proposes a systematic literature review (2015–2020) to identify the current state of research on the subject by mapping and summarising existing research efforts, as well as identifying research agendas, gaps, and opportunities for more development. Results point to that there are fourteen research opportunities, showing the potential of Industry 4.0 as an enabler of sustainable business models that changes the responsibilities of companies.

**Keywords:** Industry 4.0 · 4th industrial revolution · Sustainability · Business model · Systematic literature review

## 1 Introduction

The overlap between Sustainable Development (SD) or Sustainability and Industry 4.0 (I4.0) or also called the fourth industrial revolution has been one of the most important industrial debates in recent years [1–9]. But, although the I4.0 revolution can be described as a facilitator for sustainable development, this union remains underdeveloped in many aspects [3, 5, 6, 10–13]. One of these aspects is about Sustainable Business Models (SBM) into the I4.0 revolution [1–4, 14].

The new era of smart production can provide companies with a more aggressive attitude to embrace opportunities. As pointed by Porter and Heppelmann [14], the changing nature of products will force many companies questioning what business they doing because new products raise a new set of strategic choices related to how value is created and captured, in other words, it redefines the value chains and business models [1, 6, 8, 10, 14, 15].

© Springer Nature Switzerland AG 2021
D. A. Rossit et al. (Eds.): ICPR-Americas 2020, CCIS 1407, pp. 357–371, 2021.
https://doi.org/10.1007/978-3-030-76307-7_27

Based on this observation, the study present here was structured into three questions (Q). The set of questions increase understanding about how two of the most debated topics of the last decades are being discussed now to support future research agenda (Q1); mapping the field (Q2), and, finally, showing the chronological evolution (Q3). The questions are as follows: (1) What are the opportunities for future research agenda?; (2) Where this topic is being investigated?; (3) What is the current status of this research field?.

## 2 Conceptions Description

The study is based on three main concepts, I4.0, SD, and SBM. In this section, these concepts are presented.

### 2.1 The Connection Between Industry 4.0 and Sustainability

Sustainability does not have a unique definition [16]. Due to this, concrete implementation is considered difficult because there is a high degree of complexity regarding the depth and specifications of the fields of action [3, 16]. However, the most common definition cited was presented by the Brundtland Report, which describes sustainable development as "the ability to meet the needs of the present without compromising the ability of future generations to meet their own needs" [17]. Besides this definition, another highly referenced concept is the Triple-Bottom-Line (TBL), whose objective is to operationalize sustainability through the balance between economic profitability, respect for the environment, and social responsibility [18]. The common meaning among these different approaches is the general scope to support the increase of the companies' competitive capacity [19] and improve the use of their resource productivity [20].

Likewise, I4.0 is also a paradigm whose unique definition is not possible due to its wide range of approaches [2, 21, 22]. Most definitions consider I4.0 as a digital technological advance to make industrial production systems more intelligent, connected, and capable of decentralizing production [2]. Some practical examples of new technologies of revolution 4.0 used in the industry are: big data and analytics, Internet of Things (Internet of Things (IoT)), augmented and virtual reality, simulation, robots and autonomous vehicles, additive manufacturing, cloud, cybersecurity, among others.

According to Kuik and Diong [23], I4.0 refers to extracting real-time data and developing new and innovative information systems (IS) through the use of computer and/or sensing devices as well as information communication tools. In the same way, Piccarozzi et al. [22] state that I4.0 is based on the horizontal and vertical integration of production systems driven by real-time data interchange and flexible manufacturing to enable customized production.

Many publications on I4.0 and sustainability are dedicated to the economic and environmental dimensions [3, 21]. Although I4.0 also has enormous potential for the social dimension [6]. In the economic dimension, most of the literary contributions of I.40 to sustainability relate to the proposal of new technologies to reduce manufacturing costs through systems integration [6]. In the environmental dimension, most of the literary contributions of I.40 to sustainability are related to improving the standard of living of the society and increasing the quality of production without causing damage to the environment [6]. In the social dimension, most of the literary contributions of I.40 to sustainability are related to predicting better conditions and beneficial opportunities for workers [11].

The predictive use of production data provide by I4.0 toward more sustainable industrial value creation will change the ability to make planning production for competition between non-sustainable and sustainable products [6, 8, 23, 24]. Therefore, is also highlights the impact of this union in business management models and firms' main components and the importance of more studies with this domain [22].

### 2.2 Sustainable Business Model (SBM)

A new business mindset can generate many advantages and challenges. When sustainability is added to the core of the business model concept, it can systematically change the purpose of the company [25]. In this regard, SBM offers opportunities for companies to rethink their process of value creation, product production, pro-active multi-stakeholder management, and long-term perspective [1, 7, 26].

At the same time, SBM can be limited by the introduction of incremental or radical innovations [1]. According to Inigo et al. [27], evolutionary approaches, based on incremental sustainability innovations, adjust the creation of value to respond gradually to market needs. On the other hand, radical innovations introducing new value propositions to address a new sustainability issue. This way can result in the creation of an unprecedented market segment.

Among diverse SBM definitions, Schaltegger et al. [28] argue that a business model for sustainability "describe, analyze, manage, and communicate (i) a company's sustainable value proposition to its customers, and all other stakeholders, (ii) how it creates and delivers this value, (iii) and how it captures economic value while maintaining or regenerating natural, social, and economic capital beyond its organizational boundaries". On the other side, it is possible to find SBM definitions based on Brundtland Report as in Garetti and Taisch [29] that describes SBM as the preservation of the environment, while continuing to improve the quality of human life. Dyllick and Hockerts [30] view SBM as the meet the needs of a firm's direct and indirect stakeholders without compromising its ability to meet the needs of the future. Similarly, Bansal and DesJardine [31] argue that SBM is the ability of firms to respond to their short-term financial needs without compromising their (or others') ability to meet their future needs.

## 3 Review Methodology

To better identify the relationship between I4.0, SD, and SBM, a Systematic Literature Review (SLR) was implemented according to Tranfield et al. [32] procedures. A literature review is essential to strengthening the field of study [32]. As scientific inquiries, the SLR should be a transparent, valid, reliable, and replicable method for identifying, describing, evaluating, and synthesizing evidence related to a specific topic [32]. The SLR rigors ensure that the result found has scientific validity since it describing the information with the least possible distortion and errors; also, because it has the legitimacy to improve knowledge development and make decisions [33].

To secure the validity and transparent rigors of SLR, specific databases were selected: Scopus and Web of Science. These databases were used because contains a significant number of internationally renowned publications, like Springer, Elsevier, Emerald, and Taylor and Francis. The review process considered only formal literature that was: (i) written full-texts in English, (ii) published between May 2015 and May 2020; (iii) peer-review published papers. Then, a structured keyword search was conducted executed through a pairwise query, focusing on titles, abstracts, and keywords.

In each database, 15 different search strings were used. Table 1 reports the search strings and the resulting number of papers in each database. The search found 154 papers from both databases. To refine the results further, duplicate papers were exclusion. The total numbers of articles dropped to 36 papers after reading the abstract according to the review purpose for the final review.

**Table 1.** Search results

| | Keyword pairwise query | Scopus | Web of Science |
|---|---|---|---|
| 1 | TITLE-ABS-KEY ((sustainability AND business model AND industry 4.0)) | 8 | 4 |
| 2 | TITLE-ABS-KEY ((sustainability AND new business model AND industry 4.0)) | 4 | 2 |
| 3 | TITLE-ABS-KEY ((sustainability AND organization improvement AND industry 4.0)) | 1 | 1 |
| 4 | TITLE-ABS-KEY ((sustainability AND maturity model AND industry 4.0)) | 1 | 0 |
| 5 | TITLE-ABS-KEY ((sustainability AND maturity evaluation AND industry 4.0)) | 0 | 0 |
| 6 | TITLE-ABS-KEY ((sustainability AND dynamic capability AND industry 4.0)) | 2 | 0 |
| 7 | TITLE-ABS-KEY ((sustainability AND strateg* AND industry 4.0)) | 17 | 3 |
| 8 | TITLE-ABS-KEY ((sustainability AND competitiveness AND industry 4.0)) | 7 | 1 |
| 9 | TITLE-ABS-KEY ((sustainability AND competition AND industry 4.0)) | 9 | 2 |
| 10 | TITLE-ABS-KEY ((sustainability AND value creation AND industry 4.0)) | 8 | 4 |
| 11 | TITLE-ABS-KEY ((sustainability AND strategic planning AND industry 4.0)) | 0 | 0 |
| 12 | TITLE-ABS-KEY ((sustainability AND product AND development AND process AND industry 4.0)) | 12 | 3 |
| 13 | TITLE-ABS-KEY ((sustainability AND product AND development AND industry 4.0)) | 20 | 3 |
| 14 | TITLE-ABS-KEY ((sustainability AND product AND industry 4.0)) | 33 | 7 |
| 15 | TITLE-ABS-KEY ((sustainability AND continuous improvement AND development AND industry 4.0)) | 1 | 1 |
| | Total in each database | 123 | 31 |
| | Final result | 154 | |

## 4 Results and Analysis

This section made a descriptive analysis of the whole sample. Even at a general level, the database allowed extrapolation of interesting information. This information will be presented according to the order of the questions. Started by (Q1): What are the opportunities for future research agenda?. Not all authors present directions for future research. However, fourteen opportunities for further investigation were identified (see Table 2).

**Table 2.** Future research opportunities

| N | % | Future research directions |
|---|---|---|
| 1° | 16,70% | Apply the study in large-scale |
| 2° | 16,70% | Analyze the different functions of the company in interaction with I4.0 |
| 3° | 10,60% | TBL dimensions in the context of the I4.0 |
| 4° | 9,10% | The impact of I4.0 on supply chain network design |
| 5° | 9,10% | The impact of I4.0 on managing customer channels |
| 6° | 8% | Employee quality |
| 7° | 6% | How Industry 4.0 will change the competition |
| 8° | 6% | How incorporating the I 4.0 into theory about sustainable value propositions |
| 9° | 6% | Qualitative research |
| 10° | 4,55% | Cost-benefit analysis of Industry 4.0 solutions |
| 11° | 3,03% | Policy-making efforts to I4.0 |
| 12° | 1,52% | Transformations in the identity of a manufacturing company |
| 13° | 1,52% | Risk assessment to I4.0 implementation |
| 14° | 1,52% | Ensure data security |

Two of these were the most mentioned: apply the studies in large-scale and analyze the different functions of the company in interaction with I4.0. The first opportunities are regard to test the empirical observations found in each paper in a large-scale study of companies since very small to bigger ones; in the context within and across industries; applied to any business function; into new customized segments; diversified markets; between different cultures and nations; and different methodologies [21, 34–44]. In more, could be investigated its influence on other business models types, such as product-focused and result-focused ones [37, 38].

Analyze the different functions of the company in interaction with I4.0 is an option for future research to observe the shift industry boundaries and theories of the firm within a manufacturing context [1, 7, 34, 35, 45, 46]. More investigations are needed to analyze how can the currently highly production-oriented developments and the data they generate be included in the entire corporate information and decision-making processes [1, 46]. Additionally, future research projects should take into consideration other key variables and outcome to give a broader view on it specific characteristics and requirements that influence the implementation of I4.0, e.g., top management support, metrics of performance, environmental uncertainty, digital servitization, new technologies, absorptive capacity, and so on [6, 21, 26, 37, 38, 43, 45].

Investigations on TBL dimensions in the context of the I4.0 shows that there is still a lack of investigations about the interplaying and conflictive relationships between dimensions [1, 6, 7, 13, 26, 37]. Furthermore, current research emphasizes the need for more studies on the social impacts of I4.0 [3, 21, 26, 47].

Possible contributions to the impact of I4.0 on supply chain network design can analyze how it facilitates reverse logistics to reduce production waste, overproduction, and energy consumption; focusing on production planning, structure, and control of product recycling [4, 6, 42]. Additionally, the resource-sharing aspect of sustainable business models [49].

Future research about the impact of I4.0 on managing customer channels should try to describe it in the creation of sustainable awareness of customers, customer loyalty, and satisfaction for sustainable I4.0 products/services [4, 7, 45].

More detailed investigations should observed work quality between humans and machines regarding which job profiles emerge and which diminish because of the I4.0 [13, 21, 49]. This opportunity for future research is linked to social dimension issues. But, here further research is needed on the issue of accidents involving hazardous chemicals, new tools, and techniques by developing technologies I4.0; and, professional training [6, 42].

A research agenda is also needed to explain how I4.0 will change the competition, especially, between non-sustainable and sustainable products [4]. Is also necessary to verify how it improves the competitiveness of the company over time, i.e., in the long-term [47]. As well as the role of dynamic capabilities in resource reconfiguration and processes it changes require [45]. Moreover, future research should examine how I4.0 transforms bargaining power in different sections of manufacturers [45].

How incorporating the I4.0 into an organizational theory about shape new sustainable value propositions is a gap to further studies close [4, 48]. As well how can organizations integrate, analyze, and exploit data to generate sustainable value-propositions [48]. Novel insights are needed to explore how manufacturers shift their value creation and capture from product-centric to service-centric, and further to data-centric [1, 21, 45].

Quantitative research is another future possibility to check results presented in articles [34, 36, 50], more specific to measure social and ecological impacts [3]. Valuable contributions are needed regarding the cost-benefit analysis of Industry 4.0 solutions to supported sustainable products [4, 45]. In this cost, the transaction should be analyzed downstream and upstream interactions [45].

Literature has lacked studies, since the debate on this theme is recent, on policy-making efforts to foster the development of I4.0 by national and local governments. Potential studies can explore the regulation of political, financial, and fiscal aspects, as well as the provision of infrastructures, services, knowledge, and a skilled work-force [26, 51]. Moreover, based on the scarcity of results of the SLR, there is a need to urgently investigate how I4.0 transforms the identity of a manufacturing company [45]. In the same line, must be better understood the risk assessment to I4.0 implementation, depending on short, medium, and long-term management, different corporate and environmental contexts [50]. Finally, although ensure data security is a point commonly debated in studies about I4.0, few articles in the sample of this SLR highlighted the need to further investigations of it by a sustainable business model perspective; how it would be developed, and what companies think about it [26, 34, 46].

Data to (Q2) Where this topic is being investigated?, considering the authors' affiliations, reveals that there is a strong predominance of German (19,2%) and Italian (18,4%) authors, as shown in Fig. 1. The next predominance countries perceptual contributions are Spain (9,6%), Sweden (7,2%), and Norway (7,2%). The other 17 countries' contributions accounted for about 38,4% of the total number of works published. This shows that the major research contributions are coming from countries with few authors. This means that, despite the dominance of German and Italian authors, this field of research is of interest to several countries.

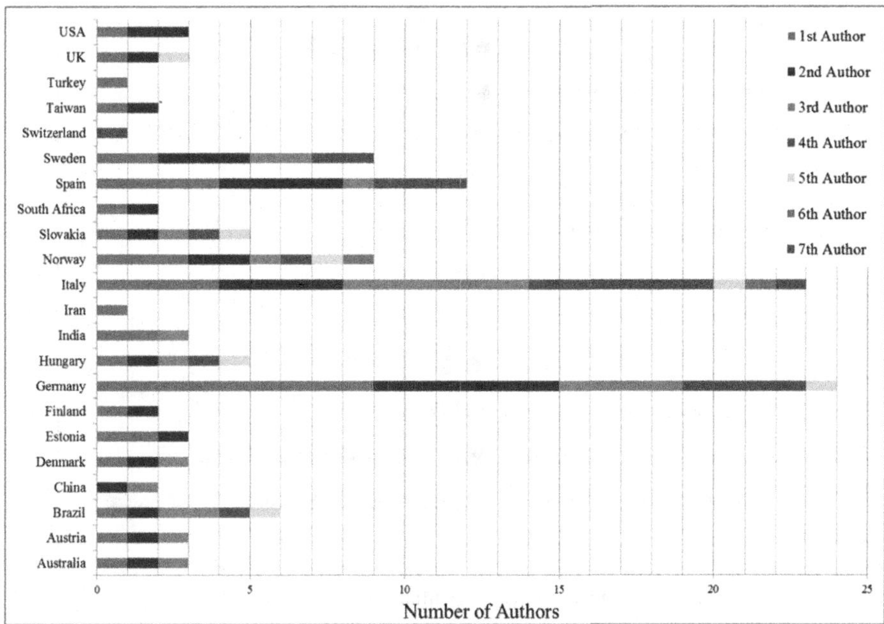


**Fig. 1.** Authors' affiliations countries.

The detail of Fig. 1 also showed that the greatest number of first authors is from Germany, Italy, and Spain. This result is equal to the second authors. Most of the third and fourth authors are form Italy and Germany. Fifth authors predominate with the diversity of countries: Slovakia, Norway, United Kingdom (UK), Italy, Hungary, Germany, and Brazil. Sixth and seventh authors are from Norway and Italy, Spain, and Italy, respectively.

According to the authors' affiliations, Fig. 2 shows from what school each author is. Most of them are researchers in business and management (31,7%), then engineering (23,5%) and mechanical engineering (9,75). Authors' affiliations from companies represent 15,44% in the sample. Most of them work in automotive companies, production systems advisory, financial, and innovation consultancies.

Going into contributions by publishers, Fig. 3 reveals that Elsevier has the highest number of scientific journal titles, with six journals, followed by Springer with three scientific journals title. However, the journal with the highest contribution (number of papers published) is Sustainability from MDPI Publisher (see Fig. 4).

Regarding the quality of these scientific journals, their classification in Scimago was verified to analyze their impact factors. As shown in Fig. 5, most journals have a great impact factor, being classified in Q1 (the best rating). Looking at how many contributions each journal had, the result pointed out that most articles are from journals with a Q2 classification. Besides, it was possible to identify the main topics of each journal according to the description of Scimago (see Fig. 6). The close link between business, I4.0, and engineering proved to be relevant because the most cited subjects were engineering and

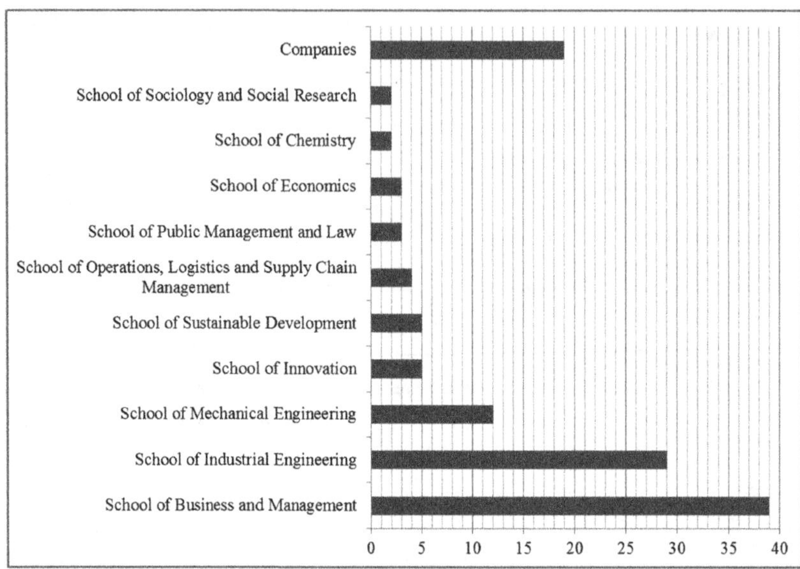


**Fig. 2.** Authors' affiliations school.

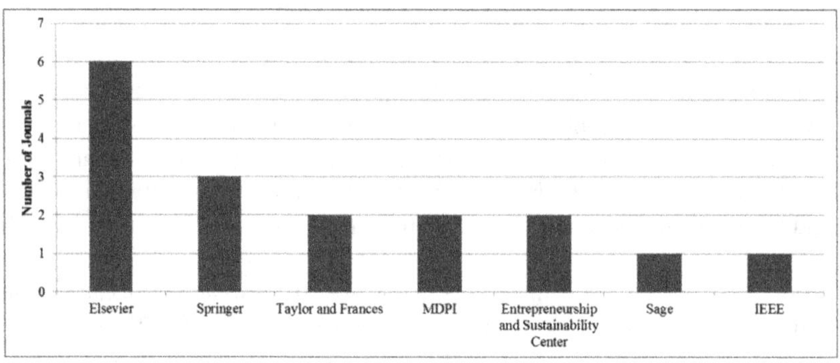


**Fig. 3.** Publisher list.

business, management and accounting; following the three dimensions of the TBL: environmental science, finance, and social sciences. In particular, this emerging link can also be verified by looking at journal titles. One-third of the entire sample deals specifically with the topic of sustainability, but not the other two (Sustainability, Entrepreneurship and Sustainability Issues, Journal of Security and Sustainability Issues, Process Safety and Environmental Protection, Journal of Cleaner Production, International Journal of Precision Engineering and Manufacturing – Green Technology).

The chronologic data to answer Q(3) What is the current status of this research field?. revels that there was an upward trend observed from the year 2015 to 2018 (see Fig. 7). However, the number of papers published from 2019 onward has decreased drastically. It is observed that 69,44% of total papers were published in the year 2015–2018. In

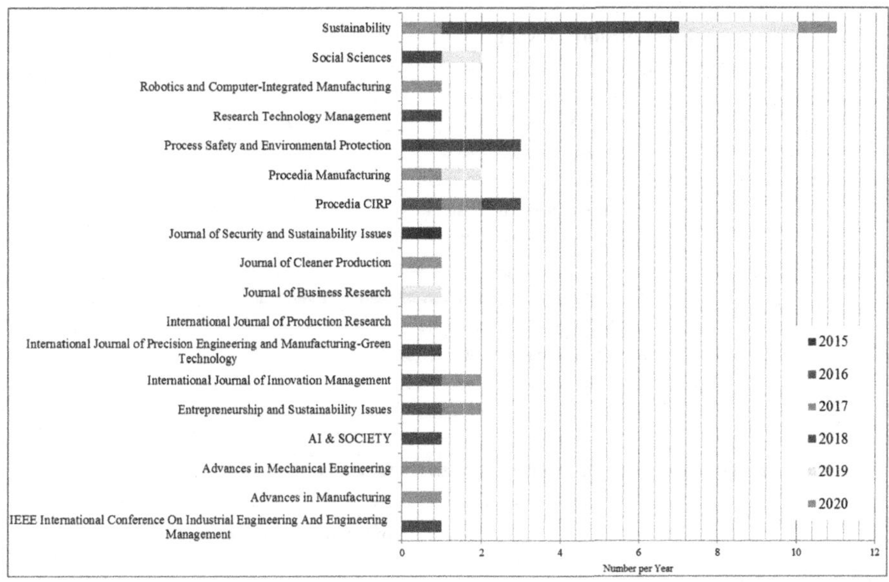


**Fig. 4.** Journal titles list.

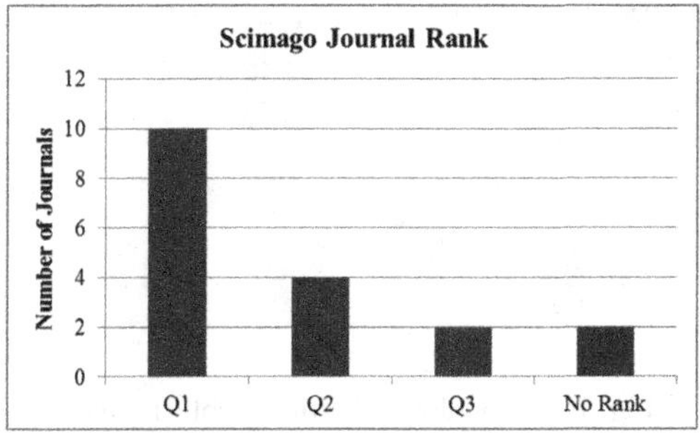


**Fig. 5.** Journal rankings by Scimago.

contrast, 2019 revealed little attention devoted to these topics by the authors. Papers' date to 2020 was published only in April and March. This outcome highlights the need for more studies covering this important link between business model, sustainability, and I4.0.

Another way to observe how is the theme currently is looking at keywords statistics (see Fig. 8). Here, were extracted the most commonly used keywords in all selected papers using the VOSviewer tool [52]. The most frequent keywords used until 2017 were: business model, research agenda, expert interviews, German industry sectors, logistics 4.0, and organizational structures. The most frequently keywords used in 2020 were business model canvas, smart manufacturing, literature review, and enabling technologies.

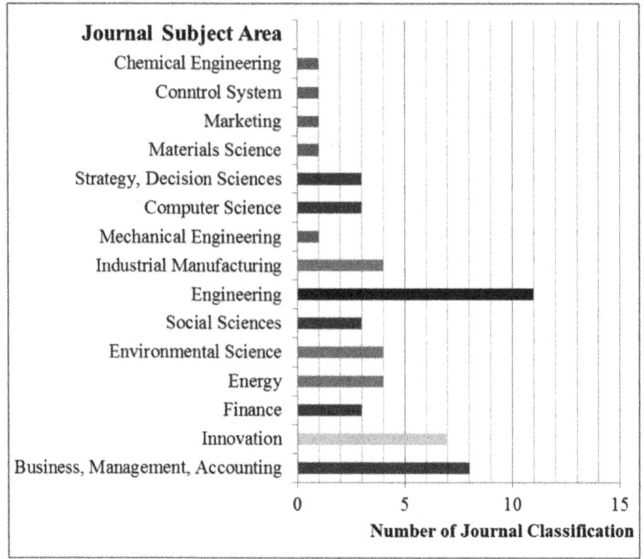


**Fig. 6.** Journal Subject Area list.

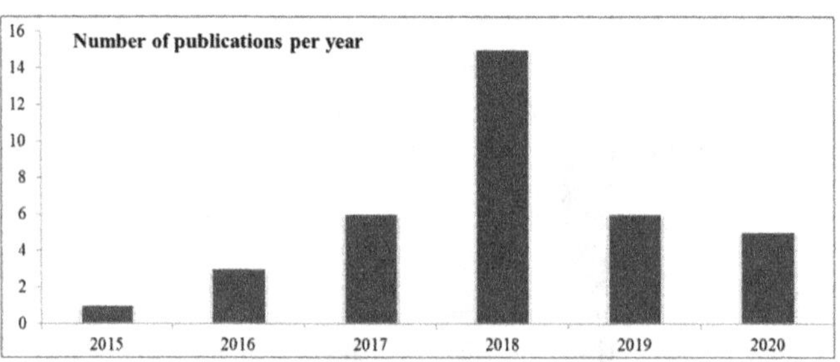


**Fig. 7.** Chronology of article publications.

Finally, Fig. 9 shows the distribution of the selected 36 papers by methodology. The sample is mostly made of surveys and/or interviews (28,5%). Then, systematic literature review (24,4%), model or framework (14,2%), case studies (14,2%), simulations or experimentation or prototypes (10,2%), and empirical contributions (8,16%). One important observation about papers' type of research was the change of type year by year. In the initial years 2015 until 2017, most of the articles were surveys or interviews, case studies, and simulations, experimentation, or prototypes. From 2018 until 2020, the main papers' type of research was a systematic literature review and model or framework. Based on this, it can be said that this recent topic was initially studied from a practical and expert point of view, while in more recent years conceptual and theoretical research has become more frequent.

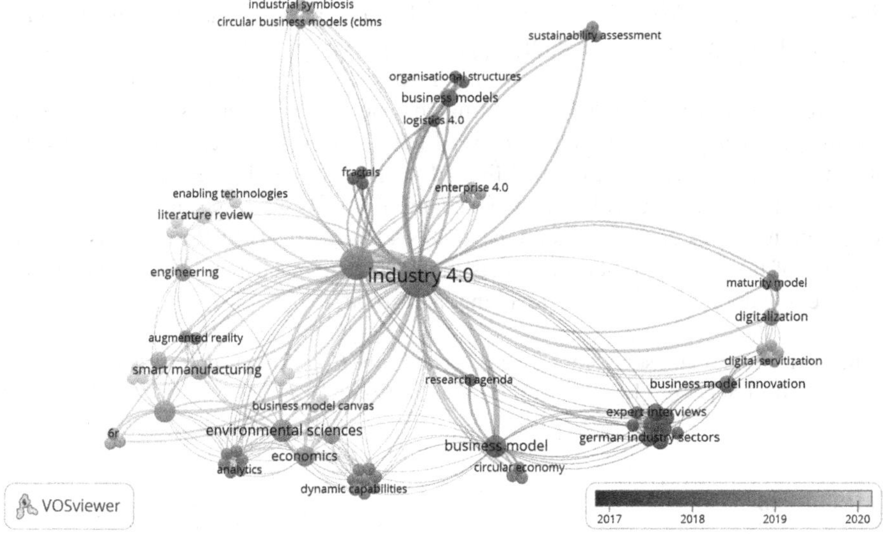


**Fig. 8.** Chronology of article publications by keywords.

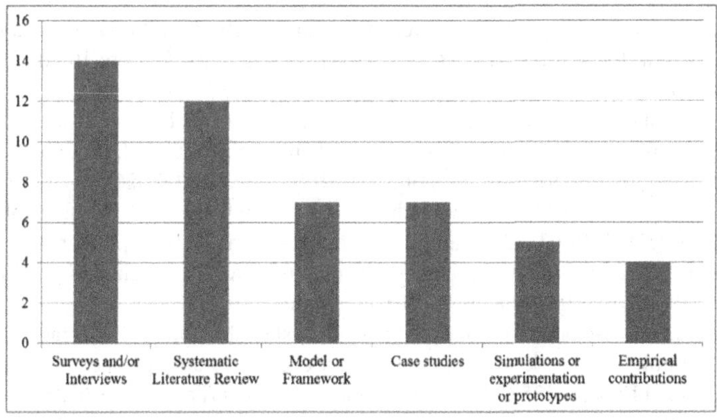


**Fig. 9.** Chronology of article publications by methodologies.

## 5 Conclusion

This study aimed to investigate, through a multidisciplinary lens, the current state of research on the theme I4.0, SD, and SBM performing an SLR. 36 articles were analyzed for this purpose. Based on the limited number of articles analyzed and the initial proposal, this study was not intended to use statistical analysis or to provide a more in-depth analysis of specific topics. However, this review was successful in demonstrating the potential of introducing I4.0 into SBM. We conclude that I4.0 is an enabler for sustainable business models that changes the responsibilities of companies, making them responsible for their entire consumption system, and not only for the production and sale of consumable goods. This crossing of borders introduced a competitive advantage of high technology that initially approached a vision of technocentric evolution for a holistic view. Nevertheless, our results have some limitations due to its qualitative nature. Selected keywords and databases may be one of the limitations of this article, as there may be articles relevant to the scope of the study outside these criteria that were not considered. Finally, the result highlighted the need for further research. A research agenda for the development of sustainable business models were presented in the results section. Future work can also extend the analysis made by this SLR and focus on how the union of the observed themes reflects in the science of business and management, such as company theory, strategic planning, and so on.

**Acknowledgments.** This study was financed in part by the Coordenação de Aperfeiçoamento de Pessoal de Nível Superior - Brasil (CAPES) - Finance Code 001.

## References

1. García-Muiña, F.E., Medina-Salgado, M.S., Ferrari, A.M., Cucchi, M.: Sustainability transition in Industry 4.0 and smart manufacturing with the triple-layered business model canvas. Sustainability **12**, 2364 (2020). https://doi.org/10.3390/su12062364
2. Rosa, P., et al.: Assessing relations between Circular Economy and Industry 4.0: a systematic literature review. Int. J. Prod. Res. **58**, 1662–1687 (2020). https://doi.org/10.1080/00207543.2019.1680896
3. Stock, T., Obenaus, M., Kunz, S., Kohl, H.: Industry 4.0 as enabler for a sustainable development: a qualitative assessment of its ecological and social potential. Process Saf. Environ. Prot. **118**, 254–267 (2018). https://doi.org/10.1016/j.psep.2018.06.026
4. de Man, J.C., Strandhagen, J.O.: An Industry 4.0 research agenda for sustainable business models. Procedia CIRP **63**, 721–726 (2017). https://doi.org/10.1016/j.procir.2017.03.315
5. Ghobakhloo, M.: Industry 4.0, digitization, and opportunities for sustainability. J. Clean. Prod. **252**, 119869 (2020). https://doi.org/10.1016/j.jclepro.2019.119869
6. Kamble, S.S., Gunasekaran, A., Gawankar, S.A.: Sustainable Industry 4.0 framework: a systematic literature review identifying the current trends and future perspectives. Process Saf. Environ. Prot. **117**, 408–425 (2018). https://doi.org/10.1016/j.psep.2018.05.009
7. Kiel, D., Müller, J.M., Arnold, C., Voigt, K.-I.I.: Sustainable industrial value creation: benefits and challenges of Industry 4.0. Int. J. Innov. Manag. **21**, 1–34 (2017). https://doi.org/10.1142/S1363919617400151
8. Stock, T., Seliger, G.: Opportunities of sustainable manufacturing in Industry 4.0. In: Seliger, G., Kohl, H., Mallon, J. (eds.) 13$^{th}$ Global Conference on Sustainable Manufacturing - Decoupling Growth from Resource Use, pp. 536–541 (2016)
9. Peukert, B., et al.: Addressing sustainability and flexibility in manufacturing via smart modular machine tool frames to support sustainable value creation. Procedia CIRP **29**, 514–519 (2015)
10. Bai, C., Dallasega, P., Orzes, G., Sarkis, J.: Industry 4.0 technologies assessment: a sustainability perspective. Int. J. Prod. Econ. **229**, 107776 (2020). https://doi.org/10.1016/j.ijpe.2020.107776
11. Varela, L., et al.: Evaluation of the relation between lean manufacturing, Industry 4.0, and sustainability. Sustain **11**, 1–19 (2019). https://doi.org/10.3390/su11051439
12. Beier, G., Niehoff, S., Xue, B.: More sustainability in industry through industrial internet of things? Appl. Sci. **8**(2), 219 (2018). https://doi.org/10.3390/app8020219

13. Müller, J.M., Voigt, K.-I.: Sustainable industrial value creation in SMEs: a comparison between Industry 4.0 and made in China 2025. Int. J. Precis. Eng. Manuf. Technol. **5**, 659–670 (2018). https://doi.org/10.1007/s40684-018-0056-z
14. Porter, M.E., Heppelmann, J.E.: How smart, connected products are transforming competition. Harv. Bus. Rev. **92**, 64 (2014)
15. Lee, J., Kao, H., Yang, S.: Service innovation and smart analytics for Industry 4.0 and big data environment. Procedia CIRP **16**, 3–8 (2014). https://doi.org/10.1016/j.procir.2014.02.001
16. Teixeira, G.F.G., Canciglieri Junior, O.: How to make strategic planning for corporate sustainability? J. Clean. Prod. **230**, 1421–1431 (2019). https://doi.org/10.1016/j.jclepro.2019.05.063
17. WCED, World Commission on Environment and Development: Report of the World Commission on Environment and Development. Our Common Futuro, Geneva, Switzerland (1987)
18. Elkington, J.: Cannibals with Forks: the Triple Bottom Line of the 21st Century. New Society, Stoney Creek (1997)
19. Brook, J.W., Pagnanelli, F.: Integrating sustainability into innovation project portfolio management e a strategic perspective. J. Eng. Technol. Manag. **34**, 46–62 (2014)
20. Esty, D.C., Porter, M.E.: Industrial ecology and competitiveness: strategic implications for the firm. J. Ind. Ecol. **2**(1), 35–43 (1998)
21. Müller, J.M., et al.: What drives the implementation of Industry 4.0? The role of opportunities and challenges in the context of sustainability. Sustain **10**, 247 (2018). https://doi.org/10.3390/su10010247
22. Piccarozzi, M., Aquilani, B., Gatti, C.: Industry 4.0 in management studies: A systematic literature review. Sustain **10**, 1–24 (2018). https://doi.org/10.3390/su10103821
23. Kuik, S., Diong, L.: A model-driven decision approach to collaborative planning and obsolescence for manufacturing operations. Ind. Manag. Data Syst. **119**, 1926–1946 (2019). https://doi.org/10.1108/IMDS-05-2019-0264
24. Erol, S., Jäger, A., Hold, P., Ott, K., Sihn, W.: Tangible Industry 4.0: a scenario-based approach to learning for the future of production. Procedia CIRP **54**, 13–18 (2016)
25. Porter, M.E., Kramer, M.R.: Creating shared value: how to reinvent capitalism - and unleash a wave of innovation and growth. Harv. Bus. Rev. **89**, 62–77 (2011)
26. Tirabeni, L., De Bernardi, P., Forliano, C., Franco, M.: How can organisations and business models lead to a more sustainable society? A framework from a systematic review of the Industry 4.0. Sustain **11**, 6363 (2019). https://doi.org/10.3390/su11226363
27. Inigo, E.A., Albareda, L., Ritala, P.: Business model innovation for sustainability: Exploring evolutionary and radical approaches through dynamic capabilities. Ind. Innov. **24**, 515–542 (2017)
28. Schaltegger, S., Hansen, E.G., Lüdeke-Freund, F.: Business models for sustainability: origins, present research, and future avenues. Organ. Environ. **29**, 3–10 (2016). https://doi.org/10.1177/1086026615599806
29. Garetti, M., Taisch, M.: Sustainable manufacturing: trends and research challenges. Prod. Plann. Control **23**(2–3), 83–104 (2012). https://doi.org/10.1080/09537287.2011.591619
30. Dyllick, T., Hockerts, K.: Beyond the business case for corporate social responsibility. Bus. Strateg. Environ. **11**, 130–141 (2002). https://doi.org/10.1002/bse.323
31. Bansal, P., DesJardine, M.: Business sustainability: it is about time. Strateg. Organ. **12**, 70–78 (2014). https://doi.org/10.1177/147612701352026
32. Tranfield, D., Denyer, D., Smart, P.: Towards a methodology for developing evidence-informed management knowledge by means of systematic review. Br. J. Manage. **14**(3), 207–222 (2003). https://doi.org/10.1111/1467-8551.00375
33. Webster, J., Watson, R.T.: Analyzing the past to prepare for the future : writing a literature review. Mis. Q. **26**, xiii–xxiii (2002). https://doi.org/10.5465/amr.1989.4308371

34. Arnold, C., Kiel, D., Voigt, K.I.: How the industrial internet of things changes business models in different manufacturing industries. Int. J. Innov. Manag. **20**, 1–25 (2016). https://doi.org/10.1142/S1363919616400156
35. Gerlitz, L.: Design management as a domain of smart and sustainable enterprise: business modelling for innovation and smart growth in Industry 4.0. Entrep. Sustain Issues **3**, 244–268 (2016). https://doi.org/10.9770/jesi.2016.3.3(3)
36. Strandhagen, J.O., et al.: Logistics 4.0 and emerging sustainable business models. Adv. Manuf. **5**, 359–369 (2017). https://doi.org/10.1007/s40436-017-0198-1
37. Bressanelli, G., Adrodegari, F., Perona, M., Saccani, N.: Exploring how usage-focused business models enable circular economy through digital technologies. Sustainability **10**, 639 (2018). https://doi.org/10.3390/su10030639
38. Bressanelli, G., Adrodegari, F., Perona, M., Saccani, N.: The role of digital technologies to overcome Circular Economy challenges in PSS Business Models: An exploratory case study. Procedia CIRP **73**, 216–221 (2018). https://doi.org/10.1016/j.procir.2018.03.322
39. Luthra, S., Mangla, S.K.: Evaluating challenges to Industry 4.0 initiatives for supply chain sustainability in emerging economies. Process Saf. Environ. Prot. **117**, 168–179 (2018). https://doi.org/10.1016/j.psep.2018.04.018
40. Munsamy, M., Telukdarie, A.: Application of Industry 4.0 towards achieving business sustainability. In: 2018 IEEE International Conference on Industrial Engineering and Engineering Management, p. 844 (2018)
41. Bal, A., Badurdeen, F.: A business model to implement closed-loop material flow in IoT-enabled environments. Procedia Manuf. **38**, 1284–1291 (2019). https://doi.org/10.1016/j.promfg.2020.01.162
42. Micieta, B., et al.: Product segmentation and sustainability in customized assembly with respect to the basic elements of Industry 4.0. Sustainability **11**, 6057 (2019). https://doi.org/10.3390/su11216057
43. Sartal, A., Bellas, R., Mejías, A.M., García-Collado, A.: The sustainable manufacturing concept, evolution and opportunities within Industry 4.0: a literature review. Adv. Mech. Eng. **12** (2020). https://doi.org/10.1177/1687814020925232
44. Cui, Y., Kara, S., Chan, K.C.: Manufacturing big data ecosystem: a systematic literature review. Robot Comput. Integr. Manuf. **62**, 101861 (2020). https://doi.org/10.1016/j.rcim.2019.101861
45. Kohtamäki, M., et al.: Digital servitization business models in ecosystems: a theory of the firm. J. Bus. Res. **104**, 380–392 (2019). https://doi.org/10.1016/j.jbusres.2019.06.027
46. Nagy, J., et al.: The role and impact of Industry 4.0 and the internet of things on the business strategy of the value chain—The case of Hungary. Sustainability **10**, 3491 (2018). https://doi.org/10.3390/su10103491
47. Garcia-Muiña, F.E., González-Sánchez, R., Ferrari, A.M., Settembre-Blundo, D.: The paradigms of Industry 4.0 and circular economy as enabling drivers for the competitiveness of businesses and territories: the case of an Italian ceramic tiles manufacturing company. Soc. Sci. **7**, 255 (2018). https://doi.org/10.3390/socsci7120255
48. Brenner, B.: Transformative sustainable business models in the light of the digital imperative—A global business economics perspective. Sustainability **10**, 4428 (2018). https://doi.org/10.3390/su10124428
49. Braccini, A.M., Margherita, E.G.: Exploring organizational sustainability of Industry 4.0 under the triple bottom line: the case of a manufacturing company. Sustainability **11**, 36 (2018). https://doi.org/10.3390/su11010036
50. Birkel, H., et al.: Development of a risk framework for Industry 4.0 in the context of sustainability for established manufacturers. Sustainability **11**, 384 (2019). https://doi.org/10.3390/su11020384

51. Lin, K., Shyu, J., Ding, K.: A cross-strait comparison of innovation policy under Industry 4.0 and sustainability development transition. Sustainability **9**, 786 (2017). https://doi.org/10.3390/su9050786
52. VOSviewer: version 1.6.15. (https://www.vosviewer.com/)

# Current Issues in Flexible Manufacturing Using Multicriteria Decision Analysis and Ontology Based Interoperability in an Advanced Manufacturing Environment

M. B. Canciglieri[1(✉)], A. F. C. S. de M. Leite[1], E. de F. Rocha Loures[1], O. Canciglieri Jr.[1], R. P. Monfared[2], and Y. M. Goh[2]

[1] Industrial and Systems Engineering Graduate Program, Pontifical Catholic University of Paraná, Curitiba, Paraná, Brazil
matheus.beltrame@pucpr.br

[2] School of Mechanical, Electrical and Manufacturing Engineering, Loughborough University, Loughborough, Leicestershire, UK

**Abstract.** The manufacturing industry is undergoing a major transformation based on the emerging industry 4.0 technologies, such as cloud computing, big data, internet of things and cyber-physical systems. These novelty technologies aim at providing central management for the user's flexible manufacturing requirements and information. Also, the advent of these technologies has transformed the process planning and became crucial for the building of knowledge-based process planning environments. However, current praxis cannot deal with all semantic issues within this new paradigm, as requirements must be clear, consistent, measurable, stand-alone, testable, unambiguous, unique and verifiable. In this context, multicriteria decision analysis models have gained focus of the scientific and industrial communities as a support tool for the decision-making process in the product development and advanced manufacturing as these processes excel in environments with numerous and conflicting alternatives, providing the optimal alternative. Therefore, the main objective of this research is to highlight the current issues and research tendencies regarding ontology-based interoperability systems, multicriteria decision analysis and their integration. To achieve this goal, it will be applied a literature review on the targeted technologies, discussing the current tendencies of the field and the main issues regarding their implementation and integration. Finally, the paper points themes for further research and indicates viable concepts that can compose a solution for the gaps in a systematic manner.

**Keywords:** Advanced manufacturing systems · Semantic interoperability · Multicriteria decision analysis · Flexible manufacturing systems process reconfiguration

© Springer Nature Switzerland AG 2021
D. A. Rossit et al. (Eds.): ICPR-Americas 2020, CCIS 1407, pp. 372–383, 2021.
https://doi.org/10.1007/978-3-030-76307-7_28

## 1 Introduction

The manufacturing industry is undergoing a major transformation based on the emerging industry 4.0 technologies, such as cloud computing, big data, internet of things and cyber-physical systems. These novelty technologies aim at providing central management for the user's manufacturing requirements and information. Also, the advent of these technologies has transformed the process planning and became crucial for the building of knowledge-based process planning environments.

Furthermore, over the last two decades, the manufacturing industry has compacted their product lifecycles through the capture and integration of lifecycle knowledge in product development. This resulted in an almost simultaneous processes, which provided efficiency and flexibility while also increasing the product's final quality [1]. This is accentuated by the interconnection of the product development with trends on integrated manufacturing systems, smart factories, and concept and technologies of Industry 4.0 [2] However, current praxis cannot deal with all semantic issues within this new paradigm, as requirements must be clear, consistent, measurable, stand-alone, testable, unambiguous, unique and verifiable [2, 3].

Flexible manufacturing systems are considered the future of manufacturing because of their adaptable nature. In these systems basic modules, such as, the working machines, the material handling system or the manufacturing control system can be rearranged, interchanged or modified adapting the production system according to the enterprise's requirements. In this context, the performance of these systems is related to the system's capability and the planning and scheduling data in agreement with the market demands. Yet, as the market demands change in time the system's capabilities need to adapt to these changes. Thus, these changes can cause problems such as increase in the production time, cost and can decrease the final product's quality.

As the product's design and its manufacturing capabilities are closely related, the manufacturing system is desired to be reconfigurable in order to easily adjust for design changes that can occur in the product.

Currently, multicriteria decision analysis models have gained focus of the scientific and industrial communities as a support tool for the decision-making process in the product development and manufacturing as these processes excel in environments with numerous and conflicting alternatives, providing the optimal alternative. In this context, the multicriteria decision analysis makes for more consistent decision making as subjectivity and possible trade-offs can be contextualized and incorporated into these models. With these characteristics the multicriteria decision analysis can be incorporated and integrated with ontological approaches in order to make the decision-making process more accurate and efficient.

## 2 Material and Methods

The research is considered to have an applied nature, with a qualitative approach. Its scientific objectives are exploratory, using as a technical procedure a literature review and qualitative analysis (Fig. 1).

The research has as its starting point a literature review on issues on the context of Reconfigurable Manufacturing Systems, such as, agility in the reconfiguration process,

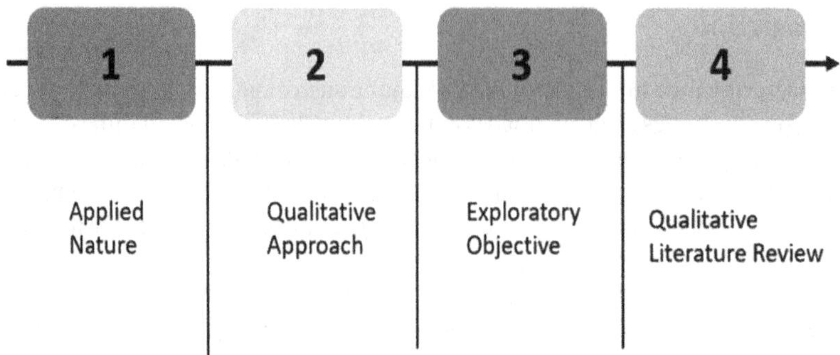


**Fig. 1.** Research methodology.

Semantic Interoperability such as Requirement Interpretation and knowledge representation, and Multicriteria Decision models. The goal is to establish a conceptual link between the studied areas as well as to improve understanding of the tools used in each theme and their intricate relations (Fig. 2). The result of this process is a discussion on the themes, represented through a schema, which serves as the starting point to expose the current issues regarding the continuous development and flexibilization of reconfigurable manufacturing models, and their impacts in Industry 4.0.

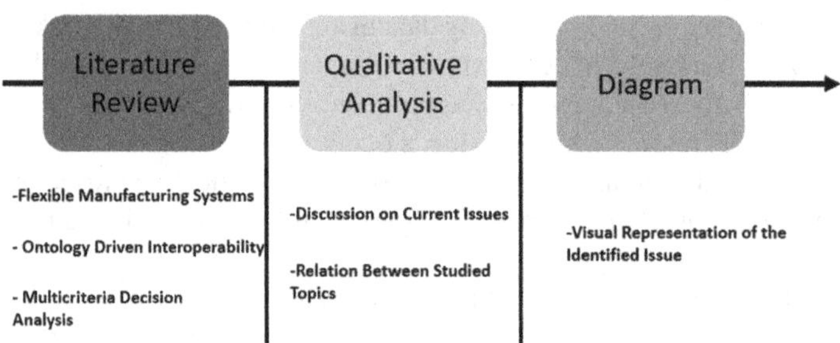


**Fig. 2.** Technical procedures.

## 3 Related Works

### 3.1 Ontology Based Interoperability

Semantic interoperability is the ability of heterogeneous computer systems to share data with clear meaning. Therefore, it is concerned with the simultaneous transmission of the meaning of the data alongside it. This is accomplished though the addition of data about the data, which links each data element to a controlled, shared vocabulary. The meaning of the data is transmitted with the data itself, in one "information package" that is independent of any information system.

Since the introduction of the Semantic Web concept by [4] there has been numerous researches and applications of the W3C (World Wide Web Consortium, W3C) standards to provide web-scale semantic data exchange, federation, and inferencing capacities. Nowadays ontologies are one of the most popular tools for representing and sharing knowledge across different systems and domains [5].

The construction of ontologies is a viable solution on the formalization of these common information models and on the sharing of the formal information throughout the stages of the product development process, which, consequently, provides increased knowledge in the domains of application [6, 7].

An Ontology is defined as "a lexicon of specialized terminology along with some specification of the meaning of terms on the lexicon" [7, 8] where the lexicon is the vocabulary of a knowledge domain. In this way, a significant differentiation can be made between ontologies by their degree of expressiveness. In this differentiation, simple ontologies, which formalizes only a taxonomy of concepts and basic relations between them are referred to as lightweight ontologies. When a lightweight ontology is enriched through the insertion of axioms in the form of constraints, they are classified as a heavyweight ontology. Nevertheless, the use of ontologies is restricted to the purpose of its application, that is, the knowledge structure formalized in an ontology has little reusability outside the scope of its application [8].

Given two different classification systems, a simple query finds all the data corresponding to a term in both information sources; however, this query can only be efficiently answered if both systems have their semantics well understood. If these systems are conceptualized in two different ontologies, the comparison of terms is a challenge due to the high variation of the detail level and logic between these ontologies [8, 9]. To solve this limitation a shared ontology approach can be adopted, which enables terminological reasoning over the definition of classes in the descriptive logic ontologies by considering the axioms, set of relations and set of class definitions defined in the shared ontology [8, 9].

Also, according to [10] and created digital links between virtual models improving the information sharing in the operation management. Similarly, [11] created a interoperable environment through the integration of different ontologies which resulted in dynamic collaborations between partners in heterogeneous and multi sited projects.

[11] affirms that with the rapid advanced and development of semantic technologies based on ontologies, there has been a shift towards more complex and better semantic modeling of manufacturing processes. This process culminates in the insertion of machining features in the semantic technologies in order to retrieve the considerable machining knowledge and shape semantics presented in these features.

[12], in the other hand, identifies that ontology based systems can be used for the optimization of machine fault diagnostics as this is a multi-disciplinary problem and the use of semantic mapping techniques and knowledge reasoning technologies can efficiently represent and organize fault knowledge with explicit and consistent semantic support. According to [12] these types of systems have the ability of supporting, sharing and scaling the fault knowledge presented in the system. [13] asserts that the ontology technology has great advantages in improving semantic consistency of multi-source heterogeneous data. However, for complex dynamic decision-making problems, there are

some limitations in the mining of ontology relations and the establishment of inference rules.

### 3.2 Multi-criteria Decision Analysis

As globalization, collaboration, and cooperation contribute to a more connected and integrated competitive environment for the manufacturing industries, they have also made the decision-making process more complex whilst reducing the time for the decision made to be relevant.

In this context, the Multi-Criteria Decision Making (MCDM) methods are gaining importance as potential tools for analyzing complex real problems. This happens due to the inherent ability of these methods to judge different alternatives through several criteria (attributes) in order to provide the most suitable alternative. These alternatives may be further explored in-depth for their final implementation (Chung & Ng 2016). A MCDM problem can be concisely expressed in a decision matrix as shown in Fig. 3.

| Alternatives | | Criteria | | | |
|---|---|---|---|---|---|
| | | $C_1$ | $C_2$ | ......... | $C_n$ |
| | $A_1$ | $X_{11}$ | $X_{12}$ | ......... | $X_{1n}$ |
| | $A_2$ | $X_{21}$ | $X_{22}$ | ......... | $X_{2n}$ |
| | ......... | ....... | ...... | ......... | ......... |
| | $A_m$ | $X_{m1}$ | $X_{m2}$ | ......... | $X_{mn}$ |

**Fig. 3.** Multi-criteria decision analysis generic decision matrix.

In this matrix each element (xij) represent the performance rating of "**m**$^{th}$" alternative in respect to "**n**$^{th}$" criteria, while considering also the significance of nth criteria [14]. This acquirement of the importance of each attribute, also called as weight of the attribute, is what boosted the introduction of numerous MCDM methods during the last decades.

Over the last decades, several MCDM methods have been proposed, the most popular of which are AHP (Analytic Hierarchy Process) [14–16], ANP (Analytic Network Process) [5], TOPSIS (Technique for Order of Preference by Similarity to Ideal Solution) [17–20], ELECTRE (ELimination and Choice Expressing REality) [21–23], and PROMETHEE (Preference Ranking Organization METHod for Enrichment Evaluations) [24–26].

These methods have been used to aid the decision making in several fields of knowledge, such as medicine, management and logistics, among others. This can be seen with the research of [26] which creates a systematic literature review on the use of MCDM in the optimization of supply chains and concludes that there are various combined approaches for supplier selection. However, integrated AHP approaches are more prevalent. The popularity of this type of approach is due to its simplicity, ease of use, and great flexibility. [26] creates a novel multi criteria model for the optimization of flexible manufacturing systems.

The MCDM methods' task is to support a decision-maker in choosing the most preferable variant from many possible options, taking into account a multitude of criteria characterizing acceptability of individual decision variants [42]. MCDM problems can be divided into 2 main categories: i) continuous problems and ii) discrete problems. The second type of MCDM problems can be solved either through utility functions or outranking methods [43]. Also, according to [43] the utility approach determines a relationship between the alternatives which can be indifferent and preference of one alternative over another, this has the consequence of the methods in this group leaving out the non-comparability of the decision criteria and assume transitivity and completeness of preference [29].

On the other hand, outranking methods expand the basic set of preferential situations with the relationships they create which can be: i) indifferent; ii) weak preference; iii) strict preference and iv) incomparability. In this method the preferential situations are combined in an outranking relation [44].

Furthermore, MCDM methods can vary according to their operational approaches: i) aggregation to a single criterion (American school); ii) aggregation by using the outranking relationship (European school) [45]. However, the mixed approach which combine elements of American and European decision-making schools have gained highlight in the scientific and industrial environments [46]. According to [46] an example of this mixed approach is the Pairwise Criterion Comparison Approach methods. Lastly MCDM methods can be differentiated according to the nature and characteristics of the used data, which will immensely affect the measurement scale. Data can be quantitative or qualitative and can be expressed in the cardinal (quantitative) or ordinal (qualitative) scale [47].

### 3.3 Flexible Manufacturing

The introduction of a new technology in a company's operations changes the organizational structures, processes and resources, factors that present themselves as obstacles in the decision phase and in the implementation of these new technologies. Currently, the choice of advanced manufacturing technologies is made through operational criteria such as productivity and performance, which can offer a competitive advantage and, therefore, must be chosen according to strategic criteria. [27]. The main characteristic of advanced manufacturing technologies is their flexibility (to produce a wide variety of products on a small scale without having extra costs or penalties) is pointed out by several authors [3, 27–29].

Advanced manufacturing systems are systems that use mechanical, electronic and computational subsystems to operate and control production, encompassing a large set of machines that execute, monitor and connect the production processes [30]. The same author emphasizes that the adoption of advanced manufacturing technologies is a key condition for maintaining the company's competitiveness in the long term. Nevertheless, many advanced manufacturing implementation projects fail to be more significant in small and medium-sized companies, as the managers of these companies are the ones who most want to implement these technologies, but they do so from instinct or information from similar companies.

Yet, [3], states in his research that small and medium-sized companies do not wish to implement, or do not understand how these technologies could assist them in improving customer satisfaction and market results. The author also states that these companies have limited financial and human resources, preventing these companies from implementing complex technological systems.

Over the past two decades, there have been reports of improvements in flexibility, quality and productivity in industries caused by the use of advanced manufacturing technologies [22, 31]. [32] claim that advanced manufacturing technologies can effectively assist the length of manufacturing objectives and simultaneously consider competitive objectives and manufacturing decision areas, being considered resources mobilized for the construction of competencies. [3] stated that with the growing emphasis on flexibility to ensure a company's long-term competitiveness, the relationship between advanced manufacturing technologies has taken on an important place for decision making. In addition, the higher the level of advanced manufacturing technologies and enterprise control technologies, the greater the level of flexibility in manufacturing.

From the many flexible manufacturing technologies additive manufacturing has been prominent in both industry and academia, as it has benefits compared to traditional manufacturing models, such as geometric flexibility as well as a potential to reduce the time and cost of manufacturing products [33].

Reconfigurable manufacturing systems (RMSs) was first proposed by the Engineering Research Center of the University of Michigan in 1999 [29]. Though the use of these systems manufacturers could make multiple variations of customized products at the price of standardized mass products [29, 34]. [29, 35] analyzed the characteristics of RMS in detail and found out that the most used approaches are biological manufacturing systems (BMSs) and holonic manufacturing systems (HMSs). The reconfiguration methods could be classified into two categories: i) knowledge-based reasoning methods; ii) artificial intelligence-based optimization methods [29]. The knowledge automation technologies of RMS have attracted considerable interest. [29, 36] proposed an ontology-based agent approach with the goal of fast reconfiguration of modular manufacturing systems. Furthermore, [29, 36–38] proposed the combination of knowledge-based reasoning methods with periodic inspection for the dynamic and automatic re-configuration of systems.

For the second kind, optimal reconfiguration of manufacturing systems needs the use of advanced artificial intelligence techniques, such as deep learning and big data [39]. However, these conventional optimization methods for RMS suffers from substantial computation complexity, which hinders rapid and online reconfiguration of the manufacturing system.

There are three significant problems in current RMS. Firstly, essential measurement factors, including the scalability and reconfigurability of the current systems, which needs to be considered during the design process of the RMSs [40, 41]. Secondly, current RMS realizes control logic validation with software testing after the mechanical structure reconfiguration is assembled, which will take a significant amount of time to do and is susceptible to reconfiguration mistakes. Thirdly, dealing with the rapid reconfiguration of the manufacturing system, requires not only a digital twin architecture as a foundation but also an optimization method. These issues mean that the complex balancing and coupling relationship between the system productivity and reconfiguration cost is difficult to model in the environment of the frequent change.

## 4 Discussion

This research works towards finding evidence that demonstrates the current issues regarding semantic interoperability and Multi-Criteria Decision-Making methods in an advanced manufacturing environment. These issues provide sustenance to the proposal of an ontology and MCDM integration model that aims to aid the real time reconfiguration and decision making in flexible manufacturing processes in order to ensure higher quality products and increase agility of the manufacturing process.

The mapping of the current issues revealed the limitations of a ontology driven interoperability system when it is used in a highly mutable environment, and, also, how the sharing of ambiguous information in such an environment can reduce the competitiveness of an enterprise as it will difficult the decision making process, raising costs and diminishing the products quality.

The MCDM models can be associated with ontological approaches as aids for the reasoning presented in them. The application of MCDM algorithms in these approaches create case-based reasoning techniques, which can aid in the flexibilization of the ontological structure and consequently promote a more accurate and efficient decision-making process. These concepts may be used integrated with ontology-based interoperability models in order to create a flexible decision-making environment that can aid in the reconfiguration of an advanced manufacturing system allowing for a more agile process and ensuring the product's quality. Figure 4 highlights the relationship of the concepts that this research approached. From the figure can be seen the decision making in a flexible manufacturing environment can be divided into three main layers. The outmost layer regards the multi-criteria analysis, while the second involves the main systems in a flexible manufacturing system and finally the decision-making layer.

The multicriteria analysis layer consists in the analysis of the many heterogeneous requirements that come from the consumer and from the enterprise. These requirements range from desired finishing qualities from the consumer to material handling and stock logistics from the enterprise. As these requirements come from various sources, they may present semantic homogeneity issues which can lead to a wrongfully interpreted information that consequently will prejudice the decision-making process and the reconfiguration of the manufacturing system.

The flexible manufacturing layer encompasses the flexible manufacturing system, from the material handling system to the working machines and the main control system. This layer consists in the capability and restrictions presented in the subsystems of the manufacturing process. These restrictions combined with the requirements of the multicriteria analysis layer make the information form that will guide the decision-making process.

However, the restrictions presented by the flexible manufacturing layer can cause misunderstanding or may present semantic conflicts with one another or even with

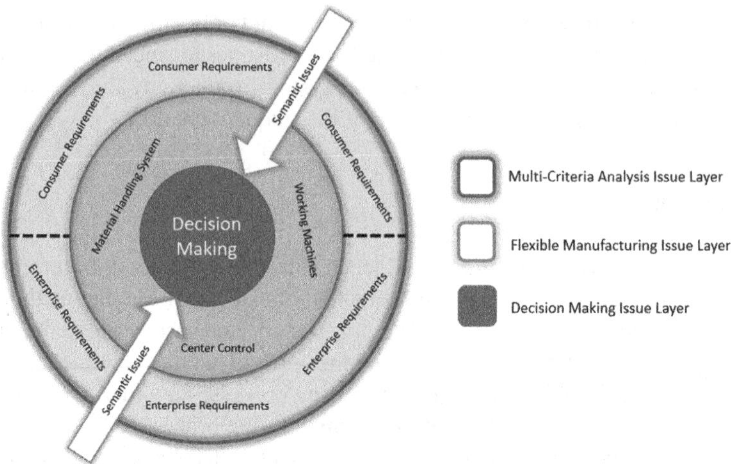


**Fig. 4.** Representation of current issues.

the requirements from the multicriteria analysis layer. These misunderstandings and conflicts may hinder the decision-making process that consequently impoverishes the manufacturing quality and thwarts the flexibility of the overall production process.

Finally, the decision making layer defines the decision making process, with the interpretation of the requirements and restrictions of the outwards layers and reconfiguring the manufacturing systems and processes in order to attend the requirements and generate the most value possible to the consumer and the enterprise. This process can be hindered with semantic heterogeneity issues from the other layer, while being susceptible to semantic errors in the decision making process which can compromise the timely reconfiguration of the manufacturing system that will in turn compromise the value generation and the attendance of the requirements and restrictions.

## 5 Conclusion

This paper presented a research that lead to the development of a model which shows some of the critical issues that hinder the development of a real time process reconfiguration environment. These areas are essential to ensure a competitive position for an enterprise in the global competitive environment it is inserted.

The standardized and formalized knowledge that is captured by an ontology driven system allows it to be retrieved, shared and reused in different stages of the reconfiguration process and, also, through the process of relating concepts, the information can be captured in its entirety as well as extended as the need arises.

Multicriteria decision analysis models have gained focus of the scientific and industrial communities as a support tool for the decision-making process in the product development and manufacturing as these processes excel in environments with numerous and conflicting alternatives, providing the optimal alternative. In this context, the multicriteria decision analysis makes for more consistent decision making as subjectivity and possible trade-offs can be contextualized and incorporated into these models. With these characteristics the multicriteria decision analysis can be incorporated and integrated with ontological approaches in order to make the decision-making process more accurate and efficient.

In order to continue this research several subjects are proposed for future works. The integration of MCDM methods with ontological applications and map the decision criteria needed for an advanced manufacturing reconfiguration process in order to develop an application ontology to aid in the reconfiguration process are some research perspectives.

## References


1. Ye, Y., Tianliang, H., Zhang, C., Luo, W.: Design and development of a CNC machining process knowledge base using cloud technology. Int. J. Adv. Manuf. Technol. **94**(9–12), 3413–3425 (2018)
2. Leite, A.F.C.S.M., Canciglieri, M.B., Szejka, A.L., Junior, O.C.: The reference view for semantic interoperability in integrated product development process: The conceptual structure for injecting thin walled plastic products. J. Indus. Inf. Integr. **7**, 13–23 (2017)
3. Canciglieri, M.B., de Moura Leite, A.F.C.S., Szejka, A.L., Junior, O.C.: An approach for dental prosthesis design and manufacturing through rapid manufacturing technologies. Int. J. Comput. Integr. Manuf. **32**(9), 832–847 (2019)
4. Berners-Lee, T., Fischetti, M.: Weaving the Web, Chapter 12. HarperSanFrancisco (1999). ISBN: 978-0-06-251587-2
5. Ceravolo, P., et al.: Big data semantics. J. Data Semant. **7**(2), 65–85 (2018). https://doi.org/10.1007/s13740-018-0086-2
6. Khan, Z.M.A., Saeidlou, S., Saadat, M.: Ontology-based decision tree model for prediction in a manufacturing network. Prod. Manuf. Res. **7**(1), 335–349 (2019). https://doi.org/10.1080/21693277.2019.1621228
7. Li, X., Zhang, S., Huang, R., Huang, B., Changhong, X., Zhang, Y.: A survey of knowledge representation methods and applications in machining process planning. Int. J. Adv. Manuf. Technol. **98**(9–12), 3041–3059 (2018)
8. Chungoora, N., Young, R.I.M.: Semantic reconciliation across design and manufacturing knowledge models: A logic-based approach. Appl. Ontol. **6**(4), 295–315 (2011)
9. Jelokhani-Niaraki, M.: Knowledge sharing in web-based collaborative multicriteria spatial decision analysis: An ontology-based multi-agent approach. Comput. Environ. Urban Syst. **72**(May), 104–123 (2018). https://doi.org/10.1016/j.compenvurbsys.2018.05.012
10. Du, Juan et al.: An ontology and multi-agent based decision support framework for prefabricated component supply chain. Inf. Syst. Front. **22**, 1467–1485 (2019)
11. Jelokhani-Niaraki, M., Sadeghi-Niaraki, A., Choi, S.M.: Semantic interoperability of GIS and MCDA tools for environmental assessment and decision making. Environ. Model Softw. **100**, 104–122 (2018). https://doi.org/10.1016/j.envsoft.2017.11.011
12. Li, X., Zhang, S., Huang, R. et al.: Structured modeling of heterogeneous CAM model based on process knowledge graph. Int. J. Adv. Manuf. Technol. **96**, 4173–4193 (2018). https://doi.org/10.1007/s00170-018-1862-8
13. Bagherifard, K., Rahmani, M., Nilashi, M., Rafe, V.: Performance improvement for recommender systems using ontology. Telematics Inf. **34**(8), 1772–1792 (2017). https://doi.org/10.1016/j.tele.2017.08.008
14. Lahdhiri, H., et al.: Supervised process monitoring and fault diagnosis based on machine learning methods. Int. J. Adv. Manuf. Technol. **102**(5–8), 2321–2337 (2019)

15. Peko, I., Gjeldum, N., Bilić, B.: Application of AHP, Fuzzy AHP and PROMETHEE method in solving additive manufacturing process selection problem. Tehnicki Vjesnik **25**(2), 453–461 (2018)
16. Almeida, D., Teixeira, A., Alencar, M.H., Garcez, T.V., Ferreira, R.J.P.: A systematic literature review of multicriteria and multi-objective models applied in risk management. IMA J. Manage. Math. **28**(2), 153–184 (2017)
17. Park, J.W., Kang, B.S.: Comparison between regression and artificial neural network for prediction model of flexibly reconfigurable roll forming process. Int. J. Adv. Manuf. Technol. **101**(9–12), 3081–3091 (2019)
18. Chourabi, Z., Khedher, F., Babay, A., Cheikhrouhou, M.: Multi-criteria decision making in workforce choice using AHP, WSM and WPM. J. Textile Inst. **110**(7), 1092–1101 (2019). https://doi.org/10.1080/00405000.2018.1541434
19. Rezaei, J.: Best-worst multi-criteria decision-making method. Omega (United Kingdom) **53**, 49–57 (2015). https://doi.org/10.1016/j.omega.2014.11.009
20. Alsina, E.F., Chica, M., Trawiński, K., Regattieri, A.: On the use of machine learning methods to predict component reliability from data-driven industrial case studies. Int. J. Adv. Manuf. Technol. **94**(5–8), 2419–2433 (2018)
21. Segreto, T., Teti, R.: Machine learning for in-process end-point detection in robot-assisted polishing using multiple sensor monitoring. Int. J. Adv. Manuf. Technol. **103**(9–12), 4173–4187 (2019)
22. Trächtler, A., Denkena, B., Thoben, K.-D.: Editorial: system-integrated intelligence – new challenges for product and production engineering. Procedia **26**, 1–3 (2016). http://dx.doi.org/10.1016/j.protcy.2016.08.001
23. Tang, D., Zheng, K., Zhang, H., Sang, Z., Zhang, Z., Xu, C., Espinosa-Oviedob, J.A., Vargas-Solar, G., Zechinelli-Martini, J.L.: Using autonomous intelligence to build a smart shop floor. Int. J. Adv. Manuf. Technol. **94**(5–8), 1597–1606 (2018)
24. Razia Sulthana, A., Ramasamy, S.: Ontology and context based recommendation system using neuro-fuzzy classification. Comput. Electr. Eng. **74**, 498–510 (2019). https://doi.org/10.1016/j.compeleceng.2018.01.034
25. Zhou, J., Yao, X.: Hybrid teaching–learning-based optimization of correlation-aware service composition in cloud manufacturing. Int. J. Adv. Manuf. Technol. **91**(9–12), 3515–3533 (2017)
26. Navarro, I.J,, Yepes, V., Martí, J.V.: A review of multicriteria assessment techniques applied to sustainable Infrastructure design. Adv. Civil Eng. **2019**, 16 p. (2019). Article ID 6134803. https://doi.org/10.1155/2019/6134803
27. Saeidlou, S., Saadat, M., Sharifi, E.A., Jules, G.D.: Agent-based distributed manufacturing scheduling: an ontological approach. Cogent Eng. **6**(1), 1–23 (2019). https://doi.org/10.1080/23311916.2019.1565630
28. Saeidlou, S., Saadat, M., Jules, G.D.: Knowledge and agent-based system for decentralised scheduling in manufacturing. Cogent Eng. **6**(1), 1–19 (2019). https://doi.org/10.1080/23311916.2019.1582309
29. Asghar, E., Zaman, U.K., Baqai, A.A., Homri, L.: Optimum machine capabilities for reconfigurable manufacturing systems. Int. J. Adv. Manuf. Technol. **95**(9–12), 4397–4417 (2018)
30. Sevinç, A., Şeyda, G., Tamer, E.: Analysis of the difficulties of SMEs in industry 4.0 applications by analytical hierarchy process and analytical network process. Processes **6**(12), 264 (2018)
31. Qu, Y.J., et al.: Smart manufacturing systems: state of the art and future trends. Int. J. Adv. Manuf. Technol. **103**(9–12), 3751–3768 (2019)
32. Wang, L., et al.: Distributed manufacturing resource selection strategy in cloud manufacturing. Int. J. Adv. Manuf. Technol. **94**(9–12), 3375–3388 (2018)

33. Wang, S., Wan, J., Li, D., Liu, C.: Knowledge reasoning with semantic data for real-time data processing in smart factory. Sensors (Switzerland) **18**(2), 1–10 (2018)
34. Widiyati, M.: “No Titleענף הקיווי: תמונת מצב.” עלון הנוטע 66: 37–39 (2012)
35. Wu, Z., et al.: Towards a semantic web of things: a hybrid semantic annotation, extraction, and reasoning framework for cyber-physical system. Sensors (Switzerland) **17**(2), 403 (2017)
36. Hamdi, F., Ghorbel, A., Masmoudi, F., Dupont, L.: Optimization of a supply portfolio in the context of supply chain risk management: literature review. J. Intell. Manuf. **29**(4), 763–788 (2018)
37. Zhang, Y., Luo, X., Zhang, B., Zhang, S.: Semantic approach to the automatic recognition of machining features. Int. J. Adv. Manuf. Technol. **89**(1–4), 417–437 (2017)
38. Zhao, Y., et al.: Dynamic and unified modelling of sustainable manufacturing capability for industrial robots in cloud manufacturing. Int. J. Adv. Manuf. Technol. **93**(5–8), 2753–2771 (2017)
39. Kumar, S., Dhingra, A.K., Singh, B.: Kaizen selection for continuous improvement through VSM-Fuzzy-TOPSIS in small-scale enterprises (2018)
40. Zhou, Q., Yan, P., Liu, H. et al.: Research on a configurable method for fault diagnosis knowledge of machine tools and its application. Int. J. Adv. Manuf. Technol. **95**, 937–960 (2018). https://doi.org/10.1007/s00170-017-1268-z
41. Liu, K., El-Gohary, N.: Ontology-based semi-supervised conditional random fields for automated information extraction from bridge inspection reports. Autom. Constr. **81**, 313–327 (2017). https://doi.org/10.1016/j.autcon.2017.02.003
42. Roy, B.: Paradigms and challenges. In: Figueira, J., Greco, S., Ehrgott, M. (eds.) Multiple Criteria Decision Analysis: State of the Art Surveys, vol. 78, pp. 3–24. Springer, New York (2005)
43. Kodikara, P.N.: Multi-objective optimal operation of urban water supply systems, Ph.D thesis. Victoria University (2008)
44. Roy, B.: Multicriteria Methodology for Decision Aiding. Springer, Boston (1996)
45. Jacquet-Lagreze, E., Siskos, Y.: Preference disaggregation: 20 years of MCDA experience. Eur. J. Oper. Res. **130**, 233–245 (2001)
46. Martel, J.-M., Matarazzo, B.: Other Outranking Approaches. Multiple Criteria Decision Analysis: State of the Art Surveys, vol. 78, pp. 197–259. Springer, New York (2005)
47. Saaty, T.L.: The Analytic Hierarchy Process: Planning, Priority Setting, Resource Allocation. McGraw-Hill International Book Co, New York, London (1980)

# Author Index

Printed in the United States
by Baker & Taylor Publisher Services